Flavio Salvati

Fundamentals of Astronomy

A guide for Olympiads

Experience suggests that it is close to impossible for the book to be error-free. If something looks amiss, please check the webpage www.astrolympiad.com for a list of typos, updates and additional problems. If you have any questions or suggestions, don't hesitate to contact me at astrolympiad@gmail.com.

ISBN: 979-8688791433

Published on the autumnal equinox, 2020.

To my parents

Contents

Preface

Chances are you too have gazed in awe and wonder at the night sky, with endless questions drifting through your mind: how did the universe begin, how did life spark on Earth, are there any other intelligent beings in the universe and, if so, why hasn't anyone visited our planet? These are questions one can hardly ignore and, while our chances of finding a definitive answer within our lifetime are slim, we can make progress and enable future generations to do so. This book will teach you the fundamentals of astronomy, passing on the discoveries of countless scientists who lived before us, giving you the tools to continue along this endless journey. Maybe you will be writing the next chapter.
I have tried to make this book accessible to a large audience, with the only prerequisites being basic high school mathematics and physics. Slightly more advanced mathematical tools are covered in the appendix, making the book self-contained. At the same time, there are plenty of exercises and problems which will help you become familiar with the concepts explained throughout the theory. The book is specifically aimed at students preparing for the regional and national phases of the Astronomy Olympiads, but will still represent a valuable resource for those chosen to represent their countries at the international competition (IOAA and IAO). While the book contains a lot of information, you are encouraged to supplement it with online resources and other textbooks that might cover some topics in more detail.
The book has been divided into 4 parts. Part I (Positional Astronomy) covers the celestial coordinate systems, the transformation and perturbation of coordinates, observation and instruments, the time systems and the Moon. Part II (Radiation Mechanisms) explains electromagnetic radiation, flux, stellar magnitudes and the cosmological ladder. Part III (Celestial mechanics) introduces gravitation and Kepler's laws, the motion of the planets, the rocket equation, orbital manoeuvres and binary systems. Finally, Part IV (Solutions) presents detailed solutions to all the exercises and problems.
I decided to distinguish between exercises, which require you to apply the concepts learned throughout the theory (aimed at sharpening your problem solving skills), and problems, which will guide you through the derivation of entirely new concepts (targeting your creative thinking). The difficulty of the exercises and problems is denoted by asterisks, from zero (easy) to four (extremely challenging). Of course, you may disagree with my judgement of difficulty, but I think that an arbitrary weighting scheme is better than none at all. Some problems are extremely hard, so it is normal you will get stuck. In this case, don't be discouraged, just set the problem aside for some time and get back to it later. You will find that thinking about the possible solutions over and over again will strengthen what you learned and resolve new doubts.

If you need some guidance, try reading the solution line-by-line (maybe covering the rest of the page with a piece of paper), pausing where you think you can continue by yourself. I have added interesting discussions at the end of many solutions, so be sure to read them after you solve a question. In the case of questions with 4 asterisks, my aim was to create an almost insurmountable challenge, that will always give you something to think about.
I have also included some exercises of various astronomy competitions, from the national to the international level. The source is stated at the beginning of each problem (see the next page for a list of references), so that you can easily find it online.
At the end of the book there are four appendices. Appendix A (Mathematics) covers the necessary concepts to understand the theory and solve most of the exercises. My advice is that you start by studying this part, if you are not familiar with high school mathematics. Appendix B (Kepler laws) contains a proof of Kepler laws from Newton's law of gravitation. I suggest you read this only after having gone through Appendix A and Part III. Appendix C (Virial theorem) contains a proof of the Virial theorem in its most general form. Finally, Appendix D (Tables and constants) has been included to speed up the process of finding known constants when solving exercises and problems.
I want to thank everyone that helped proof-read the book: Alexandra Alexiu and Amar Shah, who read the whole manuscript and made invaluable suggestions, Je Qin Chooi and Alisa Hathaway for the additional help in the more challenging chapters, and GholamReza ShahAli, who reported several corrections in the first edition. I also want to thank my parents, Lidwina and Francesco, and my siblings, Fabiana and Federico, for reading the book and encouraging me to finish this immense project. Despite scrupulous reviews by myself and many other people, it is practically impossible for the book to be error-free. If something looks amiss, visit www.astrolympiad.com for a list of typos and updates. On the website I will also publish many other problems that, for the sake of brevity, I could not include in the book. If you discover an error that has not yet been published, I would be very grateful if you could report it to me either by filling out the form on the site or by writing to the email address astrolympiad@gmail.com. If you have any questions or suggestions, don't hesitate to contact me. Finally, if you liked the book, I would be grateful if you could leave a review.
— I hope you enjoy the book!

National Olympiads

Part of the exercises were selected from the following national competitions:

- **ARAO**, the All-Russian Astronomy Olympiad;
- **ArmAO**, the Armenian Astronomy Olympiad;
- **BAAO**, the British Astronomy and Astrophysics Olympiad;
- **CAO**, the Canadian Astronomy Olympiad;
- **CzAO**, the Czech Astronomy Olympiad;
- **INAO**, the Indian National Astronomy Olympiad;
- **INT** and **NAZ**, the Italian regional and national Astronomy Olympiads;
- **MyAO**, the Malaysian Astronomy Olympiad;
- **SAO**, the Singaporean Astronomy Olympiad;
- **ONAA**, the Romanian Astronomy and Astrophysics Olympiad;
- **USAAAO**, the U.S.A. Astronomy and Astrophysics Olympiad.

You will also find exercises from the following international competitions:

- **IAO**, the International Astronomy Olympiads.
- **IOAA**, the International Olympiads on Astronomy and Astrophysics;

Part I

Positional Astronomy

1

Celestial Coordinate Systems

In this chapter we aim to study the commonly used astronomical reference systems. In the past, man believed that the universe was confined within a spherical shell and that the stars, fixed on this shell, were all equidistant from the Earth, placed at the centre of the universe. Although we know today that these assumptions are for the most part false, this simple model is still useful today, in many aspects, as it was in the past. In fact, since most of the measurements are taken from Earth, it is useful to place ourselves at the centre of the universe. Furthermore, since the distance to the stars is very large, we can neglect the orbital motion of the Earth around the Sun (which would periodically change the coordinates of the stars) and assume that all the stars are fixed on a sphere of infinite radius.
Under these hypotheses, for each coordinate system, only a fundamental plane and the direction perpendicular to it remain to be defined. We begin this chapter by defining the coordinates of a location on the surface of the Earth, and then apply a similar reasoning to the celestial vault.

1.1 Geographical Coordinate System

It is important to define the position of an observer on Earth, as the visible sky depends on the location from which it is observed. Unlike astronomical coordinates, on Earth we must also define a third coordinate, that is, the height relative to the surface.
In this coordinate system the fundamental direction is Earth's rotation axis. The fundamental plane, perpendicular to Earth's axis, is called the *equator*. The points where the axis of rotation meets the surface of the Earth are called the *north* and *south poles*. Every plane parallel to the equator is called a *parallel*; every great circle passing through the poles is called a *meridian*.

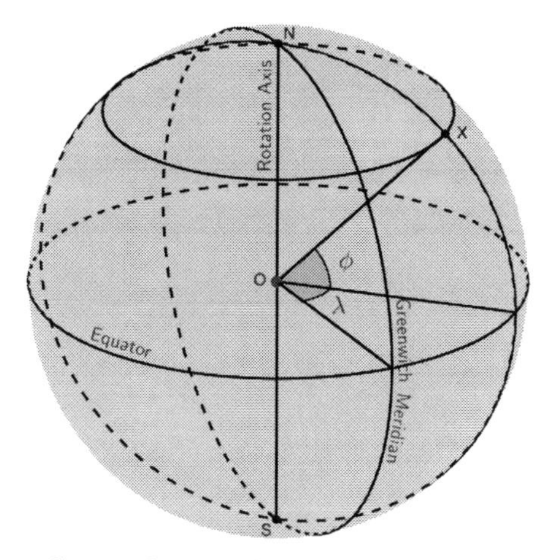

Figure 1.1: Geographic coordinates of a location X on Earth.

Let O be the centre of the Earth. Looking at Fig. 1.1, the coordinates of a location X on Earth are:

- latitude ϕ, i.e. the angle between the equatorial plane and the straight line passing through XO;
- longitude λ, i.e. the angle between the Greenwich meridian and the straight line passing through XO;
- height h, i.e. the vertical distance of X with respect to the average sea level.

The *prime* meridian is the Greenwich meridian, which passes through London. The longitude of a place to the east or west of Greenwich is denoted by $x°$ E or $x°$ W respectively, where $x°$ is the angle in degrees (normalized to 180°), between the local meridian and Greenwich. The sky visible to an observer depends on its geographical position. As we vary the longitude, the culmination time of the stars (the instant they pass through the meridian) also varies. On Earth, the sky appears to rotate from east to west (clockwise for an observer in the northern hemisphere) with an angular velocity of 15° per hour. Therefore, if a star in Rome culminates at 11:00 local time, in London it culminates about an hour later, as the location is further west by just under 15°. As we vary the latitude, the height of the stars on the horizon also varies. As will be seen more in detail in the next chapter, some stars are circumpolar (always visible), others rise and set at different times, while others permanently stay below the horizon. At the equator, all the stars rise and set over the course of a day; at the north pole, only half of the celestial vault is visible and all stars are circumpolar. Finally, varying the altitude modifies the *apparent* horizon, to be distinguished from the *astronomical* horizon, which is the line perpendicular

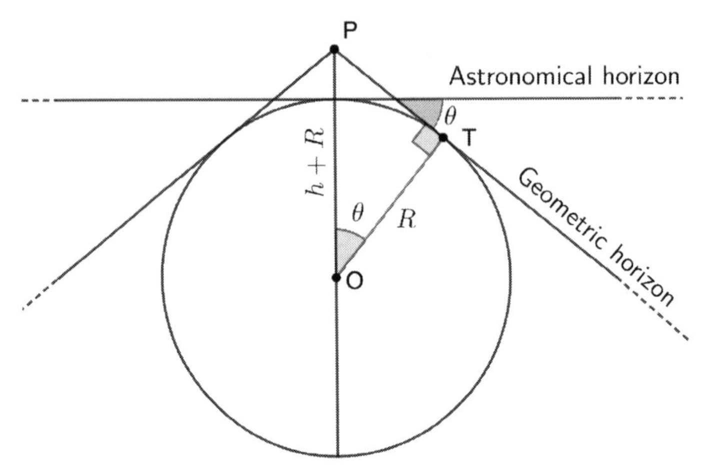

Figure 1.2: At a height h it is possible to see an additional patch of sky of angle $\theta =$ $(R/(R+h))$. For this reason, the horizon appears to be farther away when standing on a very tall building.

to the vertical at each location on Earth (Fig. 1.2). Assuming for the moment that there is no refraction, the apparent horizon coincides with the *geometric* horizon. When standing on a very tall building, we can effectively see an additional patch of sky of angle θ, and therefore the number of visible stars is greater. This phenomenon is known as the *dip* of the horizon. In the limiting case of an observer at infinite height, the Earth appears as a point, i.e. in deep space all the stars are visible. Looking back at Fig. 1.2, we see that the following relationship holds:

$$(R+h)\cos\theta = R\,,$$

Isolating θ, we find:

$$\boxed{\theta = \arccos\left(\frac{R}{R+h}\right)}\,. \tag{1.1}$$

1.2 The Horizontal System

The first and most natural astronomical reference system is the one centred on the observer. In the horizontal reference system, the fundamental direction is the vertical line at the point of observation, i.e. the direction of the plumb line coinciding with the direction of the acceleration of gravity. The intersections of the vertical with the celestial sphere to the north and south are called *zenith* and *nadir*, respectively. The plane passing through the observer and tangent to the surface of the Earth is called *horizontal plane.* The intersection of the horizontal plane with the celestial sphere marks the astronomical horizon.

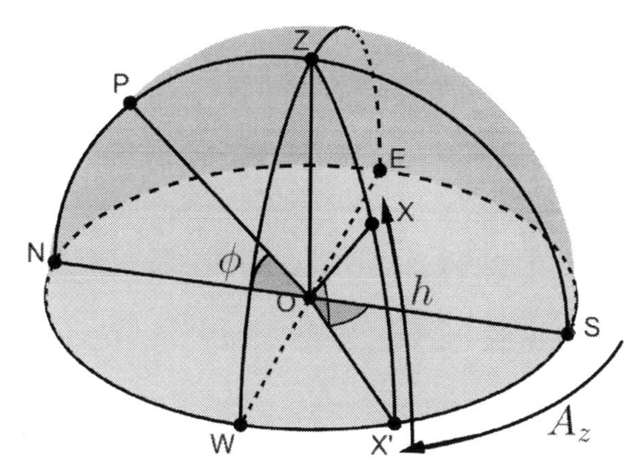

Figure 1.3: Horizontal coordinates of a star X.

Instead, we call the *true horizon* the one that is actually visible, which depends on the dip of the horizon and on the geographic landscape. Evidently, the nadir and the whole hemisphere below the true horizon are not observable. All the great circles passing through the zenith and nadir are called *verticals*. The great circle passing through the zenith (Z) and the celestial pole (P) is called the *principal vertical*. Its intersection with the horizon defines the points north (N), on the same side of the celestial pole, and south (S), at its antipode. The vertical circle at 90° from the principal vertical is called the *prime vertical*. Its intersections with the celestial horizon defines the points east (E) and west (W). The points north, east, south and west are called *cardinal points*. We call *upper* and *lower culmination*, respectively, the passage of a star on the principal vertical, to the south, or on the anti-meridian, to the north.
Let X be a star on the celestial sphere and let X′ denote its projection on the horizontal plane (Fig. 1.3). Its horizontal coordinates are:

- altitude h, i.e. the angle XX′, counted in degrees starting from the horizon, and ranging from 0° to 90° , if the object is above the horizon, and from 0° to −90°, if it is below the horizon;
- azimuth A_z, i.e. the angle SX′, counted clockwise from 0° to 360°.

In the horizontal system, each minor circle parallel to the horizon is called an *almucantar*. The altitude of the celestial pole with respect to the horizon is called *astronomical latitude*, because it is numerically equal to the geographical latitude. The horizontal system offers the advantage of being intuitive and easy to set up, but it can be impractical because both the altitude h and the azimuth A_z depend on the position and time of observation. In the next section we will see how these issues can be overcome.

1.3 The Equatorial System

There are two equatorial coordinate systems: HA-dec and RA-dec. Both use the declination δ, which can be regarded as the equivalent to the altitude h for the horizontal system. In the HA-dec equatorial system, the second coordinate is the *hour angle* H, which increases linearly with time; in the RA-dec system it is the *right ascension* α, independent of time.

The First Equatorial System, or HA-dec

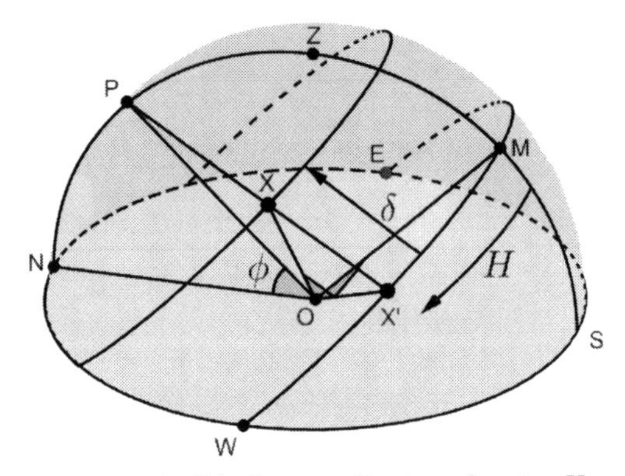

Figure 1.4: HA-dec coordinates of a star X.

As the name suggests, in this coordinate system the fundamental plane is the *celestial equator*, perpendicular to Earth's rotation axis. The great circles passing through the celestial north and south poles are called *hour circles*, as they all have a constant hour angle. Minor circles parallel to the celestial equator are called *parallels of declination*. Fig. 1.4 shows what the sky looks like to an observer at latitude ϕ who uses the first equatorial system. Let X be a point on the celestial sphere and let X$'$ denote its projection on the equator. In the first equatorial system, its coordinates are:

- declination δ, i.e. the angle XX$'$, counted in degrees starting from the celestial equator, and ranging from $0°$ to $90°$, if the object is above the equator (in the direction of the celestial north pole), and from $0\,°$ to $-90\,°$, if it is below the equator;
- hour angle H, i.e. the angle MX$'$, counted clockwise from M and measured in hours, from 0^h to 24^h.

The reason why the hour angle is counted in hours, instead of degrees, will be clarified in the next section. In this coordinate system only the hour angle H varies with time, increasing steadily because of Earth's rotation. Instead, the declination δ is constant to a first approximation, since each point on the celestial sphere moves on a parallel, and therefore at a constant distance from the celestial equator. This is true only for fixed stars, while the declination of the Sun, Moon, planets and comets is not constant.

The Second Equatorial System, or RA-dec

We now want to find a coordinate α that is completely independent of time, at least to a first approximation. In the HA-dec system, the only time-dependent coordinate was the hour angle. Since the rotation of the Earth is solely responsible for the time variation of the hour angle, it is necessary for the reference point from which we measure α to move in the sky with the same direction and velocity as the (apparent) angular velocity of rotation of the celestial sphere. Let us consider the two points of intersection between the celestial equator and the *ecliptic* (the projection in the sky of the plane of Earth's orbit). For an observer on Earth, the celestial equator is fixed while the ecliptic rotates from east to west (in the same direction as the rotation of the celestial sphere); hence their intersection points rotate from east to west, with the same direction and velocity as the angular velocity of rotation of the celestial sphere. Thus, the distances between a fixed star and the intersection points remain constant. One of these points is called *vernal equinox* or *first point of Aries*, because it used to be in the constellation of Aries (see Sec. 3.1), and is denoted by the symbol ♈ (the same symbol used for the constellation of Aries). At the antipode of the vernal equinox, we find the *autumnal equinox*, denoted by the symbol ♎ (the same symbol used for the constellation of Libra).
Let X be a point on the celestial sphere and let X′ denote its projection on the celestial equator. Its second equatorial coordinates are:

- declination δ, defined as for the first equatorial system;
- right ascension α, i.e. the angle ♈X′, counted anti-clockwise from ♈ and measured in hours, from 0^h to 24^h.

By definition, the declination of the Sun is zero when it passes through the points ♈ and ♎. The Sun, however, only transits on the celestial equator during the equinoxes. We can therefore define ♈ and ♎ as the points through which the Sun transits during the vernal and autumnal equinoxes, respectively (hence their names). At the vernal equinox, the declination of the Sun increases daily (since it is approaching the summer solstice), while it decreases at the autumnal equinox. For this reason, ♈ and ♎ are also called *ascending* and *descending nodes*, respectively.
Looking at Fig. 1.5, denoting with ST the hour angle of the vernal equinox, and with H the hour angle of the object X under consideration, the following relationship holds:

$$\boxed{\mathrm{ST} = \alpha + H}\,. \tag{1.2}$$

As will be seen in more detail in Sec. 5.1, ST is called the *sidereal time*.

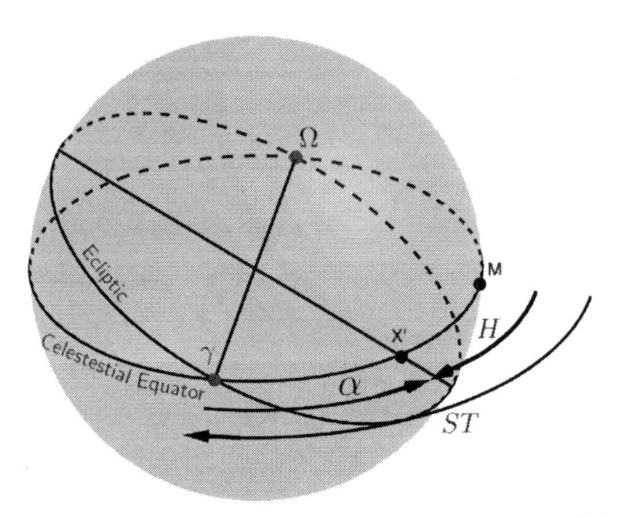

Figure 1.5: The sidereal time is equal to the sum of the hour angle and the right ascension of any object

Because it is a unit of time, the sidereal time is measured in hours; therefore it is convenient to measure H and α in hours as well. A star transiting on the meridian has an hour angle of $H = 0^h$, therefore a clock calibrated in sidereal time gives us the right ascension of every object on the meridian. To find the object in question at any other time, we need its hour angle, but this can be found from the sidereal time and the right ascension, using Eq. 1.2.
The coordinates α and δ are constant to the first order of approximation. However, in the long run, perturbation phenomena (chapter 3), in addition to the proper motion of the stars, cause them to vary.

1.4 The Ecliptic System

In the ecliptic coordinate system the fundamental direction is the normal to the plane of the ecliptic, that is, the normal to Earth's plane of revolution around the Sun. Let X be a point on the celestial sphere and let X′ be its projection on the plane of the ecliptic. Its ecliptic coordinates are:

- ecliptic longitude λ, i.e. the angle ♈X′, counted anti-clockwise starting from ♈ and measured in degrees;
- ecliptic latitude β, i.e. the angle XX′, measured in degrees starting from X′.

As will be seen in chapter 5, the ecliptic system is useful in defining the length of a *tropical year*.

1.5 The Galactic System

The galactic coordinate system is of great importance in extragalactic studies. Unlike other reference systems which are centred on Earth, the centre of the galactic system is the Sun and fundamental plane is that of the Milky Way, defined as the region of maximum emission of the neutral hydrogen 21 cm line. The galactic north pole is in the direction of the Coma galaxy cluster ($\alpha_{\rm np} = 12.8^h$, $\delta_{\rm np} = 27.4°$), while the galactic centre is located in the constellation of Sagittarius ($\alpha_{\rm gc} = 17^h 42.4^m$, $\delta_{\rm gc} = -28°55'$). Similarly to the other coordinate systems, the galactic coordinates are the galactic longitude l, measured counter-clockwise from the galactic centre, and the galactic latitude b, measured from the galactic plane and positive in the northward direction.

1.6 Exercises

1. (INT 2014, Th.S, q.4) The first accurate measurement of the size of the Earth was made by Eratosthenes of Cyrene and was obtained by measuring the difference in the altitude of the Sun at the summer solstice in two locations at a known distance. Assuming that the Earth is spherical, what is the length of the arc between two locations at latitude $+35°$ and $+45°$ (with equal longitude)?

2.* What is the angle formed by the equator with the horizon for an observer at latitude ϕ?

3.* What is the condition whereby the altitude of a celestial object remains unchanged over the course of a day?

4. What is the right ascension and declination of the vernal equinox?

5.* What is the angle formed by the ecliptic and the horizon for an observer at the north pole?

6.* What is the condition whereby the ecliptic north pole coincides with the zenith of an observer?

7.* At what latitudes are Mercury and Venus best seen in the evening or morning?

8.* Where is a star located, if its right ascension is equal to its ecliptic longitude?

9. Find the hour angle of Capella ($\alpha = 5^h 14^m 28^s$) at ST$= 4^h 20^m 13^s$.

10.* The Burj Khalifa, in the United Arab Emirates, is over 829m tall. In the absence of obstacles and absorption of light, from what distance would the skyscraper be visible?

11.* In January 2019 Uranus had an ecliptic longitude of 43°. What longitude did the planet have in 1781, the year of its discovery, and which constellation was it in?

1.7 Problems

1.** **The Seasons**
The hottest period in the northern hemisphere falls when the Earth is near aphelion. Explain why the winter in the northern hemisphere is milder than the winter in the southern hemisphere, while the summer in the southern hemisphere is slightly warmer. Attempt a quantitative analysis (Hint: read Ch. 8).

2.** **Rotation of the Earth**
An object dropped from an appreciable height does not land exactly on the vertical, but its trajectory appears displaced in the direction of the Earth's rotation. This is due to the Coriolis force, or more intuitively to the fact that the angular velocity at the launch point is the same as that at the surface of the Earth, and therefore the tangential velocity of the body is greater than the velocity of the surface. If the object is dropped from the Leaning Tower of Pisa ($h = 57\,\mathrm{m}$, $\phi = 44°$), how far does it land from where it was dropped?

3.* **Foucault's Pendulum**
Because of Earth's rotation, the plane of motion of a pendulum rotates slowly. Prove that the precession period of Foucault's pendulum is:

$$T = \frac{T_{\rm sid}}{\sin\phi}, \tag{1.3}$$

where $T_{\rm sid}$ is the time of rotation of the Earth and ϕ is the latitude.

4.* **Dip of the horizon with refraction**
So far we have ignored refraction, and drawn the line of sight to the apparent horizon as a straight line. However, in the real world, refraction is non-negligible: this causes the line of sight to the apparent horizon to be curved, not straight. Assuming we can regard the ray to the horizon as an arc of a circle, with a curvature k times the Earth's curvature, show that Eq. 1.1 still holds if we replace R by $R_{\rm eff}$, and find an expression for $R_{\rm eff}$ in terms of R and k.

2

Transformation of Coordinates

In astronomy it is very useful to change from one coordinate system to another. For example, it may be necessary to convert equatorial to horizontal coordinates, to know where to find a star. This chapter may be more challenging than others, however it is sufficient to keep in mind only those equations which will be obtained in Sec. "Upper and Lower Culmination". These can also be derived more intuitively, as will be shown in Pr. 2.1.

2.1 The Three Equations

Since all astronomical coordinate systems measure angles on a sphere, instead of the usual Cartesian system, in astronomy it is particularly convenient to use spherical polar coordinates. In the Cartesian system, the position of a point in space is determined solely by the three coordinates x, y and z. In spherical polar coordinates, however, the position of a point P can be written as a function of r, ψ and θ (see Fig. 2.1), where:

- r is the distance of P from the origin O;
- ψ is the angle between the x axis and the projection of r on the xy plane;
- θ is the angle between r and the z axis.

Starting from the definition above, it is possible to write x, y, z as a function of r, ψ, θ. Looking at Fig. 2.1, we see that $z = \overline{\mathrm{OH}_z} = \overline{\mathrm{OP}} \cos\theta = r\cos\theta$. OH is the projection of OP on the xy plane, therefore $\overline{\mathrm{OH}} = \overline{\mathrm{H}_z\mathrm{P}} = \overline{\mathrm{OP}} \sin\theta = r\sin\theta$. Thus, $x = \overline{\mathrm{OH}_x} = \overline{\mathrm{OH}} \cos\psi = r\cos\psi\sin\theta$ and $y = \overline{\mathrm{OH}_y} = \overline{\mathrm{OH}} \sin\psi = r\sin\psi\sin\theta$. To sum up:

$$\begin{cases} x = r\cos\psi\sin\theta \\ y = r\sin\psi\sin\theta \\ z = r\cos\theta \end{cases} .$$

Let xyz and $x'y'z'$ be two Cartesian systems, where $x'y'z'$ is obtained from xyz by a rotation of angle χ around its y axis.

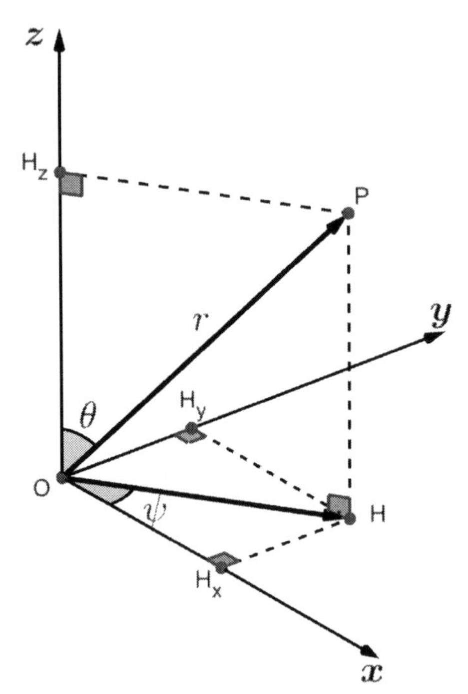

Figure 2.1: Transforming Cartesian coordinates x, y, z to spherical polar coordinates r, ψ, θ.

As can be seen in Fig. 2.2, the two coordinate systems are related by the following equations:

$$\begin{cases} x' = x\cos\chi + z\sin\chi & (2.1a) \\ y' = y & (2.1b) \\ z' = z\cos\chi - x\sin\chi \,. & (2.1c) \end{cases}$$

Writing xyz and $x'y'z'$ in spherical polar coordinates, setting $r = 1$ for simplicity:

$$\begin{cases} x = \cos\psi\sin\theta \\ y = \sin\psi\sin\theta \\ z = \cos\theta \end{cases} \qquad \begin{cases} x' = \cos\psi'\sin\theta' \\ y' = \sin\psi'\sin\theta' \\ z' = \cos\theta' \end{cases} .$$

Substituting the last two systems of equations in Eqs. 2.1, we find:

$$\begin{cases} \cos\psi'\sin\theta' = \cos\psi\sin\theta\cos\chi + \cos\theta\sin\chi & (2.2a) \\ \sin\psi'\sin\theta' = \sin\psi\sin\theta & (2.2b) \\ \cos\theta' = \cos\theta\cos\chi - \cos\psi\sin\theta\sin\chi \,. & (2.2c) \end{cases}$$

These three equations will be sufficient to determine all coordinate transformations. With this in mind, in the next sections we will identify to what angles χ, ψ, θ, ψ' and θ' correspond in the various celestial coordinate systems, and then rewrite Eqs. 2.2 accordingly.

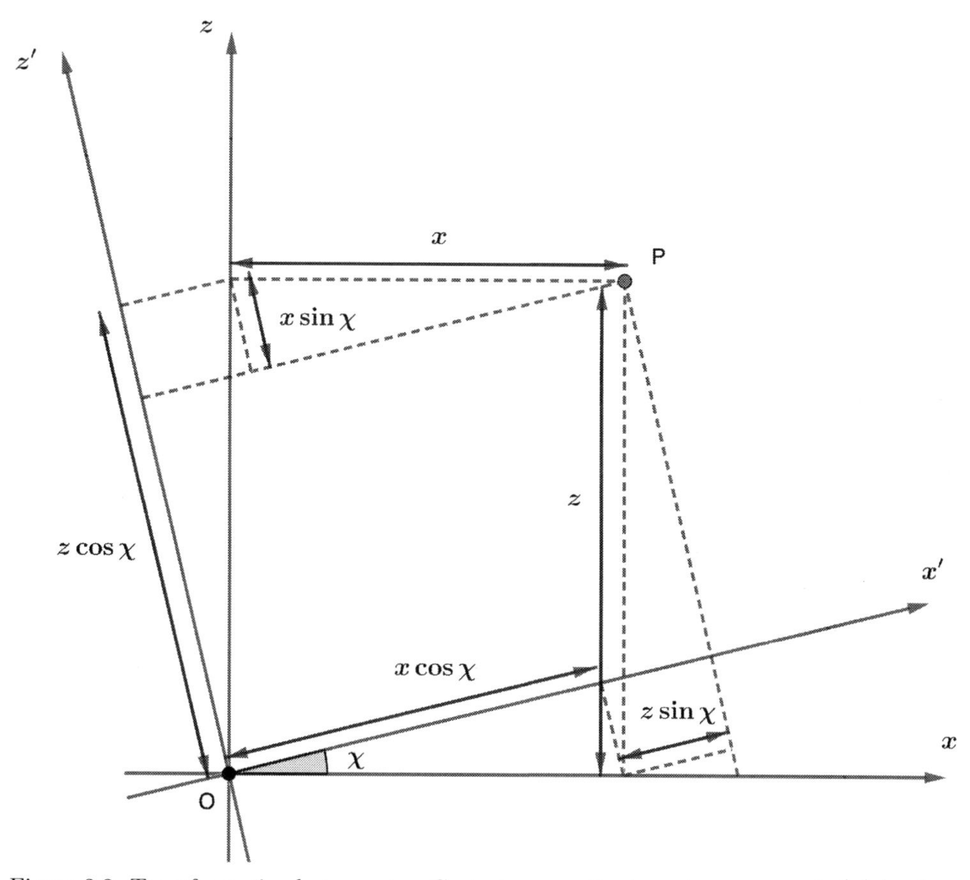

Figure 2.2: Transformation between two Cartesian coordinate systems xyz and $x'y'z'$, where the latter is obtained from xyz by a rotation of angle χ around its y axis.

2.2 From the Horizontal to the HA-dec System

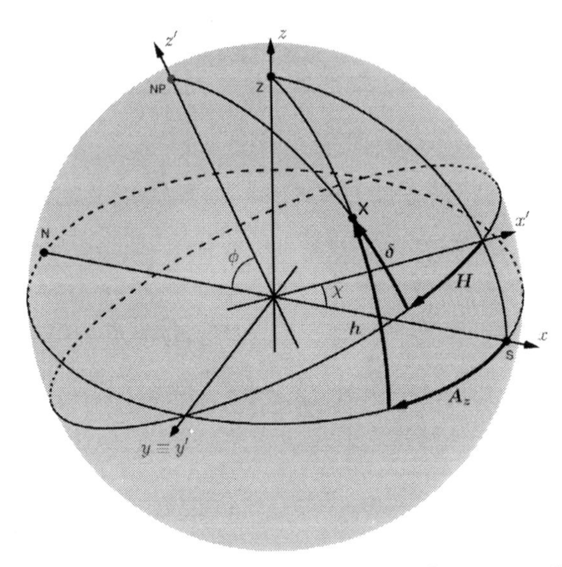

Figure 2.3: Transformation from the horizontal system to HA-dec.

We would now like to find a relationship between the angles ψ, ψ', θ, θ', χ that appear in Eqs. 2.2 , and the angles ϕ, A_z, H, h, δ in Fig. 2.3. In the horizontal and HA-dec systems, A_z and H are measured clockwise, but in the previous section we defined the angles ψ and ψ' anti-clockwise. Therefore $\psi = -A_z$ and $\psi' = -H$. The angles θ and θ' are measured starting from the z axis, while the coordinates h and δ are measured starting from the fundamental planes, so it is necessary to consider the complementaries: $\theta = 90\,^\circ - h$ and $\theta' = 90^\circ - \delta$. Finally, the angle between the two fundamental planes is $\chi = 90\,^\circ - \phi$. Summarizing:

$$\begin{cases} \psi = -A_z \\ \psi' = -H \\ \theta = 90^\circ - h \\ \theta' = 90^\circ - \delta \\ \chi = 90^\circ - \phi\,. \end{cases}$$

Substituting the above equations in Eqs. 2.2 and using the reduction formulae (Appendix A.3), we find:

$$\begin{cases} \cos H \cos\delta = \cos A_z \cos h \sin\phi + \sin h \cos\phi & \text{(2.3a)} \\ \sin H \cos\delta = \sin A_z \cos h & \text{(2.3b)} \\ \sin\delta = -\cos A_z \cos h \cos\phi + \sin h \sin\phi\,. & \text{(2.3c)} \end{cases}$$

The above equations allow us to change from the horizontal to the HA-dec system. To carry out the inverse transformation, we exchange the roles of xyz

and $x'y'z'$. In doing so, we must remember to change the sign of χ:

$$\begin{cases} \psi = -H \\ \psi' = -A_z \\ \theta = 90^\circ - \delta \\ \theta' = 90^\circ - h \\ \chi = -(90^\circ - \phi)\,. \end{cases}$$

We then obtain:

$$\begin{cases} \cos A_z \cos h = \cos H \cos\delta \sin\phi - \sin\delta\cos\phi & \text{(2.4a)} \\ \sin A_z \cos h = \sin H \cos\delta & \text{(2.4b)} \\ \sin h = \cos H \cos\delta\cos\phi + \sin\delta\sin\phi\,. & \text{(2.4c)} \end{cases}$$

Since the altitude and declination are in the range $[-90^\circ, 90^\circ]$, it suffices to know the sine of one of these angles to determine the other angle unambiguously. Azimuth and right ascension, however, can take any value from 0° to 360° (or from 0^h to 24^h), so we need both their sines and cosines to determine them completely.

Upper and Lower Culmination

For an observer in the northern hemisphere, the altitude of a celestial body is maximum when it passes through the meridian, in the southern direction. In this case, the body is said to be in upper culmination and its hour angle is $H = 0^h$. The altitude is minimum when the object is passing through the anti-meridian. Now, the body is said to be in lower culmination and its hour angle is $H = 12^h$.

In the derivation that follows, we consider only the case of an observer in the northern hemisphere, so $\phi \geqslant 0^\circ$. At the end of this section, we will generalize the result for an observer in the southern hemisphere. When $H = 0^h$, using Eq. 2.4c, we find:

$$\sin h = \cos\delta\cos\phi + \sin\delta\sin\phi = \cos(\phi - \delta) = \sin(90^\circ - \phi + \delta)\,,$$

where we used the cosine addition formula (Eq. A.16). Therefore:

$$\sin h = \sin(90^\circ - \phi + \delta)\,.$$

The previous equation has two solutions, since $\sin(180^\circ - \alpha) = \sin\alpha$. Therefore, the altitude of upper culmination of an object with declination δ, measured by an observed at a latitude $\phi > 0^\circ$, is:

$$\boxed{h_u = \begin{cases} 90^\circ - \phi + \delta, & \text{if the object culminates south of zenith;} \\ 90^\circ + \phi - \delta, & \text{if the object culminates north of zenith.} \end{cases}} \qquad (2.5)$$

The above formulae give the altitude of the object as measured from the nearest horizon (north or south). However, unless otherwise stated, the altitude of upper culmination is measured from the southern horizon by convention, therefore it is sufficient to use the first formula. When $H = 12^h$, Eq. 2.4c implies:

$$\sin h = -\cos\delta\cos\phi + \sin\delta\sin\phi = -\cos(\delta+\phi) = \sin(\delta+\phi-90^\circ)$$

Therefore:

$$\sin h = \sin(\delta+\phi-90^\circ)$$

The altitude of lower culmination, measured by convention from the northern horizon, is then:

$$\boxed{h_l = \delta + \phi - 90^\circ} \tag{2.6}$$

Eqs. 2.5 and 2.6 can be obtained using only elementary geometry, as shown in Pr. 2.1. Let us examine some special cases:

- when $h_l > 0$, i.e. $\delta > 90^\circ - \phi$, the object is *circumpolar* and it never sets;
- when $h_u < 0$, i.e. $\delta < \phi - 90^\circ$, it is never visible;
- when $\phi - 90^\circ < \delta < 90^\circ - \phi$, the object rises and sets.

For example, in Helsinki ($\phi = 60^\circ$), all stars with a declination greater than 30° never set and are therefore always visible. On the other hand, stars with a declination less than -30° can never be observed. Stars with a declination between those two values rise and set. At the north pole, all stars with $\delta > 0^\circ$ are circumpolar, while those with $\delta < 0^\circ$ are never visible. At the equator, no star is circumpolar, but all the stars rise and set. Atmospheric refraction widens the visible horizon, making stars with declination close to $\delta = 90^\circ - \phi$ circumpolar and those with a declination close to $\delta = \phi - 90^\circ$ visible.
Imagine we observe a circumpolar star and record its altitudes h_u and h_l (measured from the southern and northern horizon) during upper and lower culmination, respectively. Taking the average between the two altitudes, we obtain the declination of the star:

$$\boxed{\delta = \frac{1}{2}(h_u + h_l)}\,. \tag{2.7}$$

If, instead, we take the difference:

$$\boxed{\phi = 90^\circ - \frac{1}{2}(h_u - h_l)}\,, \tag{2.8}$$

we obtain the geographical latitude.

All the previous equations also apply to the southern hemisphere if we substitute $\delta \to -\delta$ and $\phi \to -\phi$. Indeed, by symmetry, the southern hemisphere should be equivalent to the northern hemisphere under sign exchange (the northern hemisphere has a positive sign by convention). Therefore, a star is circumpolar in the southern hemisphere if $\delta < (90° + \phi)$, while it never rises if $\delta > 90° + \phi$. For example in Sydney ($\phi = -34°$), all stars with a declination less than $-56°$ are circumpolar, while stars with a declination greater than $56°$ are never visible. At the south pole, all stars with $\delta < 0°$ are circumpolar, while stars with $\delta > 0°$ are never visible.

Rising and Setting Times

We want to obtain the rising and setting times of an object, that is, the two moments when its altitude is zero. Let's assume we know its right ascension is α. To calculate the sidereal time, with the equation $\mathrm{ST} = \alpha + H$, the only information we are missing is its hour angle. Starting from Eq. 2.4c, we isolate $\cos H$ (for $\cos\delta, \cos\phi \neq 0°$):

$$\cos H = -\tan\delta\tan\phi + \frac{\sin h}{\cos\delta\cos\phi}.$$

If we neglect atmospheric refraction, we can substitute $h = 0°$:

$$\cos H = -\tan\delta\tan\phi\,. \tag{2.9}$$

The previous equation allows us to obtain the hour angle at the moment of rising and setting, knowing the latitude of the observer and the declination of the object. Knowing its right ascension, it is then possible to obtain the sidereal time using Eq. 1.2, and therefore also the local time (see Ch. 5). Atmospheric refraction increases the visible horizon by about $-34'$. In the case of extended bodies like the Sun and the Moon, we must also consider their angular radii.

Duration of Day and Night

During the vernal and autumnal equinoxes the declination of the Sun is $\delta_\odot = 0°$, therefore, from Eq. 2.9, we find that the hour angles of rising and setting of the Sun are $H = -90°$ and $H = +90°$, which are the coordinates of east and west, respectively. Hence, during the equinoxes, the Sun rises exactly in the east and sets exactly in the west, and the day lasts 12^h everywhere on Earth. From Eq. 2.9, we see that at the equator, the Sun rises exactly in the east and sets exactly in the west every day of the year, thus the durations of day and night are always the same. At the north pole, the altitude of the Sun is always equal to its declination, hence the Sun is circumpolar and the day lasts 24^h when its declination is positive, whereas the Sun never rises and it is

permanently night when its declination is negative.
Using Eq. 2.9, we can obtain the duration of the day for a location at a latitude ϕ, when the Sun's declination is $\delta_\odot$. The Sun is visible for an hour angle of $2\arccos(-\tan\delta\tan\phi)$, or converted into hours:

$$t_{\text{day}} = 24^h \cdot \left[1 - \frac{1}{180^\circ}\arccos(\tan\delta_\odot\tan\phi)\right], \tag{2.10}$$

where we used $\arccos(-x) = 180^\circ - \arccos x$, since $\cos(180^\circ - x) = -\cos x$. In the case when $|\tan\delta_\odot\tan\phi| > 1$, the last identity cannot be used since $\arccos x$ is not defined for arguments greater than unity. In practice, this means that there is no point of rising and setting, i.e. the Sun is either circumpolar or never visible. Let us prove this statement.
If $\tan\delta_\odot\tan\phi > 1$, then $\tan\delta_\odot > 1/\tan\phi$ and, using $1/\tan\phi = \tan(90^\circ - \phi)$, we find $\tan\delta_\odot > \tan(90^\circ - \phi)$. In turn, this implies that $\delta_\odot > 90^\circ - \phi$, which is the condition for the Sun to be circumpolar in the northern hemisphere. Thus, if $\tan\delta_\odot\tan\phi > 1$, the Sun is circumpolar and the day lasts 24^h.
Similarly, if $\tan\delta_\odot\tan\phi < -1$, then $\tan\delta_\odot < -1/\tan\phi$ and, using $-1/\tan\phi = \tan(\phi - 90^\circ)$, we have $\tan\delta_\odot < \tan(\phi - 90^\circ)$. This implies that $\delta_\odot < \phi - 90^\circ$, hence the Sun never rises. Therefore, if $\tan\delta_\odot\tan\phi < -1$, the Sun never rises and the duration of the day is 0^h. To sum up, in the northern hemisphere:

$$t_{\text{day}} = \begin{cases} 24^h \cdot \left[1 - \frac{1}{180^\circ}\arccos(\tan\delta_\odot\tan\phi)\right] & \text{if} \quad \phi - 90^\circ \leqslant \delta_\odot \leqslant 90^\circ - \phi \\ 24^h & \text{if} \quad \delta_\odot > 90^\circ - \phi \\ 0^h & \text{if} \quad \delta_\odot < \phi - 90^\circ\,. \end{cases}$$

Similarly, for the southern hemisphere, $\tan\delta_\odot\tan\phi > 1$ implies $\delta > 90^\circ + \phi$, which is the condition for a star to never rise. Instead, if $\tan\delta_\odot\tan\phi < -1$, then $\delta_\odot < -(90^\circ + \phi)$, so the Sun is circumpolar. Therefore, in the southern hemisphere:

$$t_{\text{day}} = \begin{cases} 24^h \cdot \left[1 - \frac{1}{180^\circ}\arccos(\tan\delta_\odot\tan\phi)\right] & \text{if} \quad -(90^\circ + \phi) \leqslant \delta_\odot \leqslant 90^\circ + \phi \\ 24^h & \text{if} \quad \delta_\odot < -(90^\circ + \phi) \\ 0^h & \text{if} \quad \delta_\odot > 90^\circ + \phi\,. \end{cases}$$

2.3 From The Ecliptic to the RA-dec System

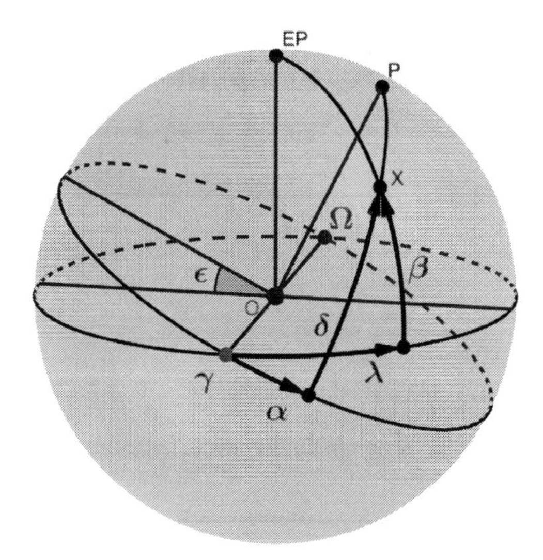

Figure 2.4: Transformation from the ecliptic system to RA-dec.

The equations for changing between the ecliptic and the RA-dec coordinate systems can be derived in a similar way to the equations converting horizontal to HA-dec coordinates. From Fig. 2.4, we see that:

$$\begin{cases} \psi = \alpha - 90^\circ \\ \psi' = \lambda - 90^\circ \\ \theta = 90^\circ - \delta \\ \theta' = 90^\circ - \beta \\ \chi = \epsilon , \end{cases}$$

where $\epsilon = 23°27'$ is the *obliquity of the ecliptic*, defined as the angle between the celestial equator and the ecliptic. Substituting in Eqs. 2.2, we find:

$$\begin{cases} \sin\lambda\cos\beta = \sin\delta\sin\epsilon + \cos\delta\cos\epsilon\sin\alpha & (2.11a) \\ \cos\lambda\cos\beta = \cos\delta\cos\alpha & (2.11b) \\ \sin\beta = \sin\delta\cos\epsilon - \cos\delta\sin\epsilon\sin\alpha\,. & (2.11c) \end{cases}$$

To obtain the inverse transformation, we swap the coordinates:

$$\begin{cases} \psi = \lambda - 90^\circ \\ \psi' = \alpha - 90^\circ \\ \theta = 90^\circ - \beta \\ \theta' = 90^\circ - \delta \\ \chi = -\epsilon\,. \end{cases}$$

Hence, we obtain:

$$\begin{cases} \sin\alpha\cos\delta = -\sin\beta\sin\epsilon + \cos\beta\cos\epsilon\sin\lambda & \text{(2.12a)} \\ \cos\alpha\cos\delta = \cos\lambda\cos\beta & \text{(2.12b)} \\ \sin\delta = \sin\beta\cos\epsilon + \cos\beta\sin\epsilon\sin\lambda\,. & \text{(2.12c)} \end{cases}$$

Declination of the Sun throughout the year

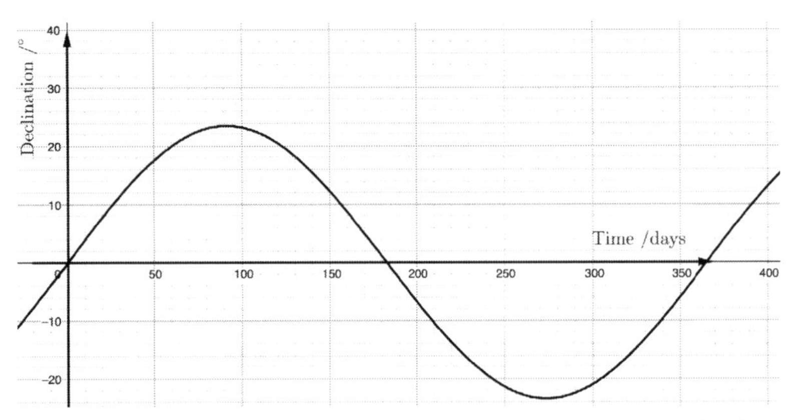

Figure 2.5: Declination of the Sun throughout the year starting from the day of vernal equinox, in the approximation that the Earth's orbit is circular.

The Sun lies on the ecliptic, hence $\beta_\odot$ is constantly equal to zero. Furthermore, neglecting the eccentricity of the Earth's orbit, its ecliptic longitude is equal to $\lambda_\odot = (2\pi/T_t)t$, where T_t is the duration of the tropical year (see Ch. 5) and t is the time elapsed since the last vernal equinox (when $\lambda_\odot$ is zero by definition). Then, using Eq. 2.12c with $\beta_\odot = 0°$, we find:

$$\sin\delta_\odot = \sin\epsilon\sin\lambda_\odot \Rightarrow \delta_\odot = \arcsin\left[\sin\epsilon\sin\left(\frac{2\pi}{T_t}t\right)\right]. \tag{2.13}$$

This equation allows us to compute the declination of the Sun at any given time of the year, in the approximation that the Earth's orbit is circular (Fig. 2.5). In particular, we can verify that Eq. 2.13 gives the correct result in the following cases:

- vernal equinox $t = 0$, $\sin(2\pi t/T_{\mathrm{yr}}) = 0$ and $\delta_\odot = 0°$;
- summer solstice $t = T_{\mathrm{yr}}/4$, $\sin(2\pi t/T_{\mathrm{yr}}) = 1$ and $\delta_\odot = \epsilon$;
- autumn equinox $t = T_{\mathrm{yr}}/2$, $\sin(2\pi t/T_{\mathrm{yr}}) = 0$ and $\delta_\odot = 0°$;
- winter solstice $t = 3/4T_{\mathrm{yr}}$, $\sin(2\pi t/T_{\mathrm{yr}}) = -1$ and $\delta_\odot = -\epsilon$.

2.4 Exercises

1. What is the condition required for a star with declination δ to pass through the zenith of an observer at latitude ϕ?

2. At what latitude can you see Sirius ($\delta = -16°42'58''$) graze the horizon during lower culmination?

3.* What can the maximum and minimum altitudes of the Sun be, in a city at latitude ϕ?

4. You are given the upper ($h_u = 53\,°54'45''$) and lower ($h_l = 34\,°34'23''$) altitudes of culmination of a star. Find the latitude of the observer and the declination of the star.

5. At what latitudes is the Sun circumpolar at least once a year?

6. (CAO 2018, Th., q.6) On a sunny afternoon on the 22nd of June, an observer standing upright on a flat surface found that his shadow had a length equal to his height. At what latitude was the observer located?

7.* (IOAA 2016, Th., q.4) An observer in the northern hemisphere noticed that the length of the shortest shadow of a 1.000 m vertical stick on a day was $l_{\text{min}} = 1.732$ m. On the same day, the length of the longest shadow of the same vertical stick was measured to be $l_{\text{max}} = 5.671$m. Find the latitude ϕ of the observer and the declination of the Sun $\delta_{\odot}$ on that day. Assume the Sun to be a point source and ignore atmospheric refraction.

8.* What must the declination of a star be and where should it be observed so that its azimuth does not change for at least half a day?

9.* What are the ecliptic coordinates of a star with right ascension $\alpha = 6^h$ and declination $\delta = 10\,°$?

10. What are the HA-dec coordinates of the four cardinal points and the zenith for an observer located in the northern hemisphere? What about an observer in the southern hemisphere?

11. Where do stars which stay above the horizon for more than 12 hours a day rise?

12. (INT 2011, Th.S, q.1) Marco and Gianna are two young astronomy enthusiasts from Rome. Gianna, who is visiting an important foreign capital, calls Marco to greet him. At one point Gianna remarks "It is a beautiful evening, the Moon is rising right now!". Marco then replies: "Great! Here the Moon is passing exactly on the meridian, so I know precisely at what longitude you are!". If Marco's longitude in Italy is $12\,°30'$ E, what is the longitude of Gianna? Can you tell which city she is in? Neglect the effect due to the finite distance of the Moon from the Earth (lunar parallax).

13.* Estimate the declination of the Sun on the 21st of April. At what latitudes is the Sun circumpolar?

14.* A star passes through the zenith at 0^h10^m of sidereal time, whereas its altitude on the horizon is $78°12'$ at 9^h2^m of sidereal time. Calculate the latitude of the observer.

15.* You want to spend a romantic evening with your partner, so you take them to the beach to see the sunset with a mechanical lift. The elevator can be raised to a height of 6 m, at a speed of v. What is the minimum v so that the sunset can be observed twice, once at sea level, and the second time at 6 m?

2.5 Problems

1.* **Upper and Lower Culmination**

Obtain the formulae for the upper and lower culmination of a star in two different ways using only elementary geometry.

In the first case, consider an inertial frame in two instants separated by 12^h. The direction of the star and of the Earth's axis are fixed, but the observation point has rotated by $180°$.

In the second case, consider the frame of the observer. Now the observer is fixed, while the star appears to be moving in a circle around the direction of the Earth's axis.

3

Perturbation of Coordinates

Even though a star can appear to be stationary with respect to the Solar System, its coordinates change over time due to various perturbation effects. Of course, the altitude and the azimuth are constantly changing, but the declination and the right ascension are also slowly varying.

3.1 Precession

The Sun and Moon, as well as the other planets in the Solar System, orbit very close to the ecliptic plane and tend to pull Earth's equatorial bulge towards this plane. The Earth reacts to this effect by rotating like a top around the axis perpendicular to the ecliptic. As a consequence, the Earth's axis traces out a pair of cones, with a period of 25765 years. This motion is called *precession of the equinoxes* since the vernal equinox, moving clockwise along the ecliptic by 50" per year, anticipates the date of the next equinox. As a result, the ecliptic longitude of all bodies on the celestial sphere increases by 50" each year.
In our century, Earth's rotation axis points very close to Polaris (α Ursae Minoris), about a degree away. In 3000 BC, Earth's axis pointed towards the fainter Thuban, in the constellation of Draco. In about $12,000$ years, it will point in the direction of Vega, which will then become the new pole star. Instead, the celestial south pole currently sits in a region of sky particularly clear of bright stars.
The equatorial coordinates are also affected by precession. To adjust for this variation, stellar catalogues must be updated every century. Currently, most maps and catalogues give the coordinates of stars at the time (or *epoch*) $J2000.0$, which corresponds to January 1, 2000.
We now derive the expressions that relate the change in declination and right ascension to the annual variation of ecliptic longitude of a celestial object, which is stationary with respect to the Solar System. We assume that only λ, δ and α vary over time (β is fixed because it is perpendicular to λ, the direction along which the precession proceeds). This is a good approximation, since ϵ changes much more slowly than the other factors involved (see Sec. 3.7).

Consider Eq. 2.12c:

$$\sin\delta = \sin\beta\cos\epsilon + \cos\beta\sin\epsilon\sin\lambda\,.$$

Differentiating both sides:

$$\cos\delta\, d\delta = \cos\beta\sin\epsilon\cos\lambda\, d\lambda\,.$$

Substituting $\cos\beta\cos\lambda = \cos\alpha\cos\delta$ (Eq. 2.12b), we find:

$$d\delta = d\lambda\,\sin\epsilon\cos\alpha\,. \tag{3.1}$$

This equation relates the variation $d\delta$ in declination to the variation $d\lambda$ in ecliptic longitude. If, instead, we differentiate Eq. 2.12b:

$$-\sin\alpha\cos\delta\, d\alpha - \cos\alpha\sin\delta\, d\delta = -\cos\beta\sin\lambda\, d\lambda\,.$$

Substituting $d\delta = d\lambda\,\sin\epsilon\cos\alpha$ (Eq. 3.1), it follows that:

$$\sin\alpha\cos\delta\, d\alpha = d\lambda\,(\cos\beta\sin\lambda - \sin\epsilon\cos^2\alpha\sin\delta)\,,$$

which, according to Eq. 2.11a, can also be written as:

$$\sin\alpha\cos\delta\, d\alpha = d\lambda\,(\sin\delta\sin\epsilon + \cos\delta\cos\epsilon\sin\alpha - \sin\epsilon\cos^2\alpha\sin\delta)\,.$$

Substituting $\cos^2\alpha = 1 - \sin^2\alpha$ (Eq. A.12, Appendix A.3) and simplifying:

$$d\alpha = d\lambda\,(\sin\alpha\sin\epsilon\tan\delta + \cos\epsilon)\,. \tag{3.2}$$

This equation relates the variation $d\alpha$ in right ascension to the variation $d\lambda$ in ecliptic longitude. Thus, given that $d\lambda/dt = 50''/\text{year}$, the annual changes $d\delta/dt$ and $d\alpha/dt$ for an object with declination δ and right ascension α are given by the equations:

$$\begin{cases} \dfrac{d\delta}{dt} = \dfrac{d\lambda}{dt}\sin\epsilon\cos\alpha & (3.3a) \\ \dfrac{d\alpha}{dt} = \dfrac{d\lambda}{dt}(\sin\alpha\sin\epsilon\tan\delta + \cos\epsilon)\,. & (3.3b) \end{cases}$$

The International Astronomical Union (IAU) divides the sky into 88 official constellations with precise boundaries, so that each point on the celestial sphere belongs to one and only one constellation. The constellations visible from northern latitudes mainly draw their origins from the Greek culture, and their names recall various mythological figures. During the vernal and autumnal equinoxes, the Sun passes through the points ♈ and ♎, respectively. At the time of the ancient Greeks this happened when the Sun was in the constellation of Aries and Libra, respectively (hence the use of these symbols). Today however, due to the precession of the equinoxes, the Sun passes through the zodiac constellations with a delay of about a month (corresponding to one constellation), compared to the convention adopted at the time of the Greeks. Hence, the points ♈ and ♎ have now moved to the adjacent constellations of Pisces and Virgo, respectively. To be precise, we should be using the symbols ♓ and ♍ for the vernal and autumnal equinoxes, but convention stuck with the tradition of the ancient Greeks. It might amuse you that astrology makes an error of about a month in associating your zodiac sign with the actual passage of the Sun through your constellation. In Tab. 3.1, the current dates of the Sun's passage through the zodiacal constellations are listed.

Capricornus ♑	20 January → 16 February
Aquarius ♒	16 February → 11 March
Pisces ♓	11 March → 18 April
Aries ♈	18 April → 13 May
Taurus ♉	13 May → 21 June
Gemini ♊	21 June → 20 July
Cancer ♋	20 July → 10 August
Leo ♌	10 August → 16 September
Virgo ♍	16 September → 30 October
Libra ♎	30 October → 23 November
Scorpius ♏	23 November → 29 November
Ophiuchus ⛎	29 November → 17 December
Sagittarius ♐	17 December → 20 January

Table 3.1

3.2 Nutation

Nutation is the oscillatory motion of Earth's rotation axis in the direction perpendicular to precession. Nutation is due to the fact that the precession angular momentum is added to that of the rotation and, consequently, the resulting angular momentum is not exactly directed along the axis of symmetry of the rotating object. The main component of nutation is associated to the regression of the Moon's nodal line and has the same period of 18.6 years. The amplitude of this motion is very small and in the order of 9.2".

3.3 Aberration

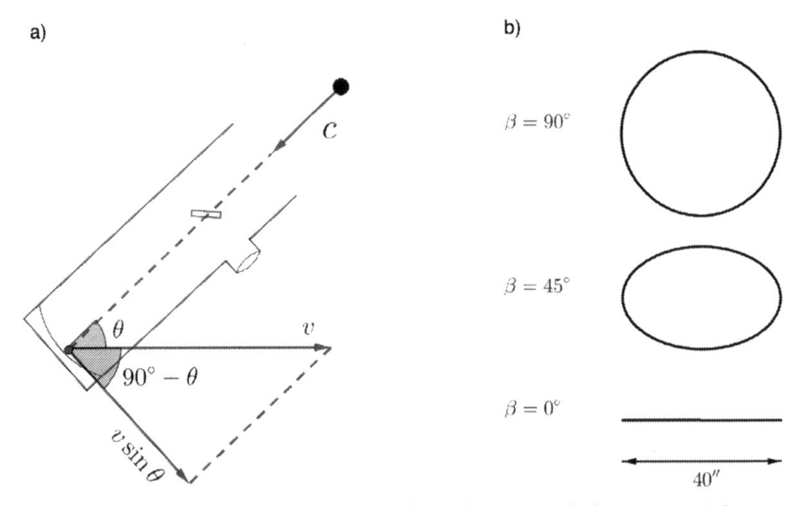

Figure 3.1: a) The angle of aberration is equal to the ratio of the tangential component of the velocity of the observer ($v \sin\theta$) to the velocity of light (c). b) Due to aberration, stars appear to be moving on ellipses whose eccentricity increases with decreasing ecliptic latitude.

Due to the finite speed of light, all stars appear slightly displaced in the direction of the observer's motion. Therefore, as a consequence of Earth's orbital motion, we see a periodic change in the apparent position of the stars, called *aberration*. In the framework of special relativity, it can be shown that, for small velocities ($v \ll c$), the aberration angle is approximately:

$$\boxed{A = \frac{v}{c}\sin\theta}, \tag{3.4}$$

where A is measured in radians and θ is the angle between the true position of the body and the direction of the observer's velocity. On Earth, the largest value of aberration is $v/c = 20''$, called the *constant of aberration.* Aberration associated with the rotation of the Earth is called *diurnal aberration*, but is usually negligible (only $0.3''$ in amplitude) compared to *annual aberration.*

Looking at Fig. 3.1 a), one can intuitively justify Eq. 3.4. Imagine a light ray propagating to a telescope. The projection of the velocity of the telescope along the direction perpendicular to the ray is $v \sin\theta$, whereas the velocity of the ray is c. Hence, stars appear displaced by an angle equal to the ratio of these velocities. Fig. 3.1 b) shows that, due to aberration, stars move on ellipses whose eccentricity increases with decreasing ecliptic latitude. Stars on the ecliptic poles describe a circle; those on the ecliptic, a segment.

3.4 Parallax

When looking at an object from different angles, we see it in different directions. This phenomenon, known as *parallax*, leads to a periodic variation of the coordinates of celestial objects during the year. The angle of parallax is related to the object's distance from Earth, therefore by measuring the angle of parallax we can obtain its distance. Parallax will be discussed in more detail in Ch. 9.

3.5 Refraction

Light is refracted by the atmosphere, therefore the apparent position of celestial objects differs from the true one. Even though refraction varies significantly with temperature and pressure, we can assume that conditions are roughly constant at the same height, i.e. that the refractive index of air depends only on the height above ground. Near the zenith, Earth's curvature is negligible and we can therefore divide the atmosphere into infinitesimal layers of air parallel to each other. Let α be the angle at which the ray enters the atmosphere and β the angle at which we observe it (measured from the zenith). The *refractive index* of a medium is defined as the ratio of the speed of light in vacuum to the speed of light in the medium. The refractive index of space (effectively a vacuum) is $n_s = 1$, whereas the refractive index of Earth's atmosphere at sea level is about $n_0 = 1.0003$. The refractive index is always greater than unity. A ray of light that travels through a medium of refractive index n_i and is incident on a surface of refractive index n_r, with angle α_i measured relative to the normal, is refracted by an angle α_r, such that:

$$\boxed{\frac{\sin\alpha_r}{\sin\alpha_i} = \frac{n_i}{n_r}}\,. \tag{3.5}$$

This equation is called *Snell's law*. Looking at Fig. 3.2, we apply Eq. 3.5 to a light ray entering the atmosphere:

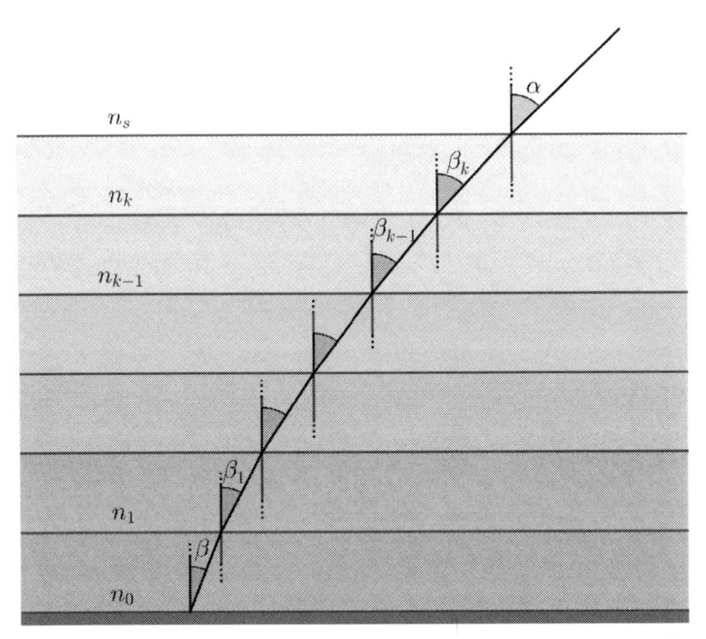

Figure 3.2: Refraction of a light ray passing through the atmosphere.

$$\begin{aligned} n_s \sin\alpha &= n_k \sin\beta_k\,, \\ n_k \sin\beta_k &= n_{k-1} \sin\beta_{k-1}\,, \\ &\dots, \\ n_1 \sin\beta_1 &= n_0 \sin\beta\,. \end{aligned}$$

Or, equivalently:

$$\sin\alpha = n_0 \sin\beta\,,$$

where we have used $n_s = 1$. Let $R = \alpha - \beta$ be the angle of deviation. Close to the zenith, $R \ll 1\,\text{rad}$, therefore $\sin R \approx R$ and $\cos R \approx 1$, giving:

$$\begin{aligned} \alpha &= R + \beta \\ \Rightarrow \sin\alpha &= \sin(R+\beta) = \sin R \cos\beta + \cos R \sin\beta \\ &\approx R\cos\beta + \sin\beta\,. \end{aligned}$$

Substituting $\sin\alpha = n_0 \sin\beta$ in the last equation, we find:

$$R = (n_0 - 1)\tan\beta\,. \tag{3.6}$$

The above equation is only valid close to the zenith, where Earth's curvature can be neglected. At the horizon $R \approx 34'$, slightly larger than the angular diameter of the Sun. When the lower part of the Sun grazes the apparent horizon, the Sun has already set on the astronomical horizon.

3.6 Proper Motion

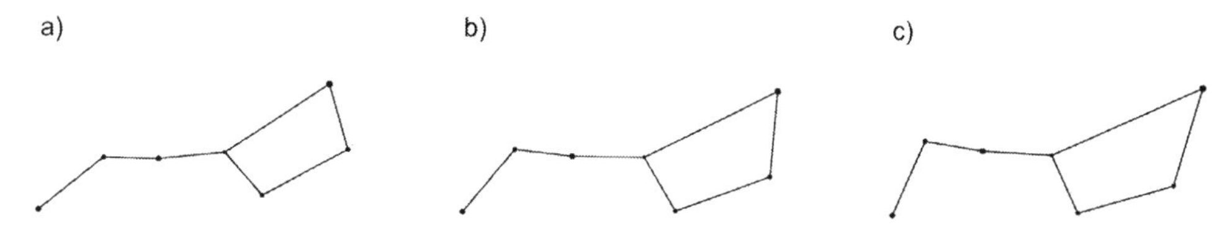

Figure 3.3: The proper motion changes the shape of the constellations. a) Big dipper today, b) in 30000 years and c) in 60000 years.

Many stars move in directions that seem to be constant in time. This effect is due to the relative motion between the Sun and the star, and is therefore called *proper motion*. Due to this motion, the shape of the constellations changes slowly with time, as shown in Fig. 3.3.
The speed of a star relative to the Sun can be divided into two components, one radial and one tangential. The tangential component is responsible for the proper motion, which can be measured by taking pictures of the star at different moments in time. The tangential motion can be further divided into the component that changes the right ascension ($u_\alpha \cos\delta$), and the component that changes the declination (u_δ). The factor $\cos\delta$ appears because a star with declination δ moves on a parallel of declination, which has a radius of $\cos\delta$. Therefore, if the rate of change of the right ascension measured on the celestial equator is u_α, the velocity on the parallel of declination is actually $u_\alpha \cos\delta$. To sum up, the total angular velocity of a star is:

$$u = \sqrt{u_\alpha^2 \cos^2\delta + u_\delta^2}\,. \tag{3.7}$$

The tangential velocity is equal to the angular velocity multiplied by the distance of the star:

$$v_\mathrm{t} = u \cdot r\,. \tag{3.8}$$

Barnard's star, with an angular velocity of about 10.3″ per year, is the star with the largest proper motion observed so far. While it is usually necessary to observe a star for a few years to measure its proper motion, the radial motion can be easily determined with a single observation, taking advantage of the Doppler effect (Sec. 7.6). The radial velocity can be determined from the displacement of the spectral lines $z = \Delta\lambda/\lambda$, according to Eq. 7.20:

$$v_\mathrm{r} = \frac{(1+z)^2 - 1}{(1+z)^2 + 1}\,c\,, \tag{3.9}$$

or, in the non-relativistic limit:

$$v_\mathrm{r} = c\,z\,. \tag{3.10}$$

Hence, the total linear velocity of the star is:

$$v = \sqrt{v_{\mathrm{r}}^2 + v_{\mathrm{t}}^2}\,. \tag{3.11}$$

3.7 Other Cycles

In addition to the relatively fast motions described above, there are other slower ones. Among these, the most important are the variation of the inclination of Earth's axis and the cycle of eccentricity.
The axis of the Earth is inclined by $23°27'$, on average, with respect to the ecliptic, but this varies from a minimum of $22.1°$, to a maximum of $24.5°$, with a periodicity of 41,000 years. This effect therefore adds to that of nutation, which instead has a much shorter period of 18.6 years. Currently, the inclination of the Earth's axis relative to the ecliptic is decreasing at a rate of $47''$ per century. Finally, the eccentricity of Earth's orbit varies from essentially zero (0.000055) to mildly eccentric (0.0679). The components of this cycle combine into a 100,000-year period. The present eccentricity is 0.0167 and is decreasing. The eccentricity varies primarily due to the gravitational attraction exerted by Jupiter and Saturn; however, the semi-major axis remains unchanged.

3.8 Exercises

1. How long does it take the Barnard to travel a distance in the sky equal to the angular diameter of the Moon?

2. Aldebaran's proper motion is $u = 20''/\mathrm{yr}$ and its parallax is $\pi_p = 0.048''$. The spectral line of iron $\lambda = 440.5\,\mathrm{nm}$ is shifted by $0.079\,\mathrm{nm}$ towards red. What are the radial and tangential velocities of the star? (Hint: use Eq. 9.3).

3.** The coordinates of Sirius at the epoch $J1900.0$ were $\alpha = 6^h40^m45^s$, $\delta = -16°35'$, and its proper motion had components $u_\alpha = -0.037$ s/a, $u_\delta = -1.12''/\mathrm{a}$. Find its coordinates at the epoch $J2000.0$. Take into account the precession of Earth's axis.

4.* The parallax of Sirius is $\pi_p = 0.375''$ and its radial velocity is -8 km/s (i.e. directed towards the Earth). By using the data from the previous exercise, find its tangential velocity. How long will it take Sirius to reach the closest point to the Sun? What will the values of its tangential and radial velocities be at that time? What about its parallax?

5.* (ArmAO 2019, Th., q.1) The atmospheric refractive index of a planet varies with the height from the surface according to:

$$n(h) = \frac{n_0}{1 + \epsilon h},$$

where n_0 and ϵ are constants. A ray emitted in the horizontal direction indefinitely circles the planet at any height (if we neglect absorption by the atmosphere). What is the radius of the planet?

6.** (IAO 2014, Th.α, β, q.4) Hydroplanet consists of a rocky "core" of radius R and a thick layer of water surrounding it from all sides. Local humans live at the bottom of this world's ocean (i.e. on the surface of the "core"), and the hydrosphere is an analogue of our atmosphere for them. Local scientists observe astronomical objects from the bottom of the ocean. Like on Earth, the duration of the day-night period on Hydroplanet is 24^h.

- Find the minimum depth h of the ocean, for which celestial bodies will be visible at the horizon.
- What will the duration of the day be for inhabitants of the planet's equator? The disk of the central star can be considered as a point source of light.
- Calculate the value of "atmospheric" refraction at the horizon on such a planet. The outer surface of the ocean is smooth, with no waves or ripples.

3.9 Problems

1.**** **Precession of the Equinoxes**

Assuming the Earth to be an ellipsoid of revolution with moment of inertia along its axis I_p and equator I_e, such that $I_e/(I_p - I_e) \approx 300$, estimate the period of precession of the equinoxes. Also prove that the influence of the Moon is approximately double that of the Sun.

4

Observation and Instruments

Until the end of the Middle Ages, the eye was the most important instrument for the observation of the sky. With the aid of different devices it was possible to record the position of celestial bodies. The telescope was invented in the Netherlands at the beginning of the 17th century, but only in 1609 was it first used, by Galileo Galilei, to conduct astronomical observations. Since then, great strides have been made in both the quantity and quality of observations. While, initially, observation was limited to within the Solar System, the telescope now allowed observations of celestial objects hundreds of parsecs away. Since Galileo, we have enjoyed 400 years of exponential improvement and growing sophistication of the tools at our disposal. In 1800, William Herschel discovered that visible light is only a small part of the electromagnetic spectrum. Soon enough, a variety of new tools were developed that would extend the study of the night sky to previously inaccessible regions of the electromagnetic spectrum.

4.1 Angular Size

Since all celestial objects appear infinitely far away for an observer on Earth, their first distinctive feature is the *angular size*. Let h_0 be the diameter of the object and d its distance from the observer. Its angular size is then:

$$\tan \alpha_0 = \frac{h_0}{d} . \tag{4.1}$$

For small α_0 (which is always the case when observing celestial objects), we can use the approximation $\tan \alpha_0 \approx \alpha_{0,\mathrm{rad}}$, so that:

$$\alpha_{0,\mathrm{rad}} = \frac{h_0}{d} , \tag{4.2}$$

where $\alpha_{0,\mathrm{rad}}$ is measured in radians. To convert this angle to degrees, we simply multiply by $360°/2\pi$ (Eq. A.11, Appendix A.3). Hence:

$$\alpha_{0,\mathrm{deg}} = \frac{180°}{\pi} \frac{h_0}{d} . \tag{4.3}$$

4.2 The Thin Lens Equation

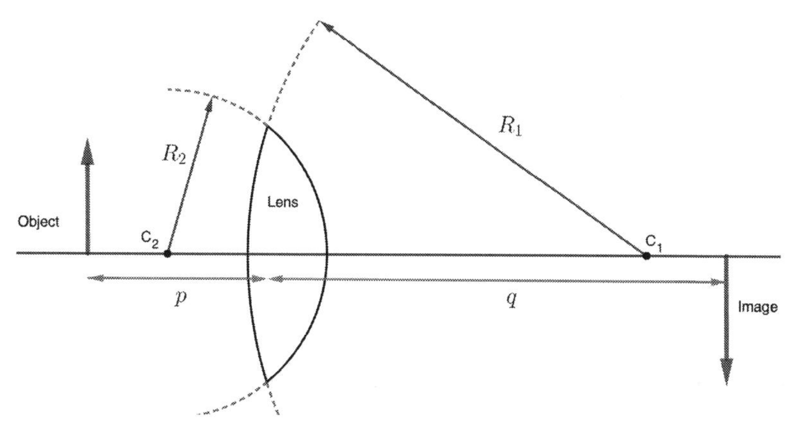

Figure 4.1: An object is placed at a distance p from a lens with refractive index n, and surfaces with radii of curvature of R_1 and R_2. The image is formed at a distance q (the thickness of the lens is exaggerated). In this case, the curvature has opposite direction on either side of the lens, hence R_1 is positive and R_2 is negative.

Let us place an object at a distance p from a lens with refractive index n, and surfaces with radii of curvature of R_1 and R_2 (Fig. 4.1). We would like to know at what distance q its image is formed. By applying Snell's law (Eq. 3.5) to a ray leaving the object and refracting through the first and second surfaces of the lens; it is possible to prove that (pr. 4.2), in the limit where the thickness of the lens tends to zero:

$$\frac{1}{p} + \frac{1}{q} = (n-1)\left(\frac{1}{R_1} - \frac{1}{R_2}\right). \tag{4.4}$$

If we define the focal length f as:

$$\frac{1}{f} = (n-1)\left(\frac{1}{R_1} - \frac{1}{R_2}\right), \tag{4.5}$$

we obtain the *thin lens equation*:

$$\boxed{\frac{1}{p} + \frac{1}{q} = \frac{1}{f}}. \tag{4.6}$$

Eq. 4.5 is called the *lens maker equation*, because it can be used to determine the values of R_1 and R_2 needed for a given refractive index and a desired focal length f. Taking $p \to \infty$, i.e. $1/p \to 0$, Eq. 4.6 reduces to:

$$\frac{1}{f} = \frac{1}{q} \Rightarrow q = f. \tag{4.7}$$

Therefore, the image of a distant object is formed in the focal plane of the lens.

Ray Diagrams for Thin Lenses

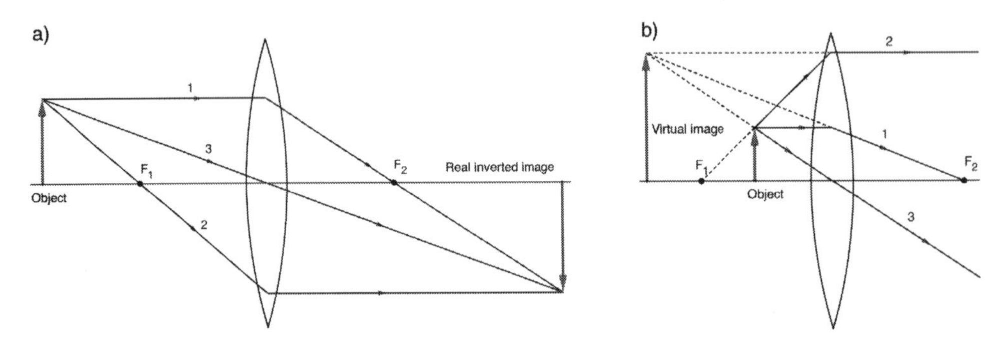

Figure 4.2: Rays are drawn to understand the type of image that is formed. In a) the image is real and inverted, in b) it is virtual and upright.

Ray diagrams are very convenient for locating images formed by thin lenses or systems of lenses. The general procedure is to draw three rays, whose paths are particularly simple to sketch (Fig. 4.2):

- ray 1 is drawn parallel to the principal axis. After being refracted by the lens, this ray passes through the focal point on the back side of the lens;
- ray 2 is drawn through the focal point on the front side of the lens and emerges parallel to the principal axis;
- ray 3 is drawn through the centre of the lens and continues in a straight line.

Properties or rays 1 and 2 follow from Eq. 4.7, since an object at infinity (whose light rays are incident on the lens parallel to the principal axis) has its image in the focal point. Property of ray 3 follows from symmetry, since both sides of the lens are identical. An example of application of this principle is shown in Fig. 4.2. In a), the image is real and inverted because it forms on the back side of the lens and it is upside-down. In b) instead, the rays would diverge. It is therefore necessary to consider their continuation (dashed lines) on the same side of the object. In this case the image is called *virtual*, since it cannot be captured on a screen.

4.3 Optical Telescopes

The term telescope usually refers to the optical telescope, operating in the visible region of the electromagnetic spectrum. However, there are many other telescopes that detect different wavelengths, from gamma rays to radio waves.

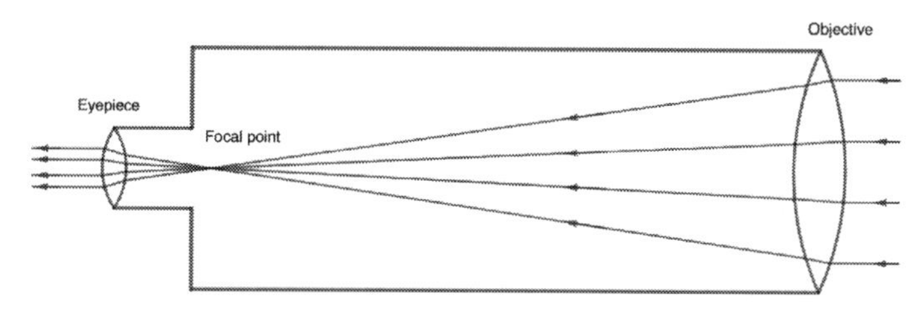

Figure 4.3: Galileo's refracting telescope.

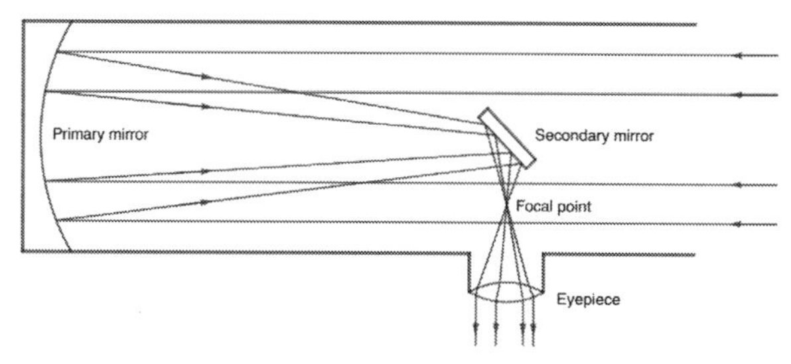

Figure 4.4: Newton's reflecting telescope.

In general, every telescope has two important functions in astrophysics:

- it collects a greater amount of light by using a larger surface than the naked eye, making it possible to observe fainter objects;
- it increases the angular size of the object and the angular resolution.

Two fundamentally different types of optical telescopes exist: the *refracting* or *Galilean* telescope and the *reflector* or *Newtonian* telescope. In order to focus the image of a celestial object, the former uses a set of lenses (Fig. 4.3), taking advantage of the phenomenon of refraction. The latter uses a set of mirrors, according to the phenomenon of total reflection (Fig. 4.4).
The operation of a Galilean telescope is shown in Fig. 4.5 (overleaf). Light from a distant object is incident on the first lens, called the *objective*, and is focused at a distance f_o, equal to the focal length of the objective (the object is essentially at infinity). The objective forms a real and inverted image close to the second lens, called the *eyepiece*. To provide the largest possible magnification, the rays leaving the eyepiece must be parallel, hence the image created by the objective must be located at the focal point of the eyepiece, at a distance f_e from the eyepiece. Thus, the distance between the lenses must be exactly $d = f_o + f_e$. This is the length of the telescope tube.
The main parameters of a telescope are:

- D, diameter of the objective;
- f_o, focal length of the objective;
- f_e, focal length of the eyepiece.

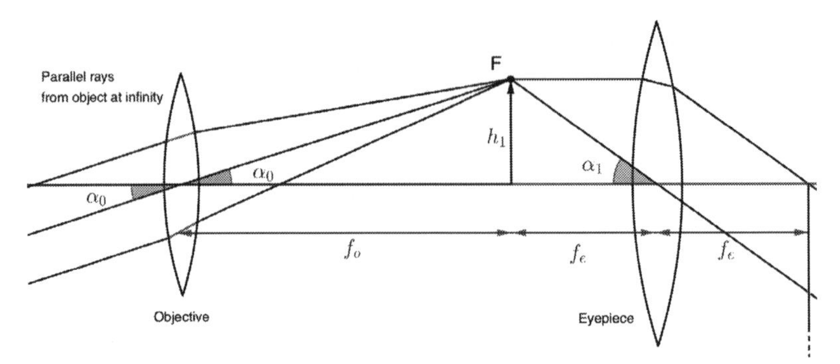

Figure 4.5: The object has angular size α_0. When viewed from the eyepiece of a telescope, its image has an angular size of α_1.

Let α_0 be the angular size of a celestial object. The linear dimension h_1 of its image formed on the focal plane of the objective is:

$$\tan \alpha_0 = \frac{h_1}{f_o} \Rightarrow h_1 = f_o \cdot \tan \alpha_0 \,.$$

In the approximation $\alpha_0 \ll 1$, then:

$$h_1 = f_o \cdot \alpha_{0,\,\mathrm{rad}} \,. \tag{4.8}$$

For example, the angular size of the Sun is approximately $\alpha_0 = 30' \approx 0.0087\,\mathrm{rad}$. Its linear size on the focal plane of a telescope with focal length $f_o = 15\,\mathrm{m}$ is $h_1 = 0.0087 \cdot 15\,\mathrm{m} = 13\,\mathrm{cm}$. If we are given the distance d and dimension h_0 of an object, we know its angular size $\alpha_{0,\mathrm{rad}} \approx h_0/d$. Substituting in Eq. 4.8:

$$\boxed{\frac{h_1}{f_o} = \frac{h_0}{d}} \,. \tag{4.9}$$

4.4 Angular Resolution

An important feature of a telescope is its angular resolution. To understand the physical mechanism that constrains the resolution of a telescope, let us consider in detail the image of a point source formed by a telescope.

When two light rays of equal amplitude interfere in phase with each other, the resulting ray has twice the amplitude of the original rays. However, when two rays interfere out of phase, the resulting amplitude is zero. Clearly, intermediate cases are also possible, so that the resultant amplitude can be any number between zero and two times the amplitude of the original rays. Because of the periodic nature of a wave, we expect two light rays to interfere constructively if the difference in their path lengths is any integer number of wavelengths (which corresponds to being completely in phase); destructively if the difference in their path lengths is any half integer (completely out of phase). Because of its wavelike property, light is diffracted by the circular aperture of a telescope, giving rise to a visible diffraction pattern (Fig. 4.6). At the centre of this pattern, all light rays have travelled the same distance from the aperture, and are therefore in phase with each other. They interfere constructively, giving rise to a bright spot. Moving to the side in any direction, light rays need to travel slightly different paths, and therefore reach the new point with slightly different phases. Hence, light intensity is reduced at the sides.

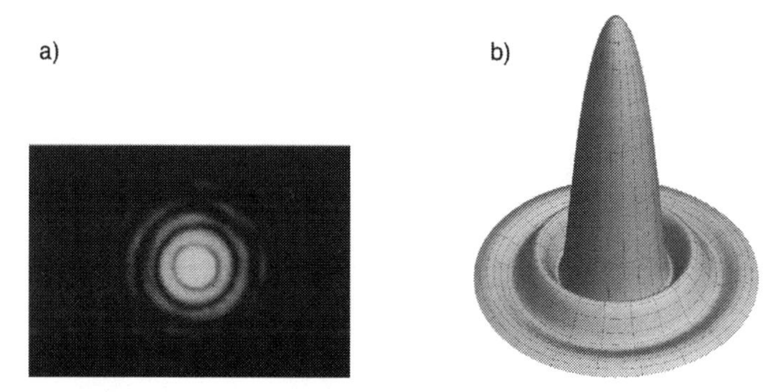

Figure 4.6: a) Image of a star as viewed with a telescope (often only the inner circle is visible). b) Intensity of the diffraction pattern.

At some particular distance from the centre, all light rays are out of phase, hence the intensity is zero. Moving further away from the centre, the phase difference will again be an integer wavelength at some other distance, leading to constructive interference and to a bright circular annulus. For this reason, point sources (like stars) produce, on the focal plane of a telescope, a diffraction pattern of alternating dark and bright zones, called *Airy disks*. In fact, this is the same phenomenon observed by Thomas Young in his double slit experiment. The general physical theory that predicts the shape of the pattern is called *Fraunhofer diffraction*. The effect of this diffraction depends on the ratio between the wavelength of light and the diameter of the telescope.
In order for two objects to be resolved, the central bright spots of their images must not overlap. In more general terms, according to *Rayleigh's criterion*, the distance between the two central maxima must be greater than the distance between the first maximum and first minimum of either image. The cross-section of the diffraction pattern of two point sources is shown in Fig. 4.7 (overleaf). In particular, in a) the two sources cannot be resolved (they appear as a single source), since the distance d between their first maxima is less than the distance d_{min}. On the contrary, in b) the sources can be *just* resolved, since d is equal to d_{min}. It can be shown (pr. 4.3) that the angular distance between the central bright spot and the first minimum is $1.22\lambda/D$, where D is the diameter of the objective and λ the wavelength. It then follows

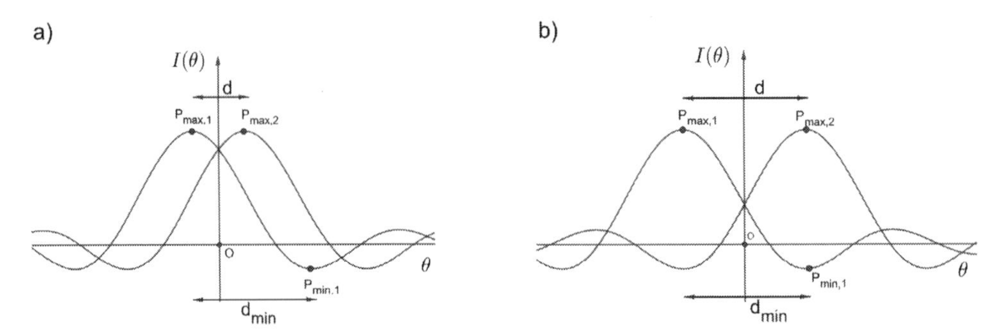

Figure 4.7: According to Rayleigh's criterion, two sources are resolved when the distance between the two central maxima is greater than the distance between the first maximum and first minimum of either image. Hence, a) cannot be resolved, whilst b) can be *just* resolved.

that the angular resolution is given by:

$$\boxed{R \approx 1.22\frac{\lambda}{D}}, \tag{4.10}$$

where R is measured in radians. This formula is valid for all light across the electromagnetic spectrum. Often, the resolution of Earth-based telescopes is limited by the conditions of the atmosphere, rather than by the instrument itself. This effect is called *seeing* and generally does not allow for a better resolution than a few arcseconds. If the telescope collects images on photographic plates, the resolution is further reduced by the finite size of the emulsion grains. These typically range between 10 to 30 micrometers. For a focal length of 1 m, the scale is 1 mm $= 206''$, so 0.01 mm corresponds to about 2" The size of a pixel in CCDs (charge-coupled devices) can be as low as $3 - 5$ micrometers, which corresponds to around one arcsecond. Therefore, the CCD is not usually the limiting factor in the resolution of an Earth-based telescope.

4.5 Magnification

The angular magnification (ω) is defined as the ratio between the angular size of the image formed by the eyepiece and the angular size of the object. Looking at Fig. 4.5:

$$\tan\alpha_0 = \frac{h_1}{f_o} \approx \alpha_{0,\text{rad}}\,,$$

$$\tan\alpha_1 = \frac{h_1}{f_e} \approx \alpha_{1,\text{rad}}\,.$$

Hence the magnification is:

$$\boxed{\omega = \frac{\alpha_{1,\mathrm{rad}}}{\alpha_{0,\mathrm{rad}}} \approx \frac{f_\mathrm{o}}{f_\mathrm{e}}} . \tag{4.11}$$

For example, if the objective has focal length $f_\mathrm{o} = 100\,\mathrm{cm}$ and we use an eyepiece with a focal length of $f_\mathrm{e} = 2\,\mathrm{cm}$, the magnification is $\omega = 100/2 = 50\mathrm{x}$. The magnification is not a fundamental property of a telescope, since it depends on the eyepiece, which can easily be changed.

Minimum Magnification

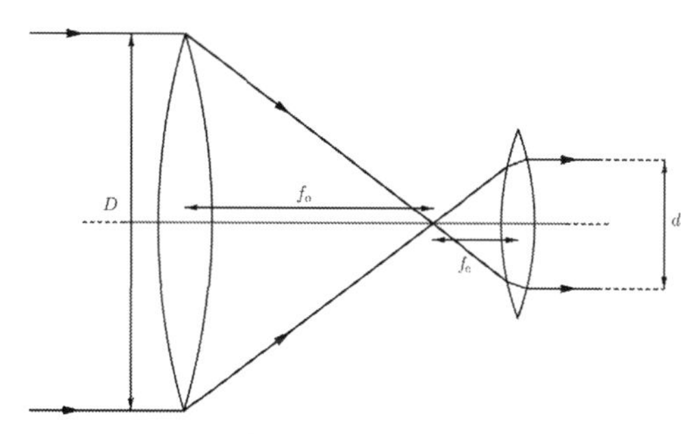

Figure 4.8: If we were to use a magnification lower than the minimum, the diameter of the outgoing light would be larger than the diameter of our pupil, which means losing photons.

The minimum magnification is obtained from the condition that the diameter of the outgoing light from the eyepiece is less than or equal to the diameter of our pupil (Fig.4.8). Thus:

$$d \geqslant D \cdot \frac{f_\mathrm{e}}{f_\mathrm{o}} = \frac{D}{\omega} .$$

Hence, the minimum magnification is:

$$\boxed{\omega_{\mathrm{min}} = \frac{D}{d}} , \tag{4.12}$$

where D and d are the diameters of the objective and of our pupil, respectively. At night, the diameter of the pupil is typically 8 mm, therefore the minimum magnification of a telescope with diameter 100 mm is approximately 13x.

Maximum Magnification

The maximum magnification ω_{max} is the largest magnification that is worth using with a telescope of diameter D. Its value is equal to the ratio of the angular resolution of the human eye ($e \approx 2' = 5.8 \cdot 10^{-4}\,\mathrm{rad}$) and the angular resolution (R) of the telescope:

$$\omega_{\mathrm{max}} = \frac{e}{R} \approx \frac{eD}{\lambda} \approx \frac{D}{1\,\mathrm{mm}} .$$

Thus:

$$\boxed{\omega_{\max} = \frac{D}{1\,\text{mm}}}. \tag{4.13}$$

For example, the maximum magnification of a telescope with aperture 100 mm is 100x. There is no point in using a magnification greater than the maximum, since the resolution of the image would then be worse than the resolution of our eyes.

4.6 Aperture Ratio

Consider an object with angular size α_0, viewed with a telescope of diameter D and focal length f_o. As we saw in Sec. 4.3, the linear size of its image on the focal plane of the telescope is:

$$h_1 = f_\text{o} \tan \alpha_0 \,.$$

Assuming the image is circular, the area it covers is:

$$A = \pi h_1^2 = \pi f_\text{o}^2 \tan^2 \alpha_0 \,.$$

Let F be the incoming light flux from the object. The power collected by the telescope is:

$$P = F \cdot \frac{D^2}{4} \,.$$

Therefore, the intensity of the image is:

$$I = \frac{P}{A} = \frac{F}{4 \tan^2 \alpha_0} \cdot \frac{D^2}{f_\text{o}^2} \,.$$

For a given object, the intensity depends only on D and f_o:

$$I \propto \frac{D^2}{f_\text{o}^2} \,.$$

We define the *aperture ratio* R_f as:

$$R_f = \frac{f_\text{o}}{D} \,. \tag{4.14}$$

Hence, the perceived brightness of an object is directly proportional to the reciprocal of the aperture ratio squared:

$$\boxed{I \propto \frac{1}{{R_f}^2}} \,. \tag{4.15}$$

Telescopes with a long focal length and a small diameter are therefore *slower* than telescopes with a short focal length and a large diameter. The lower the *speed* of the telescope, the longer the exposure time required to obtain an image of equal brightness on a CCD. For this reason, telescopes with a lower aperture ratio are preferred in photography. The aperture ratio is usually denoted by f/R_f. Therefore, a telescope with an aperture ratio of $R_f = 8$ is written as $f/8$. For fast telescopes, the focal ratio can be as low as $f/1 - f/3$, but is usually smaller and around $f/8 - f/15$.

4.7 The Role of the Atmosphere

Astronomical observation from the surface of the Earth has to deal with the atmosphere. Not only is light subject to dispersion, absorption and re-emission, but the effect of the atmosphere is strongly dependent on the wavelength, as well as the time and local conditions. The atmosphere reduces the brightness of an object (Sec. 8.4) and the maximum possible resolution. In the best conditions, the maximum possible resolution is a few arcseconds. Without any modifications, a 10 m telescope would have the same angular resolution as a 20 cm amateur telescope. Actually, the resolution of the former would be worse, because of the distortion caused by the mirror's own weight. *Active optics* are used to correct the distortion of the mirror; *adaptive optics* are used to correct atmospheric turbulence. In both cases, a thin deformable rubber layer with tens or hundreds of thousands of piezoelectric crystals is positioned under the mirror. They behave like small actuators, expanding or contracting when a potential difference is applied to their ends. To correct the effect of the atmosphere, these crystals make adjustments on the order of micrometers, a few hundred times per second. To determine the necessary corrections, the telescope automatically observes a reference star near the object to be observed. In the event that a sufficiently bright star does not exist near the object of interest, a carefully calibrated laser can be used to excite sodium atoms at an altitude of 90 km, thus allowing the required calibration.

4.8 Exercises

1. What is the smallest angle that can be resolved at a wavelength of 500 nm by the Keck telescope, which has a mirror with a diameter of 10 m? What should be the size of a radio telescope, observing at the wavelength of 20 cm, if its resolution is the same?

2. A spiral galaxy at a distance of $D = 55 \cdot 10^6$ ly, is observed perpendicular to its galactic plane and has an angular diameter of $\alpha = 400''$. What is the diameter of the galaxy in light years? How large is its image on the focal plane of a telescope with a focal length of $f = 1$ m?

3.* Calculate the apparent maximum and minimum angular sizes of the Sun from Mars. Suppose Mars has a satellite with the same orbital characteristics as the Moon. What should the minimum diameter of this satellite be in order for solar eclipses to be visible from Mars?

4. Compute the ratio of the angular diameters of Mars when seen in opposition and quadrature.

5. A telescope is pointed at a coin of diameter 16.25 mm, which is at a distance of 15.1 ± 0.1 m. The diameter of the image of the coin formed on the focal plane of the telescope is 1.35 mm. Find the focal length f_o.

6.* (NAZ 2016, Th. S, q.3) On the evening of the 20th of April 2016, in Milan, it is possible to observe three stars of equal apparent magnitudes, with coordinates: Star 1 ($\alpha_{2016} = 6^h30^m$, $\delta_{2016} = +35°20'$), Star 2 ($\alpha_{2016} = 6^h30^m$, $\delta_{2016} = 34°40'$) and Star 3 ($\alpha_{2016} = 6^h24^m$, $\delta_{2016} = +35°20'$). The stars are observed with a telescope of diameter $D = 200$ mm and an aperture ratio of f/10. A camera is placed on the focal plane of the telescope with a CCD of dimension 4096x4096 pixels. Each pixel is a square with sides $l_{\text{pix}} = 6.4\mu$m. Is it possible to obtain an image in which Star 1 and 2 appear together? What about Star 1 and 3 ?

7.* (CAO 2018, Th., q.4) Amateur astronomers use following method to estimate the *field of view* (FOV) of their telescopes: they locate a star with known declination, adjust the telescope so that the star crosses the field of view along its diameter, measure the time taken for the star to cross the FOV and, hence, estimate the FOV. For a telescope (d= 400 mm, f= 4000 mm) the crossing time of αAur ($\delta = 46°0'14.4''$) is 2.5 minutes. Can the Moon be seen in full through this telescope?

8.* (USAAAO 2020, Th.3, q.1) An astronomer used his f/5 telescope with a diameter of 130 mm to observe a binary system. He is using an eyepiece with a field of view of 45° and a focal length of 25 mm. In this system, star A and B have masses of 18.9 and 16.2 solar masses, and apparent magnitudes of 9.14 and 9.60, respectively. The period of the system is 108 days, and its distance from the Solar System is 2.29 kpc. The binary system has an edge-on orbit relative to the Solar System. Find the FOV of the telescope, its limiting magnitude, its angular resolution and the angular separation between the stars. Will the astronomer be able to observe both stars as distinct points in the telescope? (Hint: read Chs. 8 and 10).

9.* (SAO 2019, Th.1, q.3) An *achromatic* lens is used to ensure that light of different wavelengths have the same focal lengths, i.e. to correct *chromatic aberration*. An achromatic lens was made by combining plano-convex (one side flat and one side convex) and plano-concave (one side flat and one side concave) lenses, made from two different types of glasses, A and B. These two glasses have the following refractive indices: $n_{\mathrm{A,red}} = 1.48$, $n_{\mathrm{A,blue}} = 1.50$, $n_{\mathrm{B,red}} = 1.66$, $n_{\mathrm{B,blue}} = 1.70$. Find the radii of curvature R_{A}, R_{B} necessary to produce a combination equivalent to a converging lens with focal length of 600 mm (Hint: use Eq. 4.5, taking into account that A and B have different refraction indices).

4.9 Problems

1.* Combination of thin lenses

Show that if two lenses with focal lengths f_1 and f_2 are combined, they behave like a single lens with an effective focal length of:

$$\frac{1}{f_{\mathrm{eff}}} = \frac{1}{f_1} + \frac{1}{f_2}. \tag{4.16}$$

2.** Lens maker's equation

Consider two transparent media having indices of refraction n_1 and n_2, where the boundary between the two media is a spherical surface of radius R_1 and index of refraction n_2. When an object is placed at a distance p from the boundary (inside n_1), an image is formed at a distance q (inside n_2). By applying Snell's law, show that:

$$\frac{n_1}{p} + \frac{n_2}{q} = \frac{n_2 - n_1}{R_1}. \tag{4.17}$$

Now, imagine that the other side of the lens has a curvature of R_2. The image formed by the first boundary acts as the object for the second boundary. In the approximation that the thickness of the lens is much smaller than the radii of curvature, prove Eq. 4.4.

3.**** Airy Disk

Show that the amplitude $d\psi_p$ of a wavelet reaching the diffraction screen at P= (x_0, y_0), which is a distance r away from the aperture element (x, y), is proportional to:

$$d\psi_p \propto h(x,y)\frac{e^{ikr}}{r}\, dxdy\,,$$

where $h(x, y)$ is the aperture function, which is equal to unity if (x, y) belongs to the aperture and zero otherwise; and $k = 2\pi/\lambda$. In the approximation that $L \gg x_0, y_0, x, y$ (Fig. 4.9 on the next page), show that the total amplitude at P is approximately;

$$\psi_p \propto \int h(x,y)e^{-ik\frac{x_0 x + y_0 y}{R}}\, dxdy\,, \tag{4.18}$$

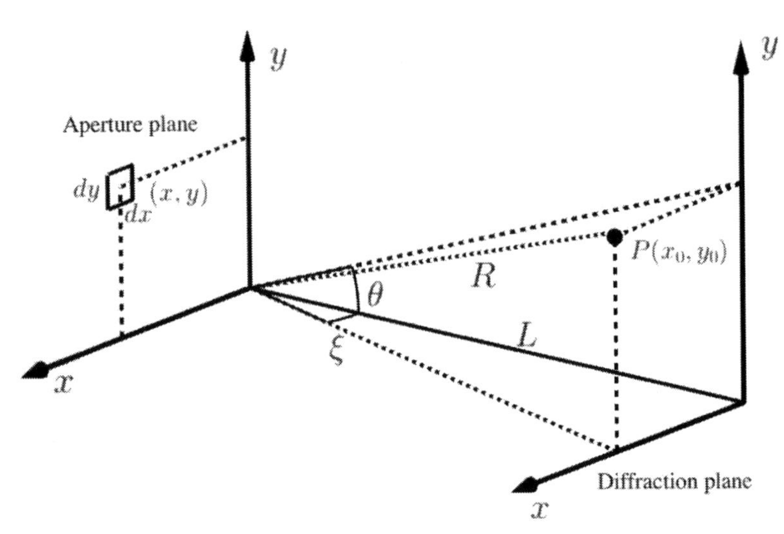

Figure 4.9: Fraunhofer diffraction

where $R^2 = L^2 + x_0^2 + y_0^2$. Now consider a circular aperture of diameter D. Using the identities:

$$\int_0^{2\pi} e^{-iq\rho\cos\phi} d\phi = 2\pi J_0(q\rho)\,, \tag{4.19}$$

and

$$\int_0^{qD/2} q\rho J_0(q\rho)\, d(q\rho) = \frac{qD}{2} J_1\left(\frac{qD}{2}\right) \tag{4.20}$$

where J_1 and J_0 are the first and zero-order Bessel functions of the first kind respectively, show that:

$$\psi_p \propto \frac{J_1(\pi D/\lambda\sqrt{\sin^2\theta + \sin^2\xi})}{\pi D/\lambda\sqrt{\sin^2\theta + \sin^2\xi}} \tag{4.21}$$

where $\sin\theta = y_0/L$ and $\sin\xi = x_0/L$. Knowing that the first zero of $J_1(x)$ is at $x = 3.8317$, prove Eq. 4.10. What is the diffraction pattern of an elliptical aperture, with major axes $2a$ and $2b$?

5

Time systems

There are two essentially different ways to measure time. The first is based on the Sun, and is therefore called *solar time*. The second is based on the rotation of the Earth, and is called *sidereal time*. They each have their own advantages. Solar time is best adopted for day-to-day time-keeping, as our lives are heavily based on the day-night cycle. In the next section we will show that the sidereal day is approximately 4 minutes shorter than the solar day. If calendars were to adopt sidereal time, after about half a year, clocks would lag behind the Sun by 12 hours: we would be working at night and sleeping during the day (as if this doesn't already happen). On the other hand, sidereal time is particularly useful in astronomy. As explained in Sec. 1.3, sidereal time allows us to find the position of any celestial object whose right ascension is known. What this also means is that the celestial vault appears to rotate clockwise (in the same direction as the apparent motion of the Sun) by about 4 minutes (or 1°) every day. This is why most stars are only visible at certain times during the year, and every season is characterized by different stars. For instance, for an observer in the northern hemisphere, the most distinctive feature of the summer sky is the *Summer Triangle*, whose defining vertices are Altair, Deneb, and Vega (in the constellations of Aquila, Cygnus, and Lyra, respectively). Instead, in winter, you should immediately recognize the constellation of Orion, or alternatively, the winter triangle, formed by Sirius, Procyon and Betelgeuse (in Canis Major, Canis Minor and Orion, respectively).

5.1 Sidereal and Solar Day

The *solar day* is the time it takes the Sun to return to the same meridian. The length of the day measured in this way is called *true solar day*. The duration of the true solar day is equal to 86400 s on average, varying from a minimum of 86378 s (mid-September), to a maximum of 86430 s (end-December).
The *sidereal day* is the time it takes a fixed star to return to the same meridian, which is also equal to the period of rotation of the Earth about its axis. In Sec. 1.3, the sidereal time was defined as the hour angle of the vernal equinox.

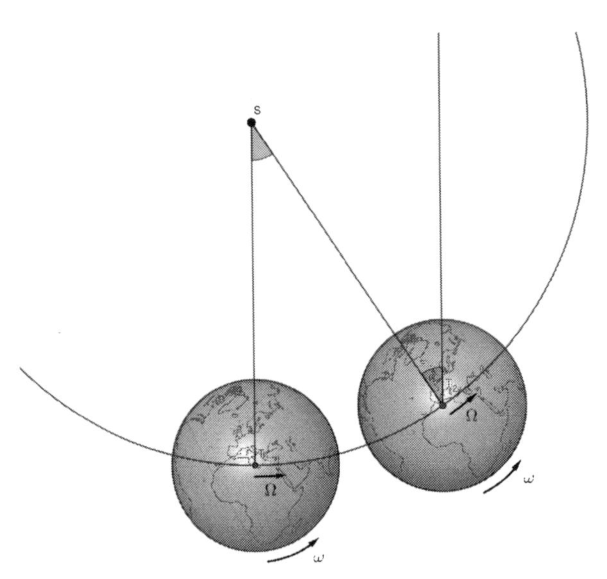

Figure 5.1: The solar day is longer than the sidereal day: after completing a revolution about its axis, the Earth needs to rotate by an additional degree (i.e. by the same amount it covered in its orbit around the Sun), in order to see the Sun at the same meridian. This additional rotation takes 4 minutes, which is the difference between the solar and sidereal day.

Since ♈ is fixed with respect to the stars, the sidereal day can also be defined as the time between two consecutive passages of the vernal equinox on the same meridian.

The lengths of the solar and sidereal days are different. After a sidereal day, the Earth has covered about one degree of its orbit around the Sun (or, more precisely, $360\,^\circ/365.25 = 0.986^\circ$). Therefore, after completing a rotation about its axis, the Sun does not appear exactly at the meridian, but the Earth needs to rotate an additional degree (Fig. 5.1). This takes around 4 minutes ($24^h/360$), therefore the solar day is about 4 minutes longer than the sidereal day. However, the difference between a solar and a sidereal day is not constant. It depends on the eccentricity of the Earth's orbit and on the declination of the Sun. In any case, the solar day is always longer than the sidereal day, since the orbital and rotational motions proceed in the same direction.

Now, let us carry out a more rigorous analysis. Let ω be the angular velocity of the Earth about its own axis and Ω the angular velocity of revolution around the Sun. Since the orbital and rotational motions proceed in the same direction, an observer on Earth sees the Sun moving clockwise with angular velocity $\omega_r = \omega - \Omega$. Let t_{sol} be the duration of the mean solar day, t_{sid} the length of the sidereal day and t_{rev} the period of Earth's orbit around the Sun (equal to a year). Since $\omega_r = 2\pi/t_{\text{sol}}$, $\omega = 2\pi/t_{\text{sid}}$ and $\Omega = 2\pi/t_{\text{rev}}$, we have:

$$\frac{1}{t_{\text{sol}}} = \frac{1}{t_{\text{sid}}} - \frac{1}{t_{\text{rev}}}\,. \tag{5.1}$$

Isolating t_{sid}, substituting $t_{\text{rev}} = 365.25\, t_{\text{sol}}$ and $t_{\text{sol}} = 86400\,\text{s}$, we find:

$$\boxed{t_{\text{sid}} = \frac{t_{\text{sol}} \cdot t_{\text{rev}}}{t_{\text{sol}} + t_{\text{rev}}} \approx 86164.1\,\text{s} = 23^h 56^m\, 4.1^s \text{ of mean solar time}}\,.$$

The mean solar day is therefore $3^m\, 56^s$ longer than the sidereal day. The sidereal day can be taken as a constant to a very good approximation, but in the long term it also varies, mainly due to the tidal force exerted by the Moon. As a consequence, the Moon gains energy, moving away from the Earth (at a current rate of 3.8 cm each year), whilst the Earth's rotation is slowed down. If the Earth–Moon system continued its motion undisturbed, after a sufficiently long time, the Earth would always show the same side to the Moon, i.e. the rotation and the orbital motions would be synchronised. There are a few examples of celestial bodies in the Solar System in which perfect synchrony has already been established. One of the most widely known examples is the dwarf planet Pluto and its largest moon, Charon.
Earlier we showed that one sidereal day, i.e. $24^h = 86400^s$ of sidereal time, is equal to $23^h 56^m\, 4.1^s = 86164.1^s$ of mean solar time. We can then write two equations to convert from sidereal to solar time and vice-versa:

$$t_{\text{sol}} = \frac{86164.1}{86400} t_{\text{sid}} \qquad \Rightarrow \qquad t_{\text{sol}} \approx 0.99727\; t_{\text{sid}}\,, \tag{5.2}$$

and:

$$t_{\text{sid}} = \frac{86400}{86164.1} t_{\text{sol}} \qquad \Rightarrow \qquad t_{\text{sid}} \approx 1.002738\; t_{\text{sol}}\,. \tag{5.3}$$

Indeed, using Eq. 5.2, we see that a sidereal day ($t_{\text{sid}} = 24^h = 86400\,\text{s}$ of sidereal time) is equal to 86164.1 s of solar time. Conversely, using Eq. 5.3 we find that a solar day ($t_{\text{sol}} = 24^h = 86400\,\text{s}$ of solar time) is equal to 86636.5 s of sidereal time.

Estimating Sidereal Time

Since the sidereal time is defined as the hour angle of the vernal equinox, it follows that ST$= 0^h$ when ♈ passes through the meridian. During the vernal equinox, the Sun coincides with ♈, therefore they culminate at the same time. Since the Sun culminates at 12 : 00 of local time, it follows that:

$$\text{ST} = T_t + 12^h\,,$$

where T_t is the *true local solar time*. This equation has the accuracy of a few minutes. Since the sidereal day is shorter than the solar day by $3^m 56^s$, after n solar days from the vernal equinox, the sidereal time will be:

$$\text{ST} = T_t + 12^h + (3^m 56^s) \cdot n\,.$$

Often we only know the value of T_m, i.e. the mean local solar time. Measuring T_t would require a sundial or a similar instrument, which I bet you don't carry in your pocket. In this case we can use the equation of time $T_t = T_m + \text{E.T.}$ (Sec. 5.2), knowing that, during the vernal equinox, E.T. $= -7^m$:

$$\boxed{\text{ST} = T_m + 11^h 53^m + (3^m 56^s) \cdot n} . \tag{5.4}$$

The vernal equinox usually falls on the 21st of March, but it can vary in either direction by about two days. For this reason, unless we are given the day of vernal equinox, it is not possible to compute the sidereal time with an accuracy greater than 5 minutes. When talking about the time at a given place, we usually refer to the (mean solar) local time, therefore it is more likely you will use Eq. 5.4.

5.2 Equation of Time

The length of the solar day varies throughout the year. Due to the eccentricity of its orbit, Earth's velocity varies from a maximum at perihelion to a minimum at aphelion. Since the daily angle covered by the Earth around the Sun is proportional to the Earth's velocity, the true solar time is longer at perihelion and shorter at aphelion. Another factor that periodically changes the length of the true solar day is the obliquity of the ecliptic. Since Earth's rotation axis is perpendicular to the celestial equator, we are only interested in the projection of the Sun's motion along this plane. Let $d\lambda_\odot/dt$ be the daily angle covered by the Sun, when its declination is $\delta_\odot$, as viewed by an observer on Earth. First, let us work out the projection of the Sun's motion on the parallel of declination $\delta_\odot$. This is just $(\cos\epsilon/\cos\delta_\odot)\, d\lambda_\odot/dt$: indeed, it is $d\lambda_\odot/dt$ during a solstice (locally the Sun moves along the parallel), and $\cos\epsilon\, d\lambda_\odot/dt$ during an equinox (when the Sun's path forms an angle ϵ with respect to the parallel, which is the equator in this case). We now need to project this motion on the celestial equator. This adds a factor of $1/\cos\delta_\odot$, i.e. the ratio of the circumference of the celestial equator to that of the parallel (see Sec. 3.6). Therefore, the projection of the Sun's motion on the equator is proportional to $\cos\epsilon/\cos^2\delta_\odot$. The eccentricity gives a maximum deviation of around 5 s per day, while the obliquity of the ecliptic is the dominant factor and gives rise to a maximum deviation of around 20 s per day.

Our clocks are based on the mean solar time, that is, on a fictitious Sun that moves on the celestial equator at a constant angular speed. By definition, the average solar time T_m is equal to the hour angle H_m of the fictitious Sun plus 12 hours (so that every day starts at midnight):

$$T_m = H_m + 12^h . \tag{5.5}$$

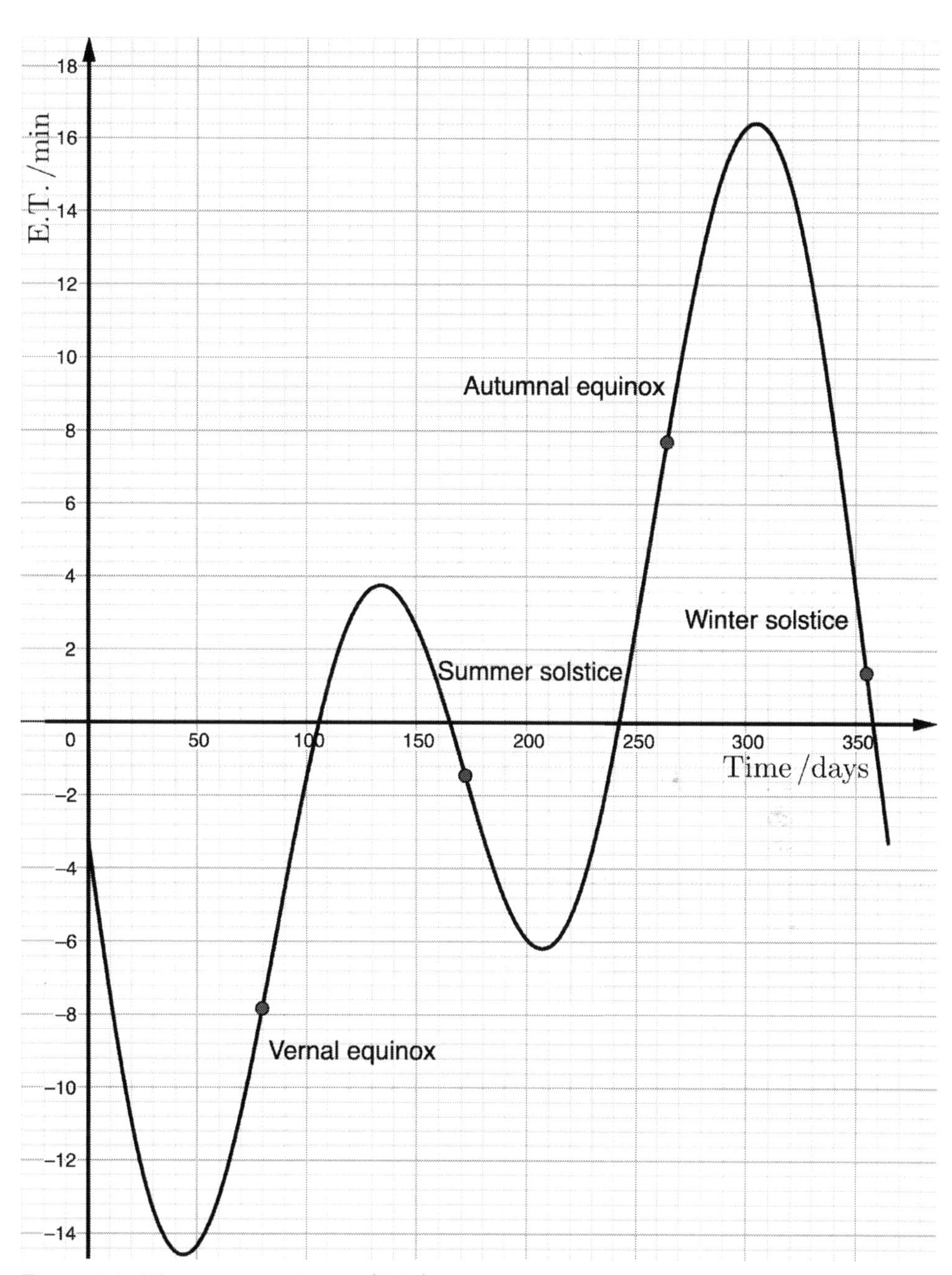

Figure 5.2: The equation of time (E.T.) gives the difference between the mean solar time and the true solar time. The value of E.T. varies throughout the year, due to the eccentricity of the Earth's orbit and the obliquity of the ecliptic.

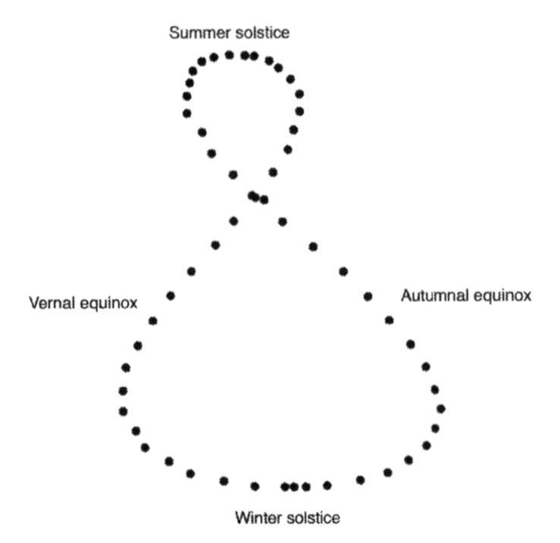

Figure 5.3: The analemma on the local meridian, as seen by an observer at an intermediate latitude in the northern hemisphere.

We can also write the same equation for the true solar time T_t and hour angle H_t of the Sun. The equation of time, denoted by E.T., is the difference between the mean solar time T_m and the true solar time T_t:

$$\boxed{\text{E.T.} = T_t - T_m}\,. \tag{5.6}$$

During a year, the small differences between each true and mean solar day sum up to significant values of E.T.: from a maximum of 16 minutes on the 2nd of November, to a minimum of -14 minutes on the 11th of February. The equation of time is zero four times per year: on the 15th of April,14th of June, 1st of September and 24th of December.
If we take a picture of the Sun every day throughout the year, at the same mean solar time, and superpose these images, we see that the Sun follows a long and slender figure-of-eight path, with one lobe much larger than the other, called the *analemma* (Fig. 5.3). The vertical component of this motion is easy to explain: since the declination of the Sun changes during the year, its height on the horizon varies accordingly, being lower in winter and higher in summer. What about the horizontal component? At the same mean solar time, the hour angle of the true Sun changes during the year in accordance with the equation of time (Eq. 5.6), therefore its horizontal position varies accordingly. For an observer in the northern hemisphere, the Sun is located in the highest part of the analemma during the summer solstice, in the lowest part during the winter solstice and in the part halfway in between (which does not correspond to the point where the lobes meet) during the vernal and autumnal equinoxes. The easternmost and westernmost points are occupied by the Sun when the equation of time is minimum and maximum, respectively. The Sun passes through the four points halfway in between the horizontal extrema when the equation of time is zero.

5.3 Local Solar Time

Both the average and true solar times are local times, i.e. they depend on the hour angle of the Sun, either real or artificial. The time on our watch is likely to be different from the true solar time (measured, for example, by a sundial). For centuries, the length of the day was directly related to the Sun's passage on the meridian. This meant that each place had its own local solar time, which could vary significantly within the same country. Clearly, our clocks must be based on a different time, one that does not vary throughout the year and that is possibly the same within the same region or country.

5.4 Greenwich Mean Time

As railways progressed across each country, the time difference between cities became problematic. The United Kingdom, the most industrialized European nation at that time, decided to standardize time within its country according to the time marked by the clock in Greenwich, central London. This time standard became known as the *Greenwich Mean Time* (GMT).
The Greenwich Mean Time was formally replaced by the *Universal Time* (UT) in 1928 (although the title has not yet entered common usage), but was essentially the same as GMT until 1967, when the definition of the second was changed. At the beginning, the universal time was obtained directly from observations (UT0). After correcting for the perturbation motions of the Earth's rotation axis (precession, nutation, motion of the poles), we obtain UT1. This system also suffers some irregularities, caused by small variations in the Earth's rotation speed. These disturbances have at least two periods, one year and half a year. By removing these effects we obtain UT2, with a precision of about 10^{-7} s. UT2 is also not exact, due to the decrease in the Earth's velocity of rotation, caused by the tidal force with the Moon. For this reason, astronomers have abandoned the rotation of the Earth as a method of measuring time. In 1967, the one-second duration was defined as 9192631770 times the period of oscillation of Caesium 133 that transits from the hyperfine level $F = 4$ to $F = 3$. Later, this definition was modified to include small relativistic effects due to the Earth's gravitational field. The precision of atomic time is approximately 10^{-12} s.

5.5 Time Zones

In 1884, the *time zone* system was proposed, according to which time is determined only for the 24 main meridians, separated in longitude by 15°. A time zone is defined as the sector on the Earth's surface delimited by the two great circles at a distance of 7.5° from a main meridian. However, the boundaries of each time zone coincide with the theoretical ones only in the open sea, while, within continents, they are modelled by borders. As a consequence of the rotation of the Earth from west to east, every time zone to the west of Greenwich is an hour of GMT earlier; to the east, an hour later. Imagine we leave Greenwich at 12:00 GMT on the 1st of January, and travel west at an arbitrarily large speed. Since we travel through 24 time zones before finally landing back in Greenwich, it looks like we need to turn our watch back by 24^h. In this case, our watch would then read 12:00 GMT, 31st of December, which is clearly a paradox, since no time has actually elapsed. To avoid this, the time zone system requires a date change at a specific meridian. Furthermore, when crossing this meridian from east to west, we need to decrease the date by one; if crossing from west to east (as in the previous example), increase it by one. By convention, this line roughly follows the Greenwich anti-meridian (the main meridian 180° from it), and is called the *International Date Line* (IDL). If you will, the IDL is a branch cut that makes time continuous. In many places, the Date Line follows exactly the Greenwich anti-meridian, in others it deviates to avoid crossing inhabited areas. The islands of the republic of Kiribati, in the archipelago of the equatorial Sporades, is the place with the farthest time zone from Greenwich (GMT + 14:00).

5.6 Daylight Saving Time

The *Daylight Saving Time* was introduced for the purpose of reducing electricity consumption, by taking advantage of daylight during summer. It is generally adopted from 2:00 a.m. on the first Sunday of spring and involves setting the clocks forward by one hour, compared to standard time. It returns to standard time at 2:00 a.m. on the first Sunday of autumn.

5.7 Measuring the Length of a Year

Sidereal Year

The *sidereal year* is the time it takes the Sun to return to the same position with respect to the fixed stars. It is equal to Earth's orbital period, given by Kepler's third law (Sec. 10.3). The duration of the sidereal year is 365.2564 average solar days (365 days, 6 hours, 9 minutes and 10 seconds). The sidereal year is not based on the relative position of the Earth and Sun; therefore, if we were to use the sidereal year in our calendars, after a sufficiently long time the seasons would no longer fall in the same months of the year, but would shift by about twenty minutes per year due to the precession of the equinoxes. The period of revolution of a celestial body is usually expressed in sidereal years.

Tropical or Solar year

The *tropical year* or *solar year* is the time it takes the Sun to return to the same position relative to the Earth. In other words, the tropical year is the time interval between two consecutive passages of the Sun through the vernal equinox. Because it is based on the motion of the Sun, the tropical year is adopted by calendars. The tropical year is equal to 365.2422 average solar days (365 days, 5 hours, 48 minutes and 46 seconds), but is subject to small perturbations, the most significant of which is due to the Moon (pr. 6.5).
The length of the tropical year can be obtained from the length of the sidereal year (which we simply call year) and the precession period of the equinoxes ($T_{\text{pr}} = 25765$ years). Since the orbital and precessional motions proceed in opposite directions, the angular velocity of the Sun with respect to the vernal equinox is given by $\omega_r = \omega_{\text{yr}} + \omega_{\text{pr}}$. Hence:

$$\frac{1}{T_{\text{tr}}} = \frac{1}{T_{\text{yr}}} + \frac{1}{T_{\text{pr}}},$$

$$\boxed{T_{\text{tr}} = \frac{T_{\text{pr,yr}}}{1 + T_{\text{pr,yr}}} \approx 0.9999612\, T_{\text{sid}} = 365.2422 \text{ mean solar days}}\,.$$

5.8 Calendars

Our calendar is the result of a long evolution. In general, a calendar must have two properties:

- the number of days in a year must be an integer, since it would be inconvenient for some years to start at night, others during the day;
- the average duration of a year must be close to that of the tropical year, so that the seasons always fall in the same period of the year.

The main problem calendars have to deal with is the incompatibility between the units of the day, month and year, since the number of days and months in a year are not integers. Our calendar draws its origins from the Roman calendar which, in its original form, was based on the phases of the moon. Around 700 BC, the calendar was based on the apparent motion of the Sun around the Earth and had a duration of twelve months. However, each month was equal to a lunar cycle, so the duration of the year was 354 days and the calendar was out of phase with the seasons. Then, under Julius Caesar, the duration of the year was extended to 365 days. To better synchronize the year with the seasons, it was also decided to add one day to every fourth year (*leap year*). Therefore, in the *Julian calendar*, the average duration of the year was 365 days and 6 hours. Even this calendar was not perfect: the tropical year is shorter by about 11 minutes and 14 seconds, so that, after 128 years, the Julian calendar was already lagging behind by one day. The difference already amounted to 10 days in 1582, when Pope Gregory XIII proposed a new calendar. In the *Gregorian calendar*, every fourth year is a leap year. However, years divisible by 100 are never leap years except when they are also divisible by 400. For example, 1900 was not a leap year, but 2000 was. The Gregorian calendar was adopted slowly, at different times by different countries, and the transition period continued until the twentieth century. As you might have guessed by now, even the Gregorian calendar is not perfect: the difference with the tropical year accumulates every year and, in about 3300 years, it will amount to one day.

5.9 Exercises

1. Is the difference in longitude between two places equal to the difference in sidereal or solar time?

2. Find the local time in Rome ($\lambda = 12°28'54.48''$ E) at 12:00 GMT. What is the difference with the time shown by a clock?

3.* When in Greenwich the local time is $10^h17^m14^s$, in Naples it is $11^h14^m15^s$. What longitude is Naples located at?

4.** Compute the hour angle of the true Sun H_t for an observer at a place with longitude $128°15'$ W on the 10th of February 2003, at $08^h46^m22.0^s$ GMT, knowing that the equation of time that day was E.T. $= -14^m13^s$.

5. On what day of the year are the solar and sidereal times the same?

6. This evening, a star rises at 22:00. At what time will it rise in a month?

7.* The sidereal time in Greenwich on the 18th of February 2003, at 0^h GMT, was $9^h\,50^m\,12^s$. At what time of GMT did a star with right ascension 18^h transit the meridian that day?

8.* What is the best month of the year to observe (around midnight) a star with right ascension $\alpha = 10^h$?

9.* What is the sidereal time in Rome ($\lambda = 12°28'54.48''$ E) at 12:00, on the 13th of April 2000? That year, the vernal equinox occurred at 7 : 35 GMT on the 20th of March.

10.* The coordinates of Arcturus are $\alpha = 14^h15.7^m$ and $\delta = 19°1.1'$. Find the sidereal time when the star sets in Boston ($\phi = 42°19'$, $\lambda = 71°$). At what (meridian) time does Arcturus set on the 10th of January?

11. What would the relation between solar and sidereal time be if the Earth rotated about its axis in the opposite direction?

12.* What is the accumulated error in the Gregorian calender after 100, 500 and 1000 years?

13.** (IAO 2012, Th.α, β, q.1) Recently, on the 6th of June 2012, an infrequent astronomical phenomenon, the transit of Venus across the solar disc, took place. The next transit of Venus will take place only in 2117. Calculate the date of that transit (Hint: read Ch. 11 first).

5.10 Problems

1.* **Effect of ecliptic obliquity on the equation of time**
Using the equations obtained in Ch. 2, prove that:

$$\frac{\mathrm{d}\alpha}{\mathrm{d}t} = \frac{\cos\epsilon}{\cos^2\delta_\odot}\frac{\mathrm{d}\lambda_\odot}{\mathrm{d}t}\,. \tag{5.7}$$

2.*** **Equation of time**
Knowing that the equation of an ellipse in plane polar coordinates with centre in the rightmost focus is (see Pr. 10.12):

$$r(\theta) = \frac{a(1-e^2)}{1+e\cos\theta}\,, \tag{5.8}$$

applying conservation of energy and angular momentum, find the tangential velocity of the Earth at each point in its orbit. Hence, prove that the daily angle covered by the Earth around the Sun is:

$$\frac{\mathrm{d}\theta}{\mathrm{d}t} = \sqrt{\frac{GM}{a^3}}\frac{(1+e\cos\theta)^2}{(1-e^2)^{3/2}}\,. \tag{5.9}$$

Therefore, compute the daily change in the Sun's right ascension $d\alpha_\odot/dt$, in terms of $\theta, \gamma, \epsilon, e$, where γ is the angle of the vernal equinox measured from the direction of perihelion, ϵ is the obliquity of the ecliptic and e is the eccentricity of Earth's orbit. Finally, using Mathematica/Matlab or similar, compute the equation of time. Sketch the analemma, observed at a mean solar time T_m, from a place at latitude ϕ (Hint: read Ch. 10).

6

The Moon

The Moon is Earth's only natural satellite. It orbits the Earth anti-clockwise (as seen from the north pole) in a slightly elliptical orbit with eccentricity $e = 0.055$ and semi-major axis $a = 384400$ km. Its orbital plane is inclined by $5°9'$ with respect to the ecliptic, which it intersects along the *node line*. The period of revolution of the Moon around the Earth is the same as the period of rotation about its axis, hence the Moon always shows the Earth its same side. However, due to its orbital inclination and eccentricity, it is possible to see 59% of its surface at one time or another. This effect is called *libration*, and will be discussed in detail in Sec. 6.5.

The Moon reflects only 8% of the light incident upon it (comparable to a lump of coal), making it one of the least reflective objects in the Solar System. Even with the naked eye, one can clearly see two distinct regions on the visible side of the Moon (the *near side*, as opposed to the *far side*): light areas, called *terrae* (or highlands), and darker areas, called *maria*. The maria were once thought to be seas and oceans, and were given beautiful names such as Oceanus Procellarum, Mare Tranquillitatis and Sinus Iridum (the Ocean of Storms, the Sea of Tranquillity and the Bay of Rainbows). When the far side of the Moon was first photographed by the Soviet probe Luna 3 in 1959, a surprising feature was its almost complete lack of maria. Today we know that the maria are made up of basaltic rocks, formed by volcanic activity, which seems to be prevalent on the near side. The surface of the Moon is covered in *lunar regolith*. It has a thickness of about 3–5 m in the maria regions and 10–20 m in the terrae.

The Moon has been studied more than any other body in the Solar System; it has been imaged from above by lunar orbiters, and its surface studied by a number of landers – the first being Lunar 9 in 1965, followed by the Russian lunar rovers and the NASA Surveyor craft. Lunar exploration culminated in the NASA Apollo programme, when six spacecraft landed men on the Moon.

6.1 Sidereal Month

The *sidereal month* is the time interval it takes the Moon to return to the same position with respect to the fixed stars. The duration of this period is 27.32166^d, i.e. $27^d\,7^h\,43^m\,12^s$. The most evident lunar period is, however, the *synodic month*, related to the so-called *lunation*.

6.2 Synodic Month

The synodic month is the time interval it takes the Moon to return to the same position relative to the Sun. After a sidereal month the Earth has travelled about 27° in its orbit around the Sun, therefore the Moon must travel that same (additional) angle around the Earth, before being aligned again with the Earth and the Sun. Since the Moon covers about 13° every day, the synodic period is about two days longer than the sidereal one.
Following the same strategy adopted in Sec. 5.1, to compute the relation between sidereal and solar days, let us consider the relative angular velocity between the Moon and the Sun. Denoting with $\omega_{☾}$ and $\omega_{\odot}$ the angular velocities of the Moon and the Sun, respectively, their relative angular velocity is $\omega_r = \omega_{☾} - \omega_{\odot}$. Substituting $\omega_{☾} = 2\pi/T_{\mathrm{sid}}$, $\omega_{\odot} = 2\pi/T_{\mathrm{yr}}$ and $\omega_r = 2\pi/T_{\mathrm{syn}}$:

$$\frac{1}{T_{\mathrm{syn}}} = \frac{1}{T_{\mathrm{sid}}} - \frac{1}{T_{\mathrm{yr}}}.$$

Solving for T_{syn}:

$$\boxed{T_{\mathrm{syn}} = \frac{T_{\mathrm{sid}} T_{\mathrm{yr}}}{T_{\mathrm{yr}} - T_{\mathrm{sid}}} = \frac{365.2564 \cdot 27.32166}{365.2564 - 27.32166} \approx 29.53059^d = 29^d 12^h 44^m 3^s}.$$

This is the mean duration of the synodic month. Due to the eccentricity of Earth's orbit, the synodic month varies from a minimum of 29.27^d, near summer solstice, to a maximum of 29.84^d, near winter solstice. A lunar year comprises 12 mean synodic months, and has a duration of $354^d\,8^h\,48^m$. In the lunar calendar, for convenience, the year is assigned an integer number of days. To better approximate the real period, one day is added every third year, so that 2 years with 354 days are followed by one with 355 days. The error is about 48 minutes every year.

Lunar Phases

The lunar phases result from the different positions occupied by the Moon, relative to the Sun, in its rotation around the Earth. Every *lunation* is characterized by four important moments: new moon, first quarter, full moon and last quarter.

New moon occurs when the Moon is in between the Earth and the Sun. During new moon, the Moon appears completely dark and it rises and sets together with the Sun. Approximately seven days after new moon, the Moon is in quadrature with the Earth and the Sun; this moment is called first quarter. Only the western part of the Moon is illuminated — the part to the east being completely dark. The Moon rises around noon and sets around midnight, lagging behind the Sun by six hours. Another seven days later, the Moon is aligned with the Sun and the Earth, on the same side of the Earth, and appears completely illuminated. During a full moon, the Moon rises at 18:00 (sunset), and sets at 6:00 (sunrise), therefore it precedes the Sun by 12 hours. Finally, in the last quarter, the Moon is again in quadrature, appearing bright to the east and dark to the west. The Moon now rises at midnight and sets at noon, making it 6 hours ahead of the Sun. The entire cycle lasts one synodic month.

6.3 Draconic Month

The *draconic month* is the time interval it takes the Moon to complete a full revolution with respect to the node line. This period is slightly shorter than the synodic month due to the precessional motion of the node line, which is in the opposite direction to the orbital motion of the Moon. The value of the draconic month (T_{dra}) can be calculated from the precessional period ($T_{\text{pr}} = 18.61$ years) and from the sidereal period (T_{sid}):

$$\frac{1}{T_{\text{dra}}} = \frac{1}{T_{\text{sid}}} + \frac{1}{T_{\text{pr}}},$$

$$\Rightarrow T_{\text{dra}} = \frac{T_{\text{pr}}\, T_{\text{sid}}}{T_{\text{pr}} + T_{\text{sid}}} \approx 27.2122^d = 27^d 5^h 5^m 34^s .$$

Instead, the *draconic year* is useful for calculating eclipses. It is defined as the time between two consecutive passages of the Sun through the same lunar node. This period can be determined from the relative angular velocity between the Sun and the node line. Let T_{tr} be the duration of the tropical year, and T_{pr} the period of the precession of the lunar nodes. Then:

$$\frac{1}{T_{\text{yr, dra}}} = \frac{1}{T_{\text{tr}}} + \frac{1}{T_{\text{pr}}},$$

$$\Rightarrow T_{\text{yr, dra}} = \frac{T_{\text{pr}}\, T_{\text{tr}}}{T_{\text{pr}} + T_{\text{tr}}} \approx 346.6201^d = 346^d 14^h 53^m .$$

6.4 Eclipses

Conditions for the Occurrence of Eclipses

If the orbital plane of the Moon coincided with the plane of the ecliptic, there would be one solar and one lunar eclipse every synodic month. However, the plane is tilted by $5°9'$; therefore, for an eclipse to occur, the Moon must be close to the node line during full or new moon. In the next sections we will show that, for a (partial or total) lunar eclipse to occur, the Sun-node distance must be smaller than $10°$ and, only in exceptional cases, can it reach $12°$. For a total lunar eclipse this distance must be less than $4.5°$. In order for a (partial or total) solar eclipse to occur, the distance of the Sun from the node must be less than $15°$, therefore solar eclipses are more frequent than lunar eclipses. Since it takes the Sun six months to travel from one node to the other, the eclipses are distributed in two groups, the so-called *eclipse seasons.*

Total Number of Eclipses in a Year

During a synodic period, the Sun moves by about $30°$ along the ecliptic. Imagine there is a new moon at the beginning of a synodic month. Half a month later, the Moon has covered half of its orbit, so there is a full moon, while the Sun has moved by $15°$. At the end of the month, there will be a new moon again, and the Sun will have moved by $30°$ relative to the starting position. Taking into account the position of the node line, it is then possible to determine the types of eclipses that can occur in a synodic month:

1. 2 solar and 1 lunar. This happens when the Sun is initially at $-15°$ from the node line. At the beginning of the month, a solar eclipse occurs. Half a month later there is a full moon, and the distance of the Sun from the node is around $0°$, therefore a lunar eclipse occurs. At the end of the month, the Sun will be at $15°$ from the node line, and a second solar eclipse will occur.

2. 1 solar and 1 lunar. In this case, the initial distance of the Sun from the node must be between $-15°$ and $-5°$. After half a month, the Sun-node distance is just right ($0° - 10°$) for a lunar eclipse to occur. At the end of the month, this distance will be greater than $15°$, therefore no other solar eclipse occurs.

3. 1 solar. The initial distance of the Sun from the node line must be less than $5°$. The first solar eclipse occurs at the beginning of the month. However, at the next full moon, the Sun's distance from the node line is greater than $10°$, so no lunar eclipse occurs. Likewise, at the end of the month, no other solar eclipse will occur.

In any case, we see that the number of solar eclipses must be greater than or

equal to the number of lunar eclipses. The maximum number of eclipses that can occur in a year is 7: 5 solar and 2 lunar, or 4 solar and 3 lunar. In the first case, at the beginning of the year there are two solar eclipses, the first one towards the beginning of January and the second during the next new moon, with a lunar eclipse between them (case 1). After half a year, in the second eclipse season, there are again two solar eclipses with a lunar eclipse in between. At the end of the year, after 354 days (or 12 synodic months) from the first eclipse, the last solar eclipse occurs.
In the second case, a lunar eclipse occurs shortly after the beginning of January, and is followed by a solar eclipse (case 2). In the second eclipse season, there are two solar eclipses with a lunar eclipse between them (case 1). Then, at the end of the year, there will be the fourth solar eclipse and the third lunar eclipse (case 2), 354 days after the first lunar eclipse. However, observing seven eclipses in a year is very rare. The most frequent event is that of 2 solar and 2 lunar eclipses in a year. The minimum number of eclipses in a year is 2: one solar eclipse (case 3), in each of the two eclipse seasons.

Lunar Eclipse

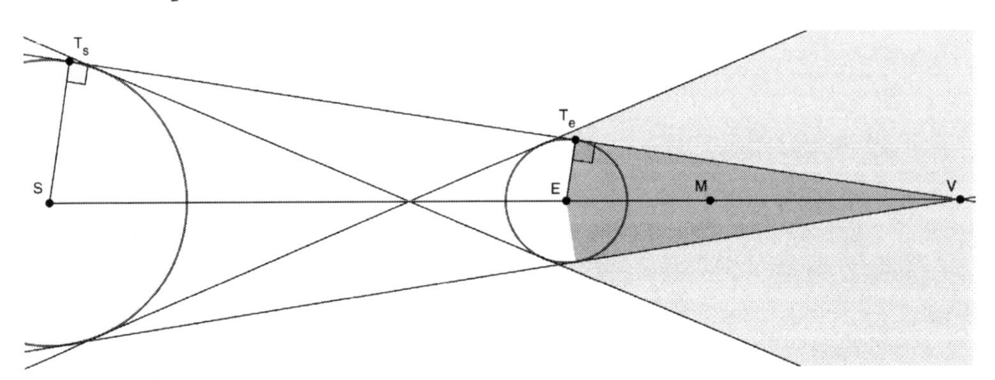

Figure 6.1: Lunar Eclipse.

Assume that the Earth, Moon and Sun are spherical bodies. Due to illumination by the Sun, the Earth casts a large cone of shadow (umbra) into space. Let V be the vertex of this cone and $l = \overline{\mathrm{VE}}$, its length (Fig. 6.1). The quantities $\overline{\mathrm{SE}} = d_\oplus$, $\overline{\mathrm{ST_s}} = R_\odot$ and $\overline{\mathrm{ET_e}} = R_\oplus$ are the Earth-Sun distance, the radius of the Sun and the radius of the Earth, respectively. Consider the similar triangles $\mathrm{T_sSV}$ and $\mathrm{T_eEV}$, then:

$$\frac{l + d_\oplus}{R_\odot} = \frac{l}{R_\oplus} \Rightarrow l = \frac{R_\oplus d_\oplus}{R_\odot - R_\oplus}.$$

Since $R_\odot = 695510\,\mathrm{km}$, $R_\oplus = 6378\,\mathrm{km}$ and $d_\oplus = 149.6 \cdot 10^6\,\mathrm{km}$, we get $l = 1.383 \cdot 10^6\,\mathrm{km}$. Hence, the length of the umbra is 3.5 times the Earth-Moon distance.

In addition to the umbra, the Earth projects a larger penumbra into space. When the Moon crosses the umbra, a *total* lunar eclipse is observed. If, instead, the Moon only crosses the penumbra, a *partial* eclipse is observed, and the brightness of the Moon is only slightly attenuated.
Let us now calculate the radius of the umbra at the distance of the Moon ($\overline{\mathrm{EM}} = d_☾$, in Fig. 6.1). Assuming that the angle in V is small, the following relation holds:

$$\frac{R_\oplus}{r} = \frac{l}{l - d_☾}.$$

Isolating r and substituting l found earlier:

$$r = R_\oplus - \frac{d_☾}{d_\oplus}(R_\odot - R_\oplus). \tag{6.1}$$

Evaluating the above equation, we find $r = 4560\,\mathrm{km}$, i.e. around 2.65 times the radius of the Moon. In fact, due to the eccentric orbits of the Moon and the Earth, this value can range from 2.6 to 2.8. The Moon covers a distance equal to its diameter every hour (see Ex. 6.6), therefore the totality of a lunar eclipse can last up to two hours.
A total lunar eclipse takes place as follows: when the Moon first enters the penumbra cone, its brightness decreases slightly. At the moment of *first contact*, the Moon enters Earth's umbra. As the eclipse continues, Earth's shadow progressively covers a greater portion of the Moon until, at the moment of *totality*, it is completely obscured. Subsequently, the Moon emerges again, its brightness slowly increasing. The moment when the Moon leaves the umbra is called *last contact*, and marks the end of the eclipse. Partial eclipses, in which the Moon only crosses the penumbra of the Earth, are very difficult to observe, because of the slight, almost imperceptible, decrease in brightness.
Let us now calculate the maximum distance between the Sun and the node line, for which a partial lunar eclipse can take place. In the limiting case, the Moon barely grazes Earth's umbra, therefore it is at a maximum distance of $r + R_☾$ from the node line. The angle from the node line is then:

$$\tan\alpha \approx \alpha_{\mathrm{rad}} = \frac{r + R_☾}{d_☾} \Rightarrow \alpha = 0.94^\circ,$$

where r is given by Eq. 6.1. Now, the maximum distance of the Sun from the node line is exactly equal to this angle, since only the relative positions of the Sun, Earth and Moon matter. However, the Moon's orbit is inclined by 5.145° relative to the plane of the ecliptic, so the maximum distance of the Sun, in ecliptic longitude, is:

$$\Delta\lambda = \frac{\alpha}{\sin 5.145^\circ} = 10.5^\circ$$

which is approximately the angle quoted at the beginning of this section.
The limit for a total eclipse is obtained by taking $r - R_☾$ as the maximum distance of the Moon from the node line. In this case, we get $\alpha = 0.42°$ and $\Delta\lambda = 4.7°$. Since the orbits of the Earth and the Moon are elliptical, these limits vary depending on the positions of the celestial bodies (Ex. 6.10).

Solar Eclipse

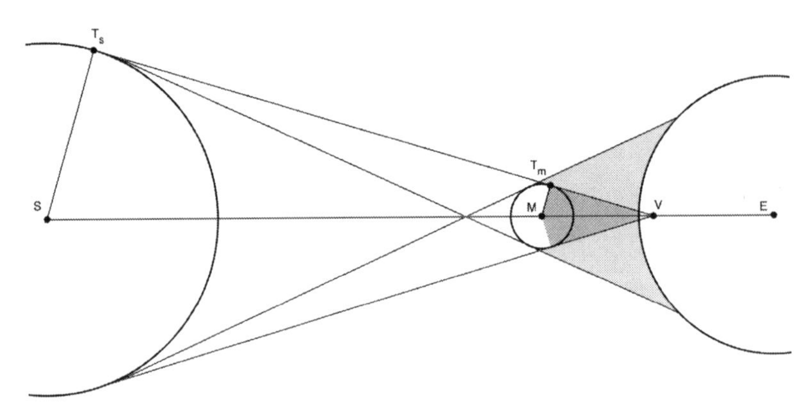

Figure 6.2: Solar eclipse: unlike lunar eclipses, visible from the entire hemisphere on Earth facing the Moon, total solar eclipses can be observed only in a region of 200 km. The penumbra is much larger, its radius being around twice that of the Moon.

When the Moon covers the disk of the Sun, a solar eclipse takes place. For a solar eclipse to be observed, there must be a new moon, and both the Moon and the Sun must be close to the node line. Since both the orbits of the Earth and the Moon are elliptical, the Moon can appear larger or smaller than the Sun. If the angular diameter of the Moon is greater than that of the Sun, a *total* solar eclipse occurs. If, instead, the angular diameter of the Moon is smaller, a very thin portion of the Sun's disk is left uncovered: in this case, an *annular* solar eclipse is observed. On average, the angular diameter of the Sun is slightly larger, hence, for a total solar eclipse to occur, either the Moon must be close to perigee, or the Earth close to aphelion.
Although solar eclipses occur more often than lunar eclipses, the former are much more difficult to observe. In fact, during a solar eclipse, the length of the umbra on the surface of the Earth is just 200 km (Ex. 6.12), which makes it visible only in very few places. On the contrary, a total lunar eclipse can be seen from the entire hemisphere on Earth facing the Moon.
What is the maximum distance of the Sun from the node line, for a (partial or total) solar eclipse to be observed? In this case, the Moon must be at point M in Fig. 6.3. The distance $\overline{\mathrm{EV}} \approx \overline{\mathrm{T_eV}}$ is just l, computed in the previous section. Also, $\overline{\mathrm{ET_e}} = R_\oplus$ and $\overline{\mathrm{ME}} = d_☾$ are the radius of the Earth and the distance Moon-Earth, respectively. Since α is small, we can approximate

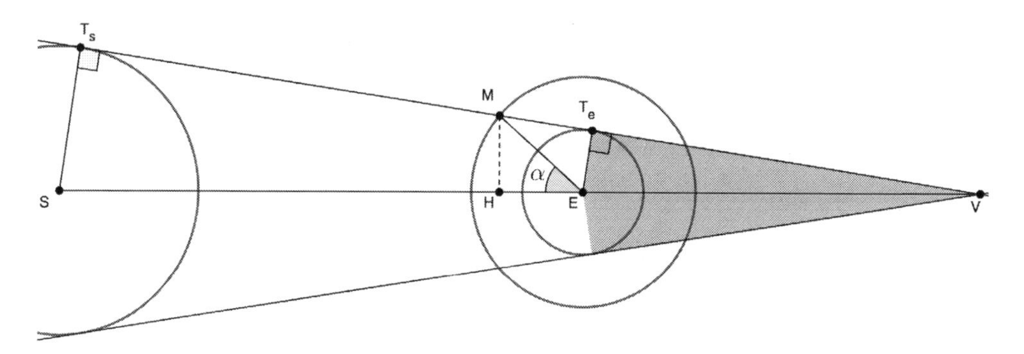

Figure 6.3: For a (partial or total) solar eclipse to occur, the distance of the Moon from the node line must be less than α.

$\overline{\text{MH}} \approx d_{☾}\ \alpha_{\text{rad}}$ and $\overline{\text{HE}} \approx d_{☾}$. The triangles EVT_e and HVM are similar, therefore:

$$\frac{d_{☾}\ \alpha_{\text{rad}}}{l + d_{☾}} = \frac{R_{\oplus}}{l} \Rightarrow \alpha_{\text{rad}} = \frac{R_{\oplus}}{d_{☾}}\left(1 + \frac{d_{☾}}{l}\right), \tag{6.2}$$

giving $\alpha \approx 1.21°$. Adding to this the angular radius of the Moon, the total angle is then $1.47°$. The maximum distance of the Sun from the node line is:

$$\Delta\lambda = \frac{1.47°}{\sin 5.145°} \approx 16°$$

which is an overestimation of the limit stated at the beginning of this section.

Saros cycle

At the time of the Chaldeans it was already known that the eclipses repeated periodically every 18 years and 11.3 days, the so-called *Saros cycle.* Indeed, eclipses repeat in the same order, when the same phase of the Moon repeats itself at a distance from the node line equal to the distance at which the first eclipse occurred. It follows that the Saros cycle is equal to the time interval necessary for the synodic month, draconic month and draconic year (29.53, 27.21 and 346.62 days, respectively) to return in phase with each other. In particular, 242 draconic months correspond approximately to 223 synodic months and 19 draconic years. More exactly:

- 242 draconic months = 6585.36 mean solar days
- 223 synodic month = 6585.32 mean solar days
- 19 draconic years = 6585.78 mean solar days

However, 223 synodic months are 0.04 days shorter than 242 draconic months, after 6585 days the Moon will return to a slightly different position (the cycle is not exact). Since the Saros cycle is an integer number of days plus approximately 1/3, the eclipses will be visible from locations shifted by around 120°

in longitude, compared to the previous cycle. For this reason, the Greeks had defined another period, known as the *exeligmos*, equal to three Saros cycles. Each Saros cycle comprises 70 eclipses: 41 solar and 29 lunar.

6.5 Libration

In addition to the motions of revolution around the Earth and precession of the node line, the Moon is subject to a rotational motion around its axis which is synchronous with the orbital motion around the Earth. As a consequence, the Moon always shows the Earth its same side (approximately). In fact, the Moon is also subject to an apparent motion, called *libration*. As a result of this motion, it is possible to see 59 % of the lunar surface at one time or another, instead of the 50 % that we would have expected.

Longitudinal Libration

While the rotation of the Moon around its axis is uniform (to a good approximation), its orbital velocity is not. For example, if the Moon is close to perigee, its speed is higher than average, and it is possible to see a thin area towards west, that wouldn't have been visible otherwise. This longitudinal motion has a period equal to the *anomalistic* month (the time it takes the Moon to return to the same position with respect to the line passing through perigee and apogee, called the *apse line*), with a length of $27^d 13^h 18.6^m$ (see Pr. 6.2). The maximum amplitude of longitudinal libration is $7°54'$ (Pr. 6.6).

Latitudinal Libration

The rotation axis of the Moon is inclined by $6°41'$ with respect to the perpendicular to its orbital plane. Since the direction of its axis stays unchanged (because of the conservation of angular momentum), every draconic month the Moon periodically shows the Earth its north and south poles. The amplitude of this motion is equal to the inclination of the Moon's axis.

Diurnal Libration

Diurnal libration is a secondary effect compared to the previous ones and is caused by parallax. Observers at different locations on Earth see the Moon at slightly different angles. As a result, they see different portions of it. The amplitude of diurnal libration is about $1°$.

6.6 Exercises

1. What planets can be eclipsed by the full moon?

2. Is the full Moon higher in summer or in winter?

3. For what range of latitudes is it possible to see the Moon at the zenith?

4. At what latitudes is the Moon circumpolar at least once a year?

5. When is the Moon closest to you: at the zenith or at the horizon?

6. How long does it take for the Moon to cover a distance equal to its diameter? (Hint: use Eq. 10.18).

7. (INT 2012, Th.S, q.2) In the morning, you look out the window of your home (in Italy) and see the moon half illuminated. Is it a first or last quarter moon? At what time does it set?

8.** (NAZ 2015, Th.S, q.5) A small outer planet travels around the Sun in a circular orbit that lies exactly on the plane of the ecliptic. It has a peculiarity: the Moon can never eclipse it. Find the radius of its orbit (Hint: use Kepler's law, Eq. 10.13).

9. On the 1st of January there was a new moon. What will be the position of the Moon, the same day of the following year?

10.* Compute again the maximum distance of the Sun from the node line for a lunar eclipse to occur. Take into account the eccentricities of the orbits.

11.* The same question as in the previous exercise, but for a solar eclipse.

12.* Compute the maximum lengths of the umbra and penumbra on Earth, during a solar eclipse. Take into account the eccentricities of the orbits.

6.7 Problems

1.* Meton cycle

In the fifth century BC, the Athenian astronomer Meton calculated how many years it would take the Moon to return to the same phase on the same day of the year. Find the length of this cycle, knowing the duration of the synodic month (29.53^d) and of the tropical year (365.24^d).

2. Anomalistic month

The apse line of the Moon completes a full rotation every 8.8504 years. Find the duration of the anomalistic month, defined as the time it takes the Moon to return to the same position with respect to the apse line.

3.* Measuring the distance of the Moon

When the Moon passes through the zenith, its angular diameter is $\alpha_z = 31'36''$. When it is on the horizon, its angular diameter is $31'4''$. Calculate the distance of the Moon from Earth. (The effects of atmospheric refraction have already been accounted for).

4. Distance of the Moon, another method**

The Moon is observed from a place of latitude $\phi = 41°55'$. When the Moon transits on the meridian and has maximum declination, its distance from the zenith is $z_u = 13°33'$; when it has minimum declination, its distance is $z_l = 71°25'$. Find the distance of the Moon from Earth. Knowing that the inclination of the ecliptic is $23°27'$, find the inclination of the Moon's orbit. (You will need to solve an equation numerically.)

5.* Lunar influence

Among the numerous factors that affect the duration of the tropical year, we want to estimate the contribution of the Moon. Assume that the centre of mass of the Earth-Moon system moves in a circular orbit around the Sun and that the Moon moves in a circular orbit around the Earth, on the plane of the ecliptic, completing (approximately) 12 and a half revolutions every year. Show that the year is longer than average if there is a first quarter moon during the spring equinox. Evaluate the difference in minutes between that year and the following year.

6. Longitudinal libration**

Estimate the amplitude of longitudinal libration. (Hint: read Ch. 10 first, and use Eq. 5.9).

Part II

Radiation Mechanisms

7

Electromagnetic Radiation

What we generally call light is only a small part of the electromagnetic spectrum, in particular the visible region between 400 and 700 nanometers in wavelength. By studying the radiation emitted by a body, it is possible to obtain a large amount of information, including its temperature, composition, surface gravity, electric and magnetic fields. In this chapter we illustrate how this can be achieved. We start by discussing the properties of electromagnetic radiation, moving on to introduce some concepts of statistical physics, which will come in handy in the study of the black body spectrum. Only a brief summary of the essential results, without delving too much into derivations, can be given here. However, some proofs will be presented in the problems. In case you wish to gain a deeper insight in this topic, you might want to consult some statistical physics and quantum mechanics textbooks (see "Suggested Resources").

7.1 Wave or particle?

Electromagnetic waves can be defined by three physical parameters:

- the *period* T, i.e. the time interval it takes a crest to reach the position previously occupied by the preceding crest;
- the *frequency* f, i.e. the number of crests that pass through a given point every second. The frequency is the reciprocal of the period, that is: $f = 1/T$;
- the *wavelength* λ, i.e. the distance between two adjacent crests or troughs, equal to the distance covered by the wave in a period.

In vacuum, the velocity of all electromagnetic radiation is equal to the speed of light, c. Since velocity is equal to space (λ) divided by time (T), it follows that:

$$\boxed{c = \frac{\lambda}{T} = \lambda f}\,. \tag{7.1}$$

When taking into consideration phenomena involving energy exchange between electromagnetic radiation and matter, the wave model becomes inadequate and unable to justify some experimental observations. In this case, a *corpuscular* model is used instead. According to this model, radiation is made up of energy packets, called *photons*. The energy carried by each photon is directly proportional to its frequency, and the constant of proportionality is called *Planck's constant*, equal to $h = 6.626 \cdot 10^{-32}\,\mathrm{J\,s}$. The energy of a photon is:

$$\boxed{E = hf}\,. \tag{7.2}$$

Since $f = c/\lambda$, we also have $E = hc/\lambda$. What distinguishes different electromagnetic waves is their energy, i.e. their frequency. High-frequency (short-wavelength) radiation is composed of high-energy photons, whereas low-frequency (long-wavelength) radiation is composed of low-energy photons. The set of all electromagnetic waves is called the *electromagnetic spectrum*. According to de Broglie's law, each photon carries a momentum of:

$$\boxed{p = \frac{h}{\lambda}}\,. \tag{7.3}$$

Since $\lambda = c/f$, we can also write $p = hf/c = E/c$. A body illuminated with light is subject to *radiation pressure*, due to the exchange of momentum between the object and the electromagnetic field. Since force equals rate of change in momentum, $p = E/c$ implies that the force is equal to the power incident on the surface, divided by the speed of light. But pressure is force divided by area, therefore, defining the *flux* (F) as the power per unit area (and therefore energy per unit area, per unit time), it follows that:

$$\boxed{P_{\mathrm{rad}} = \frac{F}{c}}\,. \tag{7.4}$$

This expression makes the implicit assumption that the body absorbs all of the incident (collimated) light. If, instead, the body is completely reflecting, the radiation pressure is double the value given by Eq. 7.4. Indeed, the photons bounce back with the same initial momentum, hence the change in momentum of the body is twice the initial momentum of the photons. The radiation force is generally too small to be noticed under everyday circumstances. However, in outer space, it is usually the main force acting on an object, besides gravity. The discovery that light carries momentum has led to the development of new propulsion methods, such as spaceships capable of deploying huge solar sails or small probes propelled by reflecting the collimated light of powerful laser beams (see Ex. 7.5).

7.2 Boltzmann Distribution Law

Boltzmann's law describes the most likely energy distribution of a system. In this and the next section, we will consider the probability distribution function $g(E)$, defined so that $g(E)\,dE$ equals the probability that the system's energy is in the range $[E,\, E+dE]$. From statistical arguments, it can be shown that the most probable distribution is:

$$\boxed{g(E) \propto e^{-E/(k_B T)}}\,, \tag{7.5}$$

where E is the energy of the state, T is the temperature and k_B is the *Boltzmann constant*, equal to $k_B = R/N_a = 1.381 \cdot 10^{-23}\,\mathrm{J/K}$.

7.3 Maxwell Distribution Law

The *Maxwell distribution law* describes the most likely velocity distribution of particles within a system. We define $f(v)$, called the Maxwell distribution function, such that $f(v)\,dv$ is equal to the probability of finding a particle with velocity in the range $[v,\, v+dv]$. This function can be derived from the Boltzmann distribution (Pr. 7.2):

$$\boxed{f(v) = 4\pi\left(\frac{m}{2\pi k_B T}\right)^{3/2} v^2 e^{-mv^2/2k_B T}}\,, \tag{7.6}$$

where m is the mass of a particle. Note the occurrence of the Boltzmann factor $e^{-E/(k_B T)}$, where E is the kinetic energy $mv^2/2$.
The most probable velocity corresponds to the maximum of $f(v)$. Equating the derivative of $f(v)$ to zero, it can be shown (Ex. 7.8) that:

$$v_{\mathrm{mp}} = \sqrt{\frac{2k_B T}{m}}\,. \tag{7.7}$$

The root mean square (rms) velocity, as the name suggests, is the square root of the average velocity squared, i.e. $\sqrt{\langle v^2\rangle}$. It is given by (Ex. 7.12):

$$v_{\mathrm{rms}} = \sqrt{\frac{3k_B T}{m}}\,. \tag{7.8}$$

It is useful for calculating the kinetic energy of a system:

$$\langle K\rangle = \left(\frac{1}{2}m\langle v^2\rangle\right)\cdot N = \frac{1}{2}mN{v_{\mathrm{rms}}}^2 = \frac{3}{2}nRT\,.$$

where we used $k_B = R/N_a$ and $n = N/N_a$, with n denoting the number of moles. Therefore:

$$\boxed{\langle K\rangle = \frac{3}{2}nRT}\,, \tag{7.9}$$

This formula is valid for a gas composed of monoatomic particles. The *equipartition theorem* states that each (quadratic) degree of freedom contributes equally to the total kinetic energy. A monoatomic gas has three degrees of freedom, one for each of the three spatial dimensions. Therefore, each degree must contribute an energy of $(1/2)nRT$. A diatomic gas has two additional degrees of freedom. These correspond to the rotation around the two axes perpendicular to the line joining the centres of the atoms (the energy associated with the rotation about the axis joining the atoms is negligible, since the atoms are very small). Hence, the kinetic energy of a diatomic gas is $(5/2)\,nRT$. Therefore, at the same temperature, a diatomic gas stores more energy than a monoatomic gas.

Planetary Atmospheres

The atmosphere of a planet is made up of several molecules: some lighter, like helium and hydrogen, others heavier, like carbon dioxide, ammonia and methane. Since the average kinetic energy of molecules in a gas is proportional to the temperature (Eq. 7.9), it follows that molecules move faster in a warmer atmosphere. At a given temperature, the average velocity of a molecule is inversely proportional to the square root of its mass. Therefore, hydrogen molecules ($m = 2\,\mathrm{u}$) move four times faster than oxygen molecules ($m = 32\,\mathrm{u}$). If a molecule located in the upper atmosphere moves at a sufficiently high speed, it can exceed the escape velocity of the planet. Therefore, small and hot planets may have easily lost all of their lighter molecules, while massive and cold planets could still retain their primordial atmosphere.
However, the situation is not that simple: Maxwell's curve (Eq. 7.6) shows that there are molecules with much higher velocities than the average. In fact, about one molecule in a million has a velocity three times higher than the rms velocity, while one molecule in 10^{16} has a velocity five times higher. It can be shown that, if the rms velocity of a molecule is less than 1/6 of the escape velocity from a planet, this molecule does not escape from the planet's atmosphere in significant quantities during the lifetime of the Solar System. For example, the mean square velocities of molecular nitrogen (506 km/s) and oxygen (473 km/s) in the Earth's atmosphere are well below 1/6 of the escape velocity from Earth (11.2 km/s). Assuming the Moon had an atmosphere with approximately the same temperature as the Earth's, the rms velocities of oxygen and nitrogen would have been only 1/5 of the escape velocity from the Moon (2.4 km/s). Therefore, no trace of such an atmosphere would remain today. The same arguments apply to the hydrogen in the Earth's atmosphere, which has a mean square velocity of 2 km/s, just over 1/6 of Earth's escape velocity. Hydrogen has had sufficient to escape, and indeed today only constitutes 0.000055 % of the Earth's atmosphere. On the other hand, the average speed of hydrogen is 1/60 of the escape velocity from Jupiter. This explains why Jupiter has been able to maintain hydrogen, which is its major constituent.

7.4 Black Body

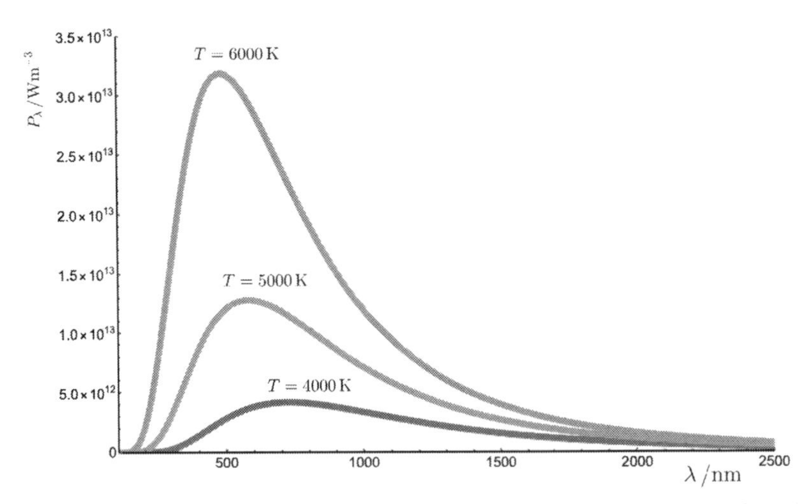

Figure 7.1: The power emitted by a black body as a function of the wavelength. With increasing temperature, the wavelength of maximum emission shifts to the left and, at the same time, the maximum power increases.

The term *black body*, first introduced by Gustav Kirchhoff in 1860, describes a theoretical body that absorbs all incident electromagnetic radiation. In equilibrium, because of energy conservation, a black body must re-emit all of the absorbed energy into space, in the form of *black body radiation.* The curve obtained by recording the amount of energy emitted by the black body as a function of the wavelength is called *black body curve* or *continuous spectrum curve* (Fig. 7.1).
Continuous spectra are produced by all bodies above absolute zero (0 K) and do not depend on the composition of the emitting body, but are a function of its temperature only. In Fig. 7.1, we see that, for long wavelengths, the power emitted by a black body increases initially, reaches a maximum at a certain wavelength $\lambda_{\max}$, then decreases again, going to zero at infinity. As the temperature is decreased, $\lambda_{\max}$ moves towards longer wavelengths and, at the same time, the maximum power decreases. Instead, as the temperature is increased, $\lambda_{\max}$ moves towards shorter wavelengths and the maximum power increases. Indeed, the area enclosed by the curve is equal to the total power emitted by the black body, which decreases with decreasing temperature. These observations are summarized in Wien's and Stefan-Boltzmann laws, respectively.

Wien's Law

The relationship between the colour of a hot body and its temperature was first noticed in 1792, by porcelain manufacturer Thomas Wedgewood. All his ovens, regardless of size, shape or construction, turned red when heated above a certain temperature. It looked like their colour was exclusively related to their temperature. Using purely thermodynamic arguments, in 1893, Wilhelm Wien showed that the wavelength (λ_{max}) at which the emission of energy is maximum is related to the temperature (T) according to:

$$\boxed{\lambda_{\text{max}} T = k}\,, \tag{7.10}$$

where k is Wien's constant:

$$k = 2.897 \cdot 10^{-3} \text{ m K}\,. \tag{7.11}$$

As the temperature is raised, a body turns first red ($620 - 750\,\text{nm}$), then orange ($590 - 620\,\text{nm}$), then yellow ($570 - 590\,\text{nm}$) and finally blue-white ($450 - 475\,\text{nm}$). In the same way, colder stars appear red and orange, whereas warmer ones appear yellow and white-blue. This does not mean that a star emits only at the wavelength corresponding to its perceived colour, but that the other wavelengths are surpassed by that of maximum emission.
For example, the surface temperature of the Sun is $T_\odot = 5778\,\text{K}$. The maximum emission wavelength is then:

$$\lambda_{\text{max}} = \frac{k}{T_\odot} = \frac{2.898 \cdot 10^{-3}\,\text{m K}}{5778\,\text{K}} = 501 \cdot 10^{-9}\,\text{m}\,.$$

Surprisingly, 501 nm corresponds to green, although it is very close to yellow 560 nm. The wavelength of maximum emission does not always correspond to the colour perceived by our eyes. This is due to the fact that the atmosphere disperses smaller wavelengths more effectively (see Sec. 8.4), and our eyes are the most sensitive to yellow.

Stefan-Boltzmann Law

The second black body law is the Stefan-Boltzmann law, experimentally obtained by Josef Stefan in 1879 and theoretically derived by Ludwig Boltzmann in 1884, using only thermodynamic arguments. The law states that the power emitted by a black body, per unit area, is directly proportional to the fourth power of its absolute temperature. The proportionality constant σ is called the Stefan-Boltzmann constant:

$$\sigma = 5.67 \cdot 10^{-8}\,\text{W/m}^2\text{K}^4\,.$$

Therefore, the power emitted by a spherical body with radius R on the entire electromagnetic spectrum is:

$$L = 4\pi R^2 \sigma T^4\,. \tag{7.12}$$

In general, a body can deviate significantly from the black body model. To improve the theoretical prediction, the *emissivity* e is introduced. Quantitatively, the emissivity is equal to the ratio of the radiation emitted by the body to the radiation emitted by a black body at the same temperature; therefore, the emissivity can also be interpreted as the ability of a body to absorb incident light. An ideal black body absorbs all incident light, so $e = 1$. On the contrary, a mirror reflects most of the incident radiation (at least the visible radiation), therefore e is close to 0. In any case, e can range from 0 to 1. In a more general form then, the Stefan-Boltzmann law can be written as:

$$\boxed{L = 4\pi R^2 e\,\sigma T^4}\,. \tag{7.13}$$

If an object is at temperature T and the space surrounding it is at temperature T_0, the body absorbs energy from the surroundings at a rate of $P_{\text{in}} = 4\pi R^2\, e\,\sigma {T_0}^4$ and emits energy at a rate of $P_{\text{out}} = 4\pi R^2\, e\,\sigma T^4$. Therefore, the total rate of exchanged energy is:

$$P_{\text{tot}} = 4\pi R^2 e\,\sigma(T^4 - {T_0}^4)\,.$$

If an object is in equilibrium (i.e. constant temperature) with the surroundings, it radiates and absorbs the same power, so that $T = T_0$. If an object is warmer than the environment, it radiates more energy than it absorbs and, in the absence of other mechanisms, its temperature decreases. The temperature of empty space, associated with the cosmic background radiation, is about $T_0 = 2.725\,\text{K}$.

7.5 Types of Spectra

There are three different kinds of spectra: *continuous*, *absorption* and *emission*. The spectrum of a star can be regarded as the superposition of a continuous spectrum, also called black body spectrum (as it depends only on the temperature of the star), and an absorption spectrum (or, more rarely, an emission spectrum) due to the absorption (or emission) of radiation by the gas in the photosphere. The continuous component of the spectrum gives information about the temperature of the photosphere, while the absorption (or emission) lines make it possible to determine the chemical composition of the star, its surface gravity and the electric and magnetic fields.
The continuous spectrum originates from a process called *bremmstrahlung*, from the German *bremsen* "to brake, decelerate" and *strahlung* "radiation", which literally translates to *braking* or *deceleration radiation.* Because of the very high temperature, matter is in the plasma state at the centre of stars, and electrons move chaotically amongst ionized atoms. An electron approaching a charged atom is scattered and accelerates away from it, consequently

emitting radiation in the form of electromagnetic waves (Pr. 7.1). Since the electron-atom distance and the electron velocity can vary continuously within a star, the resulting electromagnetic radiation is distributed continuously over all wavelengths, therefore the spectrum it produces is also continuous. Bremmstrahlung is thus characterized by a continuous radiation distribution that becomes more intense (and moves towards higher frequencies) with increasing velocity (thermal energy) of the electrons, i.e. with increasing temperature. Bremmstrahlung is sometimes referred to as *free-free* radiation: it is created by charged particles that are not bound to any ion, atom or molecule, and are therefore free, both before and after the deflection.

Another phenomenon that contributes to the continuous spectrum is the recombination of an electron with an ion. As before, the initial energy of an electron is *free* to take any value, therefore, even though the final energy must be quantized (i.e. *bound* to have the same energy as a given orbital in the atom), the energy emitted in the form of electromagnetic radiation can take any value, giving rise to a continuous spectrum. The recombination of an electron with an ion is referred to as *free-bound* radiation. As long as there is a free component, the total energy is not quantized and the resulting spectrum is continuous.

The emission spectrum is produced by the emission of energy from excited atoms. Normally, the electrons inside an atom occupy the orbitals which correspond to the lowest overall energy (*ground state*) but, when excited, they move to higher-energy orbitals. Returning to the ground state, the atom emits a photon whose energy is equal to the energy difference between the excited and the ground state. Since the energy of an electron in an atom is quantised, and the electron does not leave the atom after the transition, it follows that the energy (and therefore the wavelength) of the emitted photon must be quantized. Because the energy levels vary from atom to atom (they depend on nuclear charge and degree of ionization), each element leads to different emission lines. Therefore, the emission (and absorption) spectrum of each element is unique. In this sense, it can be regarded as the fingerprint of an element: by examining the emission spectrum of an unknown object and comparing it with other previously recorded spectra, it is possible to determine the chemical composition of the source.

The absorption spectrum is the opposite of the emission spectrum. Electrons within an atom can absorb the incident radiation and move to higher-energy orbitals, then returning to the ground state by emitting exactly the same energy. Therefore, each atom absorbs the same wavelengths that it emits. At these wavelengths, the continuous spectrum appears darker, since light is absorbed and then re-emitted in a random direction (which rarely coincides with the direction of the observer). The analysis of absorption lines, called *Fraunhofer lines*, provides information on the chemical composition of stars.

7.6 Doppler Effect

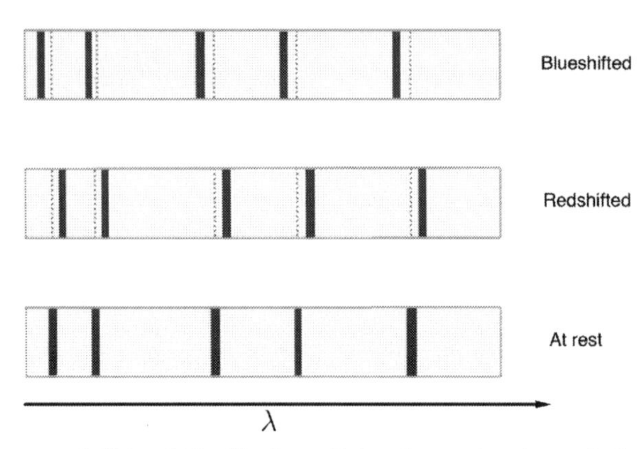

Figure 7.2: A spectrum is "blueshifted" when all the absorption (or emission) lines are shifted towards shorter wavelengths (compared to the spectrum at rest); "redshifted" when they are shifted towards longer wavelengths.

As seen in the previous section, the emission and absorption lines of atoms and molecules can be recorded in the laboratory. By comparing these with the spectra of celestial bodies, it is possible to determine their chemical composition. We often notice a shift of all these emission (or absorption) lines by a few nanometers towards longer (red) or shorter (blue) wavelengths, as in Fig. 7.2. This is due to the relative motion between the source and the observer. The law that relates the relative velocity to the shift of the spectral lines is called *Doppler effect.* From the discovery of exoplanets and spectroscopic binaries, to the expansion of the universe, this law has permanently changed the way we think of our place in the universe.

If you already studied the Doppler effect, it was probably applied to a wave that propagates in a medium, such as sound in air. You should have then learnt two different formulae, depending on whether the observer or source are in motion. However, light shows a different behaviour. Electromagnetic radiation propagates in vacuum, therefore there is no absolute reference system (such as air, for sound) against which to measure velocity. The only information we can possibly have is the *relative* velocity between two bodies. Therefore, we expect the Doppler effect for electromagnetic radiation to depend only on the relative velocity between source and observer; hence there must be only one formula that describes this effect. The hypotheses just mentioned are actually the postulates of special relativity, therefore the equation we are looking for can only be obtained within this theory. We first derive an approximate formula, using classical mechanics, and then the exact one, using special relativity.

In the classical approximation, let us only consider the case of a moving source and a stationary observer, the proof for the other case being very similar. The

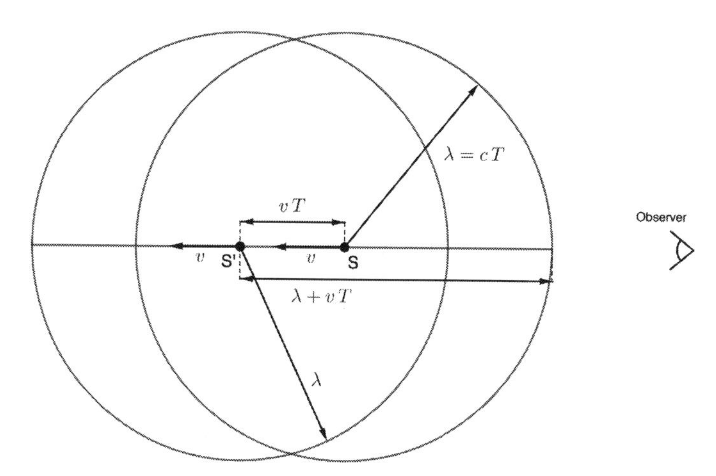

Figure 7.3: Doppler effect for a source in motion with velocity v, which emits light with wavelength λ. After a period, the source will have moved vT away from the observer, therefore the observer perceives that light is emitted with a wavelength $\lambda' = \lambda + vT$.

source moves with velocity v away from the observer. Light has wavelength λ, period T and velocity c, thus $\lambda = cT$. After a period T, the source has moved away by a distance vT, hence the next crest is emitted at a distance vT farther away from the observer. Therefore, as shown in Fig. 7.3, the perceived wavelength is slightly longer: $\lambda' = \lambda + vT = (c+v)/f$. Since light propagates with velocity c in both cases, we find $c = \lambda' f' = (c+v)f'/f$, so that:

$$f' = \frac{1}{1+v/c} f \approx \left(1 - \frac{v}{c}\right) f, \tag{7.14}$$

where we have assumed v/c to be small and used the approximation $(1+x)^\alpha \approx 1 + \alpha x$ for $x \ll 1$ (Appendix A.6), since classical physics is only valid for $v \ll c$. In the derivation that follows, we will make use of the fact that, given two inertial reference systems K and K', moving with relative velocity v, the time Δt measured by an observer in K is related to the time $\Delta t'$ measured by an observer in K' by the following equation:

$$\Delta t' = \gamma \Delta t \quad \text{where} \quad \gamma = \frac{1}{\sqrt{1-(v/c)^2}}. \tag{7.15}$$

This relation is called *time dilation*. Indeed, since $1-(v/c)^2$ is always less than unity, it follows that $\gamma \geqslant 1$. Therefore, the time measured in K' is longer than the time in K.

Let $\Delta t = 1/f$ be the emission time in the reference system of the source. The time perceived by the observer is $\Delta t' = \gamma \Delta t$, because of time dilation. In the observer's reference system, after a period $\Delta t'$, the photons have travelled a distance equal to $c\Delta t' = c\gamma\Delta t$, while the source has travelled away by $v\Delta t' =$

$v\gamma\Delta t$. Therefore, the perceived wavelength is slightly longer: $\lambda' = (c+v)\gamma\Delta t$. Then, from $f' = c/\lambda'$, we find:

$$\begin{aligned} f' &= c\frac{\sqrt{(1-v^2/c^2)}}{c+v}\frac{1}{\Delta t} \\ &= \frac{\sqrt{(1-v/c)(1+v/c)}}{1+v/c}f \\ &= \sqrt{\frac{1-v/c}{1+v/c}}f\,. \end{aligned}$$

Setting $\beta = v/c$:

$$\boxed{f' = \sqrt{\frac{1-\beta}{1+\beta}}f}\,. \tag{7.16}$$

If $\beta < 0$, i.e. if the source is moving closer to the observer, Eq. 7.16 tells us that $f' > f$. In this case, we say that light is *blueshifted*, since blue is the colour with the highest frequency. If, instead, $\beta > 0$, i.e. the source is moving away from the observer, it follows that $f' < f$. In this case, we say that light is *redshifted*, since red is the colour with the lowest frequency. The energy of a photon is proportional to its frequency, therefore redshifted light carries less energy. Conversely, blueshifed light carries more energy. Let L_{e} be the luminosity emitted by a star in its reference system. The observed luminosity is:

$$\boxed{L_{\mathrm{o}} = L_{\mathrm{e}} \cdot \frac{f'}{f}}\,. \tag{7.17}$$

In the classical limit ($\beta \ll 1$), Eq. 7.16 reduces to:

$$f' \approx \left(1 - \frac{v}{2c}\right)\left(1 - \frac{v}{2c}\right)f \approx 1 - \frac{v}{c}\,,$$

where we have used the approximation $(1+x)^\alpha \approx 1 + \alpha x$ for $x = v/c \ll 1$. This agrees with the classical result found earlier (Eq. 7.14). Eq. 7.16 can also be written as a function of the wavelength:

$$\lambda' = \sqrt{\frac{1+\beta}{1-\beta}}\lambda\,.$$

From this, we can then compute the wavelength shift $\Delta\lambda = \lambda' - \lambda$:

$$\Delta\lambda = \left(\sqrt{\frac{1+\beta}{1-\beta}} - 1\right)\lambda\,. \tag{7.18}$$

In the non-relativistic limit, this reduces to:

$$\Delta\lambda = \beta\lambda \, . \tag{7.19}$$

The term $\Delta\lambda/\lambda$ is usually denoted by z, called the *redshift parameter.* If z is known, β can be found from Eq. 7.18:

$$\boxed{\beta = \frac{(1+z)^2 - 1}{(1+z)^2 + 1}} \, , \tag{7.20}$$

which, in the non-relativistic limit ($z \ll 1$), becomes:

$$\boxed{\beta = z} \, . \tag{7.21}$$

7.7 Harvard Stellar Classification

A star can be classified based on its surface temperature, which can be estimated from the wavelength of maximum emission, using Wien's law. Another way to estimate the temperature is by studying the ionized elements in the photosphere, since the type and degree of ionization depends on the temperature. According to the Harvard stellar classification system, the stars can be divided into 7 classes: O, B, A, F, G, K and M (there are several mnemonics used to remember this sequence, the most famous one being: "Oh Be A Fine Girl/Guy, Kiss Me").

Class	Temperature / K	Colour	Luminosity / $L_\odot$	Mass / $M_\odot$
O	$\geqslant 33000$	blue	$\geqslant 30000$	$\geqslant 16$
B	10000 – 33000	light blue	25 – 30000	2.1 – 16
A	7500 – 10000	white	5 – 25	1.4 – 2.1
F	6000 – 7500	white-yellow	1.5 – 5	1.04 – 1.4
G	5200 – 6000	yellow	0.6 – 1.5	0.8 – 1.04
K	3700 – 5200	orange	0.08 – 0.6	0.45 – 0.8
M	$\leqslant 3700$	red	$\leqslant 0.08$	0.08 – 0.45

Table 7.1

However, stellar classification is not that simple. Every class is further divided into 10 subclasses, numbered from 0 to 9. In the Harvard classification system, the Sun is a G2 star. More recently, the classes L and T have been added. These correspond to brown dwarfs. Tab. 7.1 gives the range of temperatures, luminosities and masses for the 7 classes O – M.

7.8 Yerkes Stellar Classification

The Harvard scheme specifies only the surface temperature and some spectral features of the star. A more precise classification would also include the luminosity of the star, since two stars with similar temperatures, but different radii, can have very different luminosities. It is possible to obtain the luminosity by examining the spectrum of the star. The mass of giant and dwarf stars is of the same order of magnitude, but their radii are very different. Therefore, the gravitational acceleration on the surface of a giant star is much lower than the acceleration on a dwarf star, since $g = GM/R^2$. Given the lower gravity, gas pressures and densities are much lower in giant stars than in dwarf stars. It can be shown that the width of spectral lines is proportional to the pressure (Pr. 7.4), therefore measuring this width ultimately allows us to find the radius of the star. By estimating the temperature of a star using Wien's law, it is then possible to find its luminosity using Stefan-Boltzmann's equation. The Yerkes scheme divides the stars into six luminosity classes:

- Ia : Most luminous supergiants
- Ib : Less luminous supergiants
- II : Luminous giants
- III: Normal giants
- IV: Subgiants
- V : Main sequence stars (dwarfs)

For example, the Sun belongs to the luminosity class "V". Thus, the Sun would be more fully specified as a G2V-type star.

7.9 Exercises

1. At what wavelength does a star with surface temperature $T = 4000$ K emit the most energy?

2. Compute the total brightness of a star with surface temperature $T = 6800\,\text{K}$ and radius $R = 2.5R_{\odot}$. Write your answer in terms of the solar luminosity $L_{\odot}$.

3.* Assuming that dust particles ($\rho \approx 10^3\,\text{kg/m}^3$) behave like black bodies, determine the diameter of a spherical particle that is in equilibrium (at rest) at a distance of one astronomical unit from the Sun.

4. Estimate the number of neutrinos, produced by the Sun, that arrive on Earth. Each nuclear reaction produces 26.8 MeV of energy and 2 electron neutrinos. On Earth, early experiments detected only 1/3 of the expected number of neutrinos. This was known as the *solar neutrino problem*, and was finally solved by realizing that an electron neutrino can oscillate into the other two types of neutrinos (muon and tau).

5.* (ARAO 2019, Th.X/XI, q.1) A future project aims to propel small spaceships with a powerful laser beam, sending them over long distances. What speed can a spaceship be propelled at, if its perfectly reflecting base has a diameter of 1 mm, its total mass is 1 mg, and the optical laser has a power of 1 MW and divergence of $5''$? Assume that the base of the spaceship is oriented perpendicularly to the laser beam, that the beam itself is very thin when exiting the laser, and neglect the gravitational action of all the bodies surrounding the spaceship.

6.* At sea level, the Earth's atmosphere has an average temperature of $T = 14°\text{C}$. Compute:

- v_{rms} for molecular hydrogen, oxygen and nitrogen;
- the percentage of each of these molecules with velocities greater than the escape velocity from the Earth at sea level (you will need to solve an integral numerically).

7.** Prove that the probability of finding a molecule with 3 and 5 times the root mean square velocity of the gas is $5.88 \cdot 10^{-6}$ and $3.62 \cdot 10^{-16}$, respectively (you will need to solve an integral numerically).

8.* Prove that the most probable speed of molecules in a gas is $v_{\text{mp}} = \sqrt{2k_BT/m}$.

9.** Assume that the atmosphere has a uniform composition with an effective molar mass of $u = 28.9\,\text{g/mol}$ and a constant temperature of $T = 14°\text{C}$. Find the ratio of atmospheric pressure at height h and at sea level.

10.** A space station is in the shape of a cylinder with radius R_0 and is filled with air. The cylinder rotates around its axis of symmetry, providing an acceleration equal to g at the outer edge. If the temperature inside the space station is constant and equal to T, what is the ratio of the pressure P_0 at the outer edge of the station, to P_c, the pressure at the centre?

11.** Consider a spaceship with internal pressure $P_0 = 1\,\text{atm}$, constant temperature $T_0 = 285\,\text{K}$ and volume $V_0 = 100\,\text{m}^3$. Let $\mu = 28.9\,\text{g/mol}$ be the effective molar mass of air. Suddenly, the spaceship is struck by an object that makes a hole of area $A = 1\,\text{cm}^2$. Find the pressure of air inside the spaceship as a function of time. How long does it take for the pressure to halve?

12.** Prove the formula for the root mean square speed of a molecule in a gas. You may want to use the fact that the average of v^2, for the distribution $f(v)$, is given by the integral:

$$\langle v^2 \rangle = \int_0^\infty v^2 f(v)\, dv \tag{7.22}$$

7.10 Problems

1.* **Radiation emitted by accelerated charge**
The energy emitted by an accelerated charge depends exclusively on the vacuum permittivity ϵ_0, the speed of light c, the electric charge q and the acceleration a. Using dimensional analysis, derive a formula for $E(\epsilon_0, c, q, a)$. It can be shown that the numerical factor is $1/6\pi$.

2.** **Maxwell distribution**
Let $g(v_x)$ be the velocity distribution in the x direction, so that $g(v_x)\, dv_x$ is equal to the fraction of particles with velocities in the interval $[v_x, v_x + dv_x]$. Using the fact that $g(v_x)$ must be proportional to the Boltzmann factor, where $E = mv_x^2/2$, and normalizing $g(v_x)$, show that:

$$g(v_x) = \sqrt{\frac{m}{2\pi k_B T}}\, e^{-mv_x^2/2k_B T} \tag{7.23}$$

Applying a similar reasoning to v_y and v_z, find an expression for the fraction of particles with velocities in the interval $[(v_x, v_y, v_z), (v_x + dv_x, v_y + dv_y, v_z + dv_z)]$. Thus, determine the velocity distribution $f(v)$, where $v = \sqrt{v_x^2 + v_y^2 + v_z^2}$, proving the Maxwell distribution.

3.** **Natural width of emission lines**
Excited atoms spontaneously emit photons when relaxing back to the

ground state. Consider a hydrogen atom that emits visible light. By equating the difference in energy between two energy levels with the energy radiated by an accelerated electron (Pr. 7.1), estimate the time Δt necessary for the transition. Using the Heisenberg uncertainty principle applied to energy and time ($\Delta E \Delta t \geqslant \hbar/2$), find the uncertainty in the energy ΔE. This is the natural thickness of the emission lines for a hydrogen atom that emits visible light.

4.* **Doppler broadening**

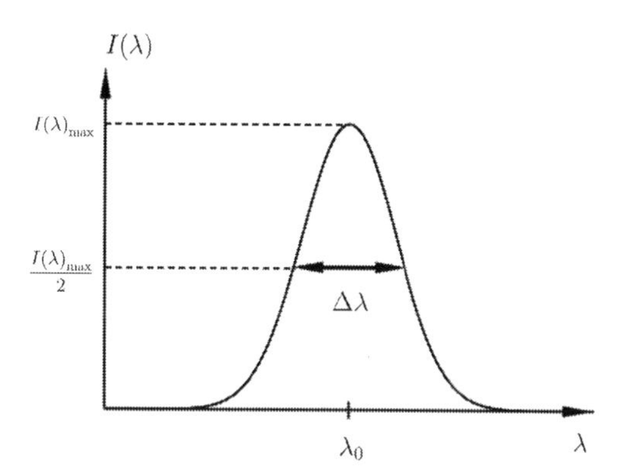

Figure 7.4: Emission lines from stellar spectra are widened because of the (thermal) motion of atoms on the surface.

In addition to the natural width of emission lines, the random motion of atoms on the surface of a star causes, due to the Doppler effect, a further increase in width (Fig. 7.4). Let x be the axis of observation. Starting from the Boltzmann distribution law, show that the fraction of atoms with mass m, emitting between λ and $\lambda + d\lambda$, is:

$$I(\lambda) \propto \exp\left\{-\frac{mc^2(\lambda-\lambda_0)}{2k_BT\lambda_0^2}\right\}, \tag{7.24}$$

where λ_0 is the wavelength of emission at rest. Therefore, show that the width for which the intensity is at least half the maximum value ($I(\lambda) > I(\lambda)_{\max}/2$) is:

$$\frac{\Delta\lambda}{\lambda} = 2\sqrt{2\ln 2\frac{k_BT}{mc^2}}\,. \tag{7.25}$$

5.* **Wien and Stefan-Boltzmann laws**
The blackbody distribution law can be obtained from the Boltzmann distribution, where the energy is that of a quantum harmonic oscillator

$E = (\frac{1}{2} + n)\hbar\omega$, since electromagnetic waves inside the body form standing waves. It is then possible to obtain the following expression for the power emitted in the frequency interval $[f,\, f + df]$:

$$P_f\, df = \frac{2\pi h}{c^2} \frac{f^3}{e^{\frac{hf}{k_B T}} - 1}\, df\,. \tag{7.26}$$

Compute the expression for $P_\lambda\, d\lambda$, defined as the energy emitted in the interval $[\lambda,\, \lambda + d\lambda]$. Prove that the maximum of P_λ is at $\lambda T = 0.0029\,\mathrm{K\,m}$ (Wien's law). Also prove that the maximum of P_f is at $fT = 0.0051\,\mathrm{K/s}$. By integrating over all frequencies, prove that $P = \sigma T^4$ (Stefan-Boltzmann law), where σ is given by:

$$\sigma = \frac{2\pi^5 k_B^4}{15 c^2 h^3}\,. \tag{7.27}$$

You can use the following integral:

$$\int_0^\infty \frac{x^3}{e^x - 1} dx = \frac{\pi^4}{15}\,.$$

6. **Ultraviolet catastrophe**

Obtain an approximate expression for P_λ, in the limiting case of long wavelengths. Show that, in this approximation:

$$P_\lambda = \frac{2\pi c k_B T}{\lambda^4}\,. \tag{7.28}$$

This is the Rayleigh-Jeans law. It is possible to derive this equation using only classical physics (in fact, Planck's constant does not appear in Eq. 7.28). Before quantum mechanics, this was believed to be the distribution law governing black body radiation. Explain why people talked about the "ultraviolet catastrophe".

7.** **Why do stars emit energy?**

Understanding the physical process that allows stars to emit energy is essential for estimating their lifetime. Over time, numerous models have been proposed. Explain, with reasoning supported by calculations, why the first two proposals cannot be correct, while the third is believable.

- 1841, Mayer's hypothesis. The Sun takes energy from the impact of asteroids on its surface. (Hint: first solve Ex. 10.13)
- 1854, Kelvin-Helmholtz hypothesis. According to the so-called contraction theory, the Sun emits energy at the expense of its gravitational potential energy. Over time, the star contracts, until all of its potential energy is exhausted (Hint: first solve Pr. 10.4).

- In the nucleus of the star, mass is converted directly into energy through a process known as *nuclear fusion*, according to $E = mc^2$. Furthermore, only 10 % of the initial mass of hydrogen in the star can be transformed into helium. In the hydrogen-helium conversion, 0.72 % of the hydrogen mass is converted into energy.

8.* **Inside a star**

It is possible to describe the interior of a star with four differential equations that govern the distribution of pressure, mass, energy production and temperature gradient. Assume throughout that the star is spherically symmetric.

- The equation of *hydrostatic equilibrium* governs the pressure inside a star. Assuming that the star is in equilibrium with its internal pressure $P(r)$ and gravitational force, prove that:

$$\frac{\mathrm{d}P}{\mathrm{d}r} = -\frac{Gm(r)\rho(r)}{r^2}, \tag{7.29}$$

where $\rho(r)$ is the density at a distance r from the centre of the star and $m(r)$ is the mass contained within that distance (Hint: first solve Pr. 10.2).

- The second equation governs the mass distribution. Prove that:

$$\frac{\mathrm{d}m(r)}{\mathrm{d}r} = 4\pi r^2 \rho\,. \tag{7.30}$$

- The third equation governs the production of energy. Let ϵ be the coefficient of energy production per unit mass and unit time, and let L_r the luminosity produced at a distance r from the centre of the star. Prove that:

$$\frac{\mathrm{d}L_r}{\mathrm{d}r} = 4\pi r^2 \rho\,\epsilon\,. \tag{7.31}$$

- The fourth equation governs the temperature gradient inside the star. Let $k(r)$ be the coefficient of energy absorption, such that $k\,\rho\,dr$ is equal to the energy absorbed by a layer dr of the star, with density ρ. Given that the radiation pressure of a photon gas is $P_{\mathrm{rad}} = 1/3\,aT^4$, where a is the *radiation constant*, show that:

$$\frac{\mathrm{d}T}{\mathrm{d}r} = \frac{3}{4ac}\frac{k\rho}{T^3}\frac{L_r}{4\pi r^2}\,. \tag{7.32}$$

It can be proven that $a = 4\sigma/c$.

8

Flux and Magnitude

In this chapter we introduce the basic concepts that characterize electromagnetic radiation.

8.1 Flux

The flux is the amount of energy incident on a surface per unit time and unit area. Assuming that a star radiates isotropically with power L (*intrinsic brightness*), at a distance d from the star its energy is distributed on a sphere of radius d, with surface area $4\pi d^2$. Therefore, the flux at this distance is:

$$\boxed{F = \frac{L}{4\pi d^2}}\,. \tag{8.1}$$

The flux is directly proportional to the intrinsic brightness and inversely proportional to the square of the distance. Consider, for example, the Sun observed from Earth. Its brightness is $L_\odot = 3.828 \cdot 10^{26}\,\mathrm{W}$, while the Sun-Earth distance is $d_\oplus = 149.6 \cdot 10^9\,\mathrm{m}$. Hence, the solar flux on Earth is $k = L_\odot/(4\pi d_\oplus^2) = 1367\,\mathrm{W}$. This is known as the *solar constant.*
If we know the radius and the temperature of a star, we can replace L in Eq. 8.1 with the expression given by the Stefan-Boltzmann law. The previous equation then becomes:

$$F = \frac{\sigma e R^2 T^4}{d^2}\,. \tag{8.2}$$

8.2 Albedo

The albedo (A) of a body is the ratio of the reflected (L_r) to the incident (L_i) power:

$$\boxed{A = \frac{L_\mathrm{r}}{L_\mathrm{i}}}\,.$$

Only a part of the radiation incident on a body is again reflected into space, while the rest is absorbed. As a result, the body warms up, emitting the absorbed energy as thermal radiation. The difference between the incident and the reflected power represents the absorbed power: $L_\mathrm{i} - L_\mathrm{r} = L_\mathrm{i}(1 - A)$. A body that reflects most of the incident radiation has a high albedo, while a darker body is characterized by a low albedo. Snow has an albedo of 0.9, the oceans — 0.06. On land, the albedo varies from 0.1, in forests, to 0.4, in deserts. Because the Earth is mainly covered by water, the average albedo of its surface is about $A_s = 0.08$. The albedo of the atmosphere is instead $A_\mathrm{atm} = 0.24$, due to the high reflectivity of the clouds. The average albedo of the Earth is about $A_\oplus = 0.31$ (a little less than the sum of A_s and A_atm).

8.3 Stellar magnitudes

Since ancient times astronomers have recorded the positions of stars on the celestial sphere and measured their brightnesses. The first stellar catalogue was compiled in 160 BC by the Greek astronomer Hipparchus. He divided the stars into six classes of brightness, from the first magnitude, formed by the brightest stars, to the sixth, formed by the stars barely visible to the naked eye. However, when more sensitive instruments became available, it was clear that two stars with a magnitude difference of one had a flux ratio of approximately 2.5. In 1854, Norman Pogson replaced the Hipparchus scale with a more precise one, while remaining as faithful as possible to the old classification method. According to this new scale, it was established that the ratio between the fluxes of a first and sixth magnitude star should be 100. Therefore, two stars with a magnitude difference of one would have a flux ratio of $\sqrt[5]{100} = 2.512$. Adopting the convention used by Hipparchus, we can write:

$$\frac{F_x}{F_0} = (\sqrt[5]{100})^{-\Delta m}\,, \tag{8.3}$$

where Δm represents the difference in magnitude between an object with flux F_x and another with flux F_0. The minus sign in front of Δm signifies that higher magnitudes correspond to lower fluxes.

In the following derivation, we will use the properties of the logarithm given in Appendix A.5. In Eq. 8.3, we take the logarithm of both sides, isolating Δm:

$$\begin{aligned}\log\frac{F_x}{F_0} &= \log 100^{-\frac{\Delta m}{5}}\\ &= \log 10^{-\frac{2\Delta m}{5}}\\ &= -\frac{2}{5}\Delta m\,,\\ \Rightarrow \Delta m &= -2.5\log\frac{F_x}{F_0}\,.\end{aligned}$$

In general, for two stars with magnitudes m_1 and m_2:

$$\boxed{m_1 - m_2 = -2.5 \log \frac{F_1}{F_2}}. \tag{8.4}$$

While the inverse formula is:

$$\boxed{\frac{F_1}{F_2} = 10^{-0.4(m_1 - m_2)}}. \tag{8.5}$$

Sirius, the brightest star in the night sky, has an apparent magnitude of -1.5. The full Moon has a magnitude of -12.74; the Sun of -26.74. If we know the radii, temperatures and distances of two stars from Earth, we can substitute Eq. 8.2 for the fluxes:

$$m_1 - m_2 = -2.5 \log \frac{{R_1}^2 {T_1}^4 {d_2}^2}{{R_2}^2 {T_2}^4 {d_1}^2},$$

which simplifies to:

$$m_1 - m_2 = -5 \log \frac{R_1 {T_1}^2 d_2}{R_2 {T_2}^2 d_1}. \tag{8.6}$$

If the stars share some parameters, a simpler equation can be obtained. Note that Eq. 8.6 compares the *bolometric* magnitudes ($m_{\rm bol}$), that is, the fluxes integrated over the whole spectrum. Since our eyes record light in a small window of the electromagnetic spectrum, the bolometric magnitude is, in general, different from the visual magnitude $m_{\rm v}$ (see "Magnitude Systems").
The flux, recorded by light-sensitive devices, is proportional to the brightness of the body. Our eyes perceive brightness in a slightly different way, and it is for this reason that we use the magnitude scale. The magnitude is directly proportional to the logarithm of the brightness (i.e. to the logarithm of the flux).
The fact that the human eyes perceive light in a logarithmic way can be clarified with the following argument. Let us imagine we have a large number of light bulbs, all of the same brightness. At first we only turn on one, then another, and compare the overall brightnesses in the two cases. You will agree that the perceived brightness doubles. Now, let us turn on 100 light bulbs, and see how the brightness changes by adding another one. For a photosensitive device, the increase in brightness is the same as it was in going from one to two bulbs. However, for the human eye, the brightness does not noticeably change. We would perhaps need to turn on another 100 bulbs to notice the same increase in brightness. In other words, what our eyes perceive is not the absolute change in flux ΔF, but rather the relative change $\Delta F/F$. Hence the logarithmic relation, which is just the integral of $\int dF/F$.

Absolute Magnitude

The flux of a star depends on both its intrinsic brightness and on its distance from Earth. Therefore, an intrinsically brighter star, placed at a greater distance than an intrinsically fainter star, may appear less luminous. Let us imagine moving both stars to the same distance from Earth. If we now find that the first star is brighter than the second, we conclude that the former is also *intrinsically* brighter. For this reason, it is convenient to introduce the quantity called *absolute magnitude*, equal to the magnitude that a star would have if it were placed at the conventional distance of 10 parsec. Therefore, by comparing the absolute magnitudes, we are effectively comparing the intrinsic luminosities of the stars, rather than their fluxes.
Starting from this definition, let us derive the formula that relates the absolute magnitude of a star to its apparent magnitude and distance from Earth. If a star was placed at a distance of 10 pc, its flux would be:

$$F_M = \frac{L}{4\pi(10\,\mathrm{pc})^2}\,.$$

However, the actual distance of the star (in parsec) is d, therefore the flux is:

$$F_m = \frac{L}{4\pi d^2}\,.$$

Denoting with M the absolute magnitude of the star, and applying Eq. 8.4:

$$\begin{aligned} M - m &= -2.5\log\frac{L/(4\pi\cdot 10\,\mathrm{pc})^2}{L/(4\pi d^2)} \\ &= -2.5\log\left(\frac{d}{10}\right)^2 \\ &= -5\log d + 5\,. \end{aligned}$$

Therefore, the absolute magnitude is:

$$\boxed{M = m - 5\log d + 5}\,, \tag{8.7}$$

where d is measured in parsec. If, instead of the distance, we know the parallax of the star, it is possible to use the formula $d = 1/\pi_p{}''$ (Eq. 9.5):

$$M = m - 5\log\frac{1}{\pi_p{}''} + 5\,.$$

Using the logarithm property $\log 1/x = -\log x$, the last equation becomes:

$$\boxed{M = m + 5\log\pi_p{}'' + 5}\,. \tag{8.8}$$

Magnitude of a Composite System

We want to find the overall magnitude of multiple light sources that appear to the observer as a single, brighter object. Since the magnitude scale is logarithmic, to obtain the magnitude of a composite system we cannot simply add up the magnitudes of each source. We can, however, add up their fluxes, which are proportional to their luminosities. Using Eq. 8.5 for each source:

$$F_1 = 10^{-0.4(m_1 - m_0)} \cdot F_0 \,,$$
$$F_2 = 10^{-0.4(m_2 - m_0)} \cdot F_0 \,,$$
$$\dots ,$$
$$F_n = 10^{-0.4(m_n - m_0)} \cdot F_0 \,;$$

where m_0 and F_0 are the magnitude and the flux of a reference object (introduced for convenience, but will simplify at the end). The flux of the system is obtained by summing the fluxes of the components. Therefore, applying Eq. 8.4 to the composite system and to the reference object:

$$m_{\mathrm{sys}} - m_0 = -2.5 \log \frac{F_1 + F_2 + \dots + F_n}{F_0} \,.$$

Substituting the fluxes as a function of the magnitudes:

$$m_{\mathrm{sys}} - m_0 = -2.5 \log \left[\frac{10^{-0.4(m_1 - m_0)} + 10^{-0.4(m_2 - m_0)} + \dots + 10^{-0.4(m_n - m_0)}}{F_0} F_0 \right] .$$

Simplifying F_0:

$$m_{\mathrm{sys}} - m_0 = -2.5 \log \left[10^{-0.4(m_1 - m_0)} + 10^{-0.4(m_2 - m_0)} + \dots + 10^{-0.4(m_n - m_0)} \right] .$$

Dividing by the factor $10^{0.4\, m_0}$:

$$m_{\mathrm{sys}} - m_0 = -2.5 \log \left[10^{0.4\, m_0} \cdot (10^{-0.4\, m_1} + 10^{-0.4\, m_2} + \dots + 10^{-0.4\, m_n}) \right] .$$

Finally, using the logarithm property $\log(x \cdot y) = \log x + \log y$:

$$m_{\mathrm{sys}} - m_0 = -2.5 \log 10^{0.4\, m_0} - 2.5 \log [10^{-0.4\, m_1} + 10^{-0.4\, m_2} + \dots + 10^{-0.4\, m_n}] .$$

But $-2.5 \log 10^{0.4\, m_0} = -m_0$, therefore the last equation simplifies to:

$$\boxed{m_{\mathrm{sys}} = -2.5 \log [10^{-0.4\, m_1} + 10^{-0.4\, m_2} + \dots + 10^{-0.4\, m_n}]} \,. \qquad (8.9)$$

Integrated Magnitude

The relationships given above refer to point objects, such as stars and planets. The total brightness of an extended source, such as a galaxy or a star cluster, can be expressed in terms of the *integrated* magnitude, equal to the magnitude that the object would have if all of its light were concentrated into a point source. The integrated magnitude can be obtained from the *surface brightness* and the area of the object under consideration. The surface brightness, with units of mag/arcsec^2, is a measure of the brightness of a portion of the surface with an area equal to one arcsecond squared.
Let F_s be the flux from one arcsecond squared of the object and A its total area (expressed in arcsecond squared). Let m_{int} and m_s be the integrated magnitude and surface brightness respectively. From Eq. 8.4, it follows that:

$$m_{int} - m_s = -2.5 \log \frac{F_{int}}{F_s} .$$

Since $F_{int} = A \cdot F_s$, the last equation can be written as:

$$\boxed{m_{int} = m_{sup} - 2.5 \log A} . \quad (8.10)$$

The apparent magnitude of an astronomical object is generally quoted as an integrated value. Therefore, if a galaxy has a magnitude of 12.5, it means we receive the same total amount of light from the galaxy as we would from a star with magnitude 12.5. However, the star is so small that it can be effectively considered as a point source (the largest angular diameter, that of R Doradus, is 0.057 ± 0.005 arcsec), whereas the galaxy may extend over several arcseconds or arcminutes. Therefore, the galaxy will be more difficult to see than the star. Surface brightness can be useful to understand the visibility of a certain object. It can be compared to the surface brightness of the night sky, which is 21.9 mag/arcsec^2 at the zenith, in the V band.

Magnitude Systems

Usually, when talking about brightness, we refer to the bolometric brightness, i.e. the brightness of the object summed over all wavelengths. However, practically speaking, the bolometric brightness cannot be measured directly, since each measurement only records light in a narrow region of the electromagnetic spectrum. For this reason, it is customary to specify the brightness of an object in a particular region of the electromagnetic spectrum. The sensitivity of the human eye varies with the wavelength. In daylight, it has a maximum at the wavelength of 550 nm (yellow), while it decreases towards red and violet. The magnitude that corresponds to the part of the electromagnetic spectrum visible to the naked eye is called *visual* magnitude, and is denoted by m_v. Photographic plates are usually more sensitive to blue and violet and can also record radiation not visible to the naked eye. The photographic magnitude is usually denoted by $m_{\rm pg}$, and is generally slightly different from the visual magnitude. However, the sensitivity of the eye can be simulated using a yellow filter and plates sensitive to yellow and green light. If, ideally, we were able to measure the magnitude of the star over its entire spectrum, we would obtain the bolometric magnitude, denoted by $m_{\rm bol}$. In practice, this is difficult since a large part of the electromagnetic spectrum is absorbed by the atmosphere, and different wavelengths require different sensors. Bolometric magnitude and visual magnitude are related by:

$$\boxed{m_{\rm bol} = m_v + {\rm BC}}, \tag{8.11}$$

where BC is called the *bolometric correction*. By definition, the bolometric correction is zero for Sun-type stars (G2V spectral class). In all other cases, for both warmer and colder stars, the bolometric magnitude is greater than the visual magnitude, therefore BC is always negative.
One of the most common systems used to classify the magnitude of stars, in the region of the spectrum close to visible light, is the UBV system, developed by Harold L. Johnson and William W. Morgan in 1950. In this system, magnitudes are measured using three different filters: U=Ultraviolet, B=Blue and V=Visual, hence the name UBV. In the UBV system it is usual to give the visual magnitude of the star, together with the quantities U−B and B−V, called the *colour indices*. The colour indices were chosen to be zero for an A0-type star (like Vega). Therefore, Vega has V= 0.03, B−V=U−B=0. Instead, the Sun has V= −26.8, B−V= 0.62 and U−B= 0.10. Later, this system was expanded to UBVRI, in order to include the red R and infrared I.

8.4 Extinction and Optical Depth

All the above equations are valid if the space between the source and the observer is completely empty, i.e. in the absence of a medium (such as interstellar dust or the atmosphere), which absorbs and scatters light. The interstellar dust is composed of grains of various sizes, the smallest of which are planar molecules, such as aromatic hydrocarbons (PAHs), formed by a few dozen atoms. Then, there are small three-dimensional grains (SG, between 100 − 2000 nm) and big grains (BG, over 2000 nm) mostly composed of silicates and graphite. If the wavelength is less than or comparable to the grain size, light is mostly absorbed. If, instead, the wavelength is significantly greater, it can pass through the interstellar medium without appreciable absorption. Therefore, PAHs absorb ultraviolet radiation, while BG can also absorb light emitted in the red band by relatively old stars. Since the interstellar medium alters the colour of the stars, it is usually possible to trace back the value of the *extinction* (absorption and scattering), by comparing the observed colour with the expected one (from the spectral class of the star). Since blue light is scattered more than red light, the B–V difference increases in the presence of an interstellar medium.

Additionally, light is scattered and absorbed by the atmosphere. The most important wavelength range for which the atmosphere is transparent is the so-called *optical window*, as it corresponds to the sensitivity range of the human eyes (400 – 700 nm). In the visible region, light is mainly absorbed and scattered by fine dust particles. At wavelengths below 300 nm, ozone is the main absorber, and it prevents ultraviolet light from reaching Earth's surface. Ozone is concentrated in a small band at a height of about 20 – 30 km, protecting the Earth from harmful, high-energy radiation. At shorter wavelengths, the molecules mostly responsible for absorption are molecular oxygen and nitrogen. Almost all radiation below 300 nm is absorbed by the atmosphere. At wavelengths greater than visible light, however, the atmosphere is quite transparent up to 1.3 μm. At this wavelength, electromagnetic radiation reaches the Earth in only a few thin windows. Wavelengths between 20 μm-1 mm are completely absorbed by the atmosphere. The *radio window* corresponds to the range 1 mm-20 m. Beyond 20 m (actually the upper limit depends on the daily conditions of the atmosphere), the ionosphere absorbs all radiation.

Light scattering by air molecules is inversely proportional to the fourth power of the wavelength (*Rayleigh scattering*). Therefore, the sky appears blue during the day because this is the colour with the shortest wavelength (and, hence, the most scattered by the atmosphere). At sunset, light from the Sun has to travel a greater distance through the atmosphere before reaching our eyes. In this case, both blue and red light are highly scattered, but blue light is scattered too early to be observed this far, therefore the intensity of red light reaching our eyes is greater.

Let us now derive a relationship between the extinction of light and the distance it travels through an opaque medium. It is reasonable to assume that, in a small interval $[r, r + dr]$, the extinction is proportional to the incident brightness L and to the distance dr. Therefore:

$$dL = -\kappa L \, dr \, ,$$

where κ is called the *opacity*, and is a measure of the amount of radiation absorbed by the medium per unit length. The opacity of vacuum is zero, while it is greater when the medium is more opaque or turbid. If the opacity is constant along the line of sight, it is possible to integrate both sides of the previous equation, while bringing κ outside the integral:

$$\int_{L_0}^{L(r)} \frac{dL}{L} = -\kappa \int_0^r dr \, ,$$

$$\ln \frac{L(r)}{L_0} = -\kappa r \,.$$

We then obtain:

$$\boxed{L = L_0 e^{-\kappa r}} \,. \tag{8.12}$$

The above equation gives the brightness of a star at a distance r, in the presence of a medium with opacity κ. The ratio L/L_0 is called the *transmittance*, and is often denoted by the letter e, followed by the wavelength at which it is measured. For instance, the transmittance of the Earth's atmosphere for yellow light is approximately 0.8, and it is smaller for ultraviolet and infrared. The plot of the transmittance as a function of wavelength is just the absorption spectrum of the gas constituting the medium. The quantity κr is called the *optical depth*, and is often denoted by the Greek letter τ.
In order to determine the effective magnitude of a star in the presence of an opaque medium, it is necessary to add a correction factor to the equations in the previous sections. Let F_0 be the light flux on the surface of the star (with radius R) and F_d the flux at a distance d. Isolating the luminosity in Eq. 8.1:

$$L_d = 4\pi d^2 F_d \,, \qquad L_0 = 4\pi R^2 F_0 \,.$$

Using Eq. 8.12 to write L_d/L_0, taking the ratio of the above equations:

$$F_d = F_0 \frac{R^2}{d^2} e^{-\kappa d} \,.$$

To compute the absolute magnitude, we need the flux (measured without extinction) at a distance of 10 pc:

$$F_{10\,\mathrm{pc}} = F_0 \frac{R^2}{(10\,\mathrm{pc})^2} \,.$$

Therefore, measuring d in parsecs:

$$\begin{aligned} M - m &= -2.5 \log \frac{F_{10\,\mathrm{pc}}}{F_d} \\ &= -2.5 \log \frac{F_0 R^2/(10\,\mathrm{pc})^2}{F_0 R^2 e^{-\kappa d}/d^2} \\ &= -5 \log \frac{d}{10} - 2.5 \log e^{\kappa d} \\ &= -5 \log d + 5 - 2.5 \log e^{\kappa d} \,. \end{aligned}$$

Taking extinction into account, the equation that relates the apparent and absolute magnitudes is:

$$\boxed{M = m - 5 \log d + 5 - 2.5 \log e^{\kappa d}} \,. \tag{8.13}$$

Compared to the formula without extinction, we note the additional factor $A(d) = -2.5 \log e^{\kappa d}$. Let us examine some limiting cases. For κ equal to zero, the term $e^{\kappa d}$ is equal to unity, therefore the logarithm and the correction term are zero. This situation corresponds to the absence of an interstellar medium, in which case Eq. 8.13 reduces to Eq. 8.7. For κ that tends to infinity, $e^{\kappa d}$ tends to infinity, therefore the correction factor is minus infinity. Now, consider two possibilities. If we fix the absolute magnitude of the star, it follows that the apparent magnitude is infinite, i.e. the apparent brightness is zero. On the other hand, if we fix the apparent magnitude, we find an absolute magnitude of minus infinity. Indeed, the intrinsic brightness of the object must be infinite to give a finite apparent magnitude in the presence of a medium that lets no light through.

Let us now compute the variation of the luminosity of a star as a function of its altitude. Let $R_\oplus$ be the radius of the Earth, h the height of the atmosphere and $90° - z$ the altitude of the star (where z is the zenith distance). Since most of the air is concentrated in the first 10 km of the atmosphere, $h \approx 10\,\text{km} \ll R_\oplus$. From Fig. 8.1, it can be seen that, for stars close to the horizon, the distance which light travels through the atmosphere is $h/\sin z$. Therefore:

$$m_{\text{space}} = m_{\text{surface}} - 2.5 \log e^{\kappa_{\text{atm}} h/\sin z} .$$

Taking into account extinction by the interstellar medium and the atmosphere, we obtain the most general formula:

$$\boxed{M = m_{\text{surface}} - 5 \log d + 5 - 2.5 \log e^{\kappa_{\text{int}} d} - 2.5 \log e^{\kappa_{\text{atm}} h/\sin z}} . \tag{8.14}$$

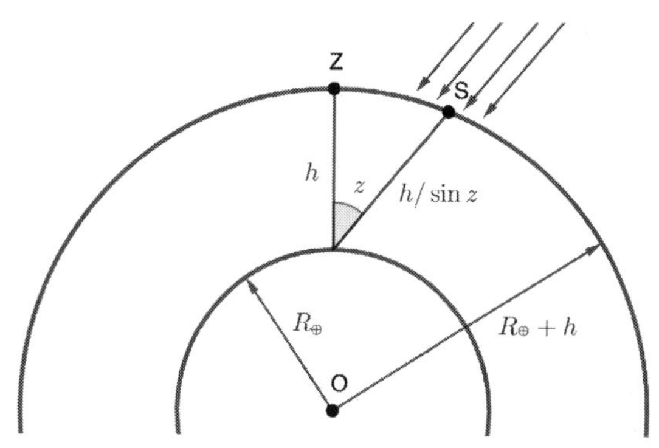

Figure 8.1: For small zenith distances (z), the distance which light travels through the atmosphere is $h/\sin\theta$, where $h \approx 10\,\text{km}$ is the height of the atmosphere with significant density.

8.5 Exercises

1. (NAZ 2015, Th.S, q.4) Approximating the Sun as a blackbody, with a temperature of $T_e = 5778\,\text{K}$, compute its luminosity $L_\odot$, the light flux on Earth and the fraction of energy that is incident on the Earth.

2.* (ONAA 2019, Th.S2, q.10) The decay of radioactive elements inside the Earth gives rise to an average heat flux of $5 \cdot 10^{-2}\,\text{W/m}^2$ on the surface of the Earth. Express this flux as a fraction of the thermal radiation absorbed from the Sun, corresponding to a temperature of 300 K. If these radioactive decays were the only source of energy, what would the temperature of the Earth be?

3. From what distance do we need to observe a 100 W lamp so that it appears as bright as the Sun?

4. With a magnitude of $m_a = -1.43$, Sirius is the brightest star in the night sky. Actually, Sirius has a small companion star (a white dwarf), with an apparent magnitude of $m_b = 8.68$. How many times is Sirius brighter that its companion?

5. The magnitude of the faintest star that can be observed with the naked eye is $m_b = 6$, that of the brightest, $m_t = -1.5$. What is the ratio of the fluxes of the two stars?

6. The apparent magnitude of the Sun, as seen from Earth, is -26.8. Compute the apparent magnitudes of the Sun as seen from Mercury, Venus, Mars, Jupiter and Saturn.

7.* (NAZ 2015, Th.S, q.3) After a geomagnetic storm, the Sun's brightness temporarily increases by 10 %. At this moment, Jupiter is in opposition. Find the apparent magnitudes of the Sun and Jupiter (assume the normal magnitudes are $m_s = -26.74$ and $m_{♃} = -2.2$). What are the time intervals between the occurrence of the storm and the increase in brightness of the Sun, and between the observed increase in brightness of the Sun and the increase in brightness of Jupiter? Assume that the orbits of Earth and Jupiter are circular.

8. The difference in apparent magnitudes of two identical stars is $\Delta m = 3.5$. Compute the ratio of their distances from Earth.

9. A star of magnitude $m = -1.45$ is $d = 8.6\,\text{ly}$ away from Earth. What is its absolute magnitude? Express its luminosity in terms of $L_\odot$.

10. Arcturus (α Bootes) has apparent and absolute magnitudes of $m = -0.05$ and $M = -0.3$, respectively. Compute its distance from Earth. What is

its parallax?

11. The absolute magnitudes of the Sun and Sirius are $M_\odot = 4.83$ and $M_s = 1.4$, respectively. At what distance from the Earth should Sirius be, to appear as bright as the Sun?

12.* Estimate the number of photons that reach your eyes every second, having been emitted by a star with apparent magnitude $m = 6$. Assume that the star only emits energy at the wavelength of $\lambda = 550\,\text{nm}$.

13.* Find the albedo of the Moon. You are given the apparent magnitudes of the full moon $m_☾ = -12.74$ and the Sun $m_\odot = -26.74$, and the angular diameter of the Moon $\alpha = 32'$. Assume the Moon only reflects light coming from the Sun.

14.* (IAO 2019, Th.β, q.3) Supergiant UY Scuti is the largest and one of the fastest burning stars currently known, with a volume of around 5 bilion times that of the Sun. The mass lost per unit time, due to radiation, is only 0.04% of the total mass lost over the same time, and only 0.5% of the light passes through the upper layers of the star (that is, reaches the observer). Estimate the absolute magnitude, the temperature, and the remaining life time of UY Scuti.

15.* (IAO 2013, Th.β, q.2) Gliese 581g is the most Earth-like planet found outside the Solar System, and the exoplanet with the greatest recognized potential for harbouring life. The parallax of Gliese 581 is $\pi_p = 0.13''$, its magnitude $m = 8.0$ and its mass $M = 0.31 M_\odot$. Estimate the orbital period of the planet.

16.** Altair is a main sequence star with mass $M = 1.7 M_\odot$, magnitude $m = 0.77$ and parallax $\pi_p = 0.195''$. Estimate its density.

17. A system is composed of three stars with apparent magnitudes of 3.67, 4.65 and 5.12. What is the magnitude of the composite system?

18. A spectroscopic binary is comprised of two stars which periodically eclipse each other: one is large and cold, the other is small and hot. If the maximum and minimum magnitudes of the system are $m_{\text{max}} = 3.70$ and $m_{\text{min}} = 4.85$, respectively, what are the magnitudes of the two stars? What is the ratio of the maximum and minimum fluxes on Earth?

19.* (IAO 2015, Th.β, q.4) According to an ancient legend of the Middle Volga there was a constellation called White Leopard (Pardus Album) in the sky in the very distant past, in which the number of stars was exactly equal to the number of letters in the Greek alphabet, and the stars had magnitudes $m_{\alpha\text{PaA}} = 0.10$, $m_{\beta\text{PaA}} = 0.20$, $m_{\gamma\text{PaA}} = 0.30$,

$m_{\delta\mathrm{PaA}} = 0.40$ and so on, adding 0.1 every time until ωPaA. Calculate the total magnitude of the stars of this constellation.

20. Two stars of magnitudes 3 and 8 are observed with a telescope. A picture is taken of the first star, with an exposure time of 10 seconds. What should the exposure time for the second star be, if we want it to appear as bright as the first star?

21. On an especially foggy day, the Sun appears as bright as the full moon (observed on a clear day). Compute the optical thickness τ.

22.** (INAO 2017, Th., q.3) Human life can survive on Earth only if the equilibrium temperature remains below 333 K. Assume that the Milky Way contains 10^{11} stars uniformly distributed across the galaxy, with and average number density of $0.14\,\mathrm{pc}^{-3}$, and that there is a supernova every 30 years, with luminosity $10^{11}\,L_\odot$. What is the probability that a supernova wipes out life on Earth, in the total life span of the Sun?

8.6 Problems

1.* **Olbert's paradox**
Assuming that the stars are uniformly distributed in the universe, show that the amount of energy which arrives on Earth from a spherical shell (of universe) of radius r and thickness dr is the same as that from a spherical shell of radius R and thickness dr, with $r \neq R$. What is the total brightness of the sky, as observed from the Earth? How can you explain this result?

2.* **Earth's temperature**
Compute the mean temperature on Earth, as a function of the solar radius ($R_\odot$) and temperature ($T_\odot$), the Earth-Sun distance ($d_\oplus$) and the albedo of the Earth ($A_\oplus$). Compute the numerical value for $A_\oplus = 0.31$, neglecting the greenhouse effect of the atmosphere.

3.*** **Earth's temperature, another model**
As can be seen from the previous problem, the greenhouse effect is non-negligible. Let e_v and e_i be the transmittances of the atmosphere for the solar (visible) and terrestrial (infra-red) radiations, and let A_{atm} and A_s be the albedos of the atmosphere and of the surface of the Earth, respectively. Assume that the atmosphere is a thin envelope surrounding the Earth, with radius $R_\oplus$ and uniform temperature T_{atm}. Compute, as a function of $R_\odot$, $T_\odot$, $d_\oplus$, A_s, A_{atm}, e_v, and e_i, the new value for the Earth's temperature. Estimate the temperature for $A_s = 0.08$, $A_{\mathrm{atm}} = 0.24$, $e_v = 0.8$, $e_i = 0.20$.

9

Cosmic Distance Ladder

In this chapter we will examine the techniques that astronomers use to measure the distances of celestial bodies. Many different methods exist, however each of them is only applicable to a specific length scale. In fact, a direct distance measurement is only possible for up to a few hundred parsecs, using the method of diurnal parallax, for objects within our Solar System, and annual parallax, for the closest stars. At greater distances, several methods rely on standard candles (i.e. objects of known intrinsic brightness), such as Cepheid variables or Type Ia supernovae. Even binary systems, where the orbital characteristics can be used to calculate the masses of the component stars (and hence their luminosities), can be used as standard candles. At even greater distances, it becomes impossible to resolve each star, therefore it is necessary to estimate the intrinsic brightness of whole galaxies. For example, the Tully-Fisher relation links the luminosity of a spiral galaxy with its rotational velocity, while the Faber-Jackson method relates the luminosity of an elliptical galaxy to the dispersion of the velocities of stars around its centre. At extremely large scales, we can measure the redshift of a galaxy to obtain its recession velocity, and therefore its distance, from Hubble's law. The cosmic distance ladder is the succession of these, and many other methods. The ladder analogy arises because no single technique can measure distances at all ranges encountered in astronomy. Each rung of the ladder provides information that can be used to determine the distances at the next, higher, rung.

9.1 Parallax

Parallax is a very common phenomenon that manifests itself as the change in the angular position of an object as seen by an observer, due to the relative motion between them. Try placing one of your fingers in front of your eyes: first, look at it with your right eye only, then with your left eye only. Your finger will appear in different directions, with respect to the background. In this example, the distance between your eyes is the *parallax baseline*. The closer your finger is to your eyes, or the greater the parallax baseline (for instance, you

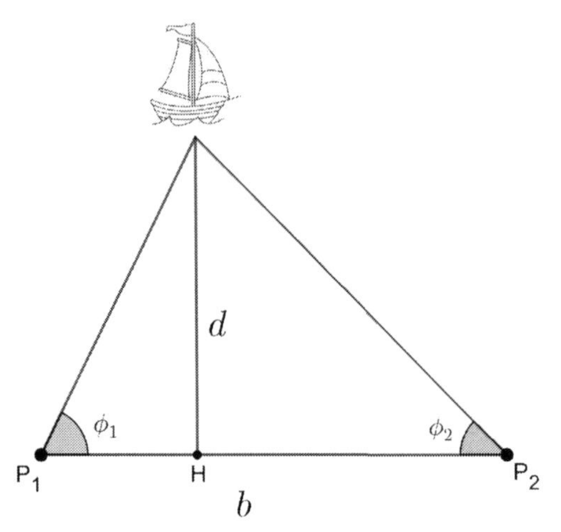

Figure 9.1: Triangulation method: the distance of an object (such as a ship) from the baseline, can be obtained by recording the angles ϕ_1 and ϕ_2 it forms with the baseline at points P_1 and P_2, which are a distance b apart.

could move your head sideways before observing the second time), the farther apart the two images appear. It is clear that the parallax angle must depend on the ratio between the parallax baseline and the distance of the object.
This simple concept proves immediately useful. The parallax baseline and the parallax angle are quantities easily measurable from our position, therefore the distance of the object can be easily obtained. In addition to being an essential tool in astronomy, the parallax method is also used in terrestrial triangulation. Let $\overline{\mathrm{P_1H}} = x$ and $\overline{\mathrm{HP_2}} = b - x$ in Fig. 9.1. Then:

$$\begin{cases} \dfrac{d}{b-x} = \tan\phi_2 & (9.1a) \\ \dfrac{d}{x} = \tan\phi_1 & (9.1b) \end{cases}$$

Dividing the first equation by the second:

$$\frac{x}{b-x} = \frac{\tan\phi_2}{\tan\phi_1}$$
$$x\left(1 + \frac{\tan\phi_2}{\tan\phi_1}\right) = b\,\frac{\tan\phi_2}{\tan\phi_1}$$
$$\Rightarrow x = b\,\frac{\tan\phi_2}{\tan\phi_1 + \tan\phi_2}\,.$$

Substituting x in Eq. 9.1b, we get d:

$$\boxed{d = b\,\frac{\tan\phi_1 \cdot \tan\phi_2}{\tan\phi_1 + \tan\phi_2}}\,. \tag{9.2}$$

This equation can be used to find the distance of an object from the baseline, given the angles ϕ_1 and ϕ_2 that it forms with the parallax baseline at points P_1 and P_2, which are a distance b apart. In astronomy, usually $\overline{P_1H} = \overline{HP_2} = b/2$, therefore $\phi = \phi_1 = \phi_2$. Eq. 9.2 then simplifies to:

$$d = \frac{b}{2} \tan \phi \,.$$

For celestial objects $d \gg b$, therefore $\phi \approx 90°$. Since the position of a star is measured with respect to the background of fixed stars, it is useful to define the parallax angle π_p as the complementary of ϕ: $\pi_p = 90° - \phi$, where $\pi_p \ll 1$ (hence, the parallax angle is just the angle subtended by the baseline). Since $\tan(90° - \pi_p) = 1/\tan \pi_p \approx 1/\pi_{p,\text{rad}}$, we find:

$$\boxed{d = \frac{b/2}{\pi_{p,\text{rad}}}} \,. \tag{9.3}$$

In the next sections, we will look at two different choices for the parallax baseline: Earth's diameter, and the diameter of the Earth's orbit around the Sun. These give the *diurnal* and *annual* parallaxes, respectively.

Diurnal Parallax

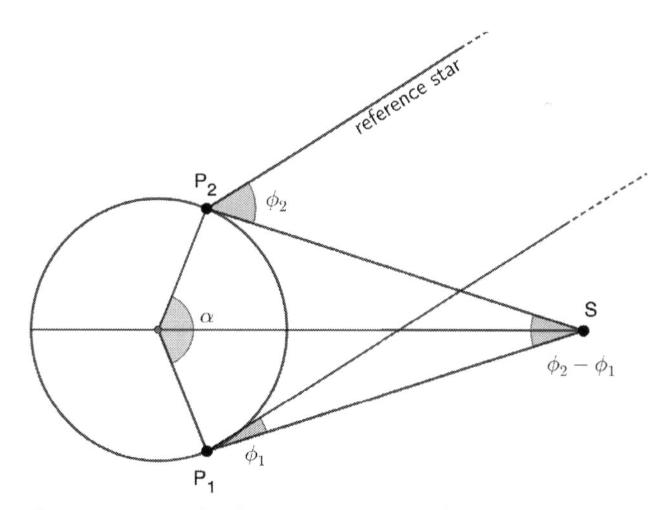

Figure 9.2: In order to measure the distance to a celestial object using the method of diurnal parallax, we record the angles ϕ_1 and ϕ_2 between the object under consideration and the reference object, for two observation sites P_1 and P_2 on the equator, at a distance α in longitude. From the parallax angle $\pi_p = (\phi_1 - \phi_2)/2$, it is then possible to obtain the distance, using Eq. 9.4.

The diurnal parallax of an object (within the Solar System) can be measured by simply performing two observations 12^h apart, or by observing the same

object from two different locations on Earth at the same time. In the first case, after twelve hours, the Earth has completed half a rotation, and the observation site (preferably at the equator) has moved by a distance equal to the Earth's diameter. Either way, if the observation site is at the equator, the parallax baseline is Earth's diameter. Half of the angle between the two images of the object is called the *diurnal parallax*. This method allows us to determine the distances of most bodies in the Solar System, but the diurnal parallax is too small to measure the distances of stars, which appear fixed.
To find the diurnal parallax, we take a fixed star, and measure the angles ϕ_1 and ϕ_2 it forms with the object under consideration, at two observation sites on the equator separated in longitude by α (Fig. 9.2). The angle of diurnal parallax π_p is then $\pi_p = (\phi_1 - \phi_2)/2$. Using Eq. 9.3, where $b/2 = R_\oplus \sin(\alpha/2)$:

$$\boxed{d = \frac{R_\oplus \sin(\alpha/2)}{\pi_{p,\mathrm{rad}}}}. \tag{9.4}$$

Since the Moon is the celestial body closest to us, lunar parallax measurements began rather early compared to those for the Sun and the planets. One of the most famous measurements was the one conducted by Hipparchus. This was carried out during a total solar eclipse in Syene and during a partial solar eclipse in Alexandria. At the same time, an observer in Syene saw the solar disk completely eclipsed by the Moon, while one in Alexandria was able to observe 1/5 of the solar disk, that is, 1/5 of 30 arcminutes. Assuming the Sun to be at an infinite distance, Hipparchus estimated the Earth-Moon distance to be in the range of 59 to 67 Earth radii: an impressive result, considering that the accepted value today is 60.2. The reason for this uncertainty lies in the fact that Hipparchus had not been able to determine the parallax of the Sun. In fact, the solar parallax was more difficult to measure, since the Sun is so bright that it is impossible to record its position relative to the stars. One of the first methods used to calculate the distance of the Sun was the transit of Venus on the solar disk. An alternative way is estimating the parallax of Venus at the moment of maximum elongation (Pr. 9.1), or estimating the parallax of Mars in opposition.

Annual Parallax

Calculating the distances of celestial objects outside the Solar System requires a larger baseline than the Earth's diameter. The widest basis we can use is the diameter of the Earth's orbit, which can be calculated using the diurnal parallax, as described in the previous section (in this sense, the diurnal parallax allows us to climb one rung up the ladder). In this case, we can simply carry out two observations of an object 6 months apart (Fig. 9.3). The *annual* parallax is then half of the angle between the two images of the object against the background of fixed stars. This method allows us to record parallax angles for stars at most a few hundred parsecs away. Distant stars appear fixed and constitute the background on which we observe the succession of all celestial phenomena.

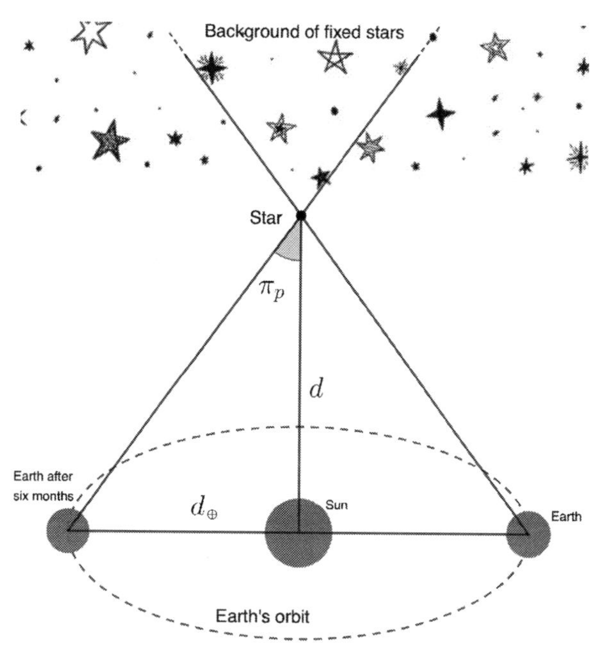

Figure 9.3: With the method of annual parallax, the distance to a celestial object can be obtained by carrying out two observations six months apart. The parallax baseline is the diameter of Earth's orbit.

Looking at Fig. 9.3, we see that $b/2 = d_\oplus$. From Eq. 9.3, it follows that:

$$\frac{d_\oplus}{d} = \pi_{p,\text{rad}} \,.$$

Since π_p is a small angle, it is best to measure it in arcseconds. In a radian there are $(180/\pi)\cdot 60\cdot 60 = 206264.8062 \approx 206265$ arcseconds. Therefore:

$$\frac{d_\oplus}{d} = \frac{\pi_p''}{206265}$$
$$\Rightarrow \frac{1}{d/(206265 \cdot d_\oplus)} = \pi_p'' \,.$$

What prevents us from defining a new unit? Since we are dealing with the *par*allax of an arc*sec*ond, why not call it *parsec*! To get rid of that annoying numerical factor, we then decide that 1 pc= 206265 au (where au is the astronomical unit, equal to the Earth-Sun distance, $d_\oplus$).
Therefore, $d/(206265 \cdot d_\oplus)$ is equal to the distance d expressed in parsec:

$$\boxed{\pi_p{}'' = \frac{1}{d_{\rm pc}}}. \tag{9.5}$$

The parsec is therefore defined as the distance of a star with an annual parallax of one arcsecond. We also note that 1 pc= 3.262 ly. Proxima Centauri, the closest star to the Solar System, has a parallax of 0.768 arcseconds. Applying Eq. 9.5, we see that its distance from Earth is about 1.3 pc, or 4.23 ly.
The annual parallax is a tiny effect and was only measured in 1838, by Friedrich Wilhelm Bessel. With the Hipparcos satellite (1989 – 1993), the European Space Agency (ESA), was able to measure the parallax of 118,000 stars with an accuracy of 0.001", significantly better than the resolution allowed by telescopes on Earth. Despite the impressive precision of the Hipparcos satellite, the distances it can measure are still small compared to the distance of the galactic centre, about 8kpc away from the Solar System. A more recent satellite, Gaia, measured the parallax of 1 billion stars with a precision of 10 μarcsec.

9.2 Spectroscopic Parallax

The spectroscopic parallax does not have much in common with the trigonometric parallax, but the use of the term "parallax" refers to the fact that this method is also used to measure the distances to celestial bodies.
Consider Eq. 8.13:

$$m - M = 5 \log d - 5 + A(d)\,,$$

where d is the distance measured in parsec and $A(d)$ is the correction term for extinction. If we know $m - M$, we can obtain d, in principle.
While m can be directly calculated from observation, finding M is a much harder task. The absolute magnitude can be estimated from the spectral class of the star, but the relation that links spectral class and intrinsic brightness is only a rough approximation. Indeed, the intrinsic brightness depends on several factors, such as the percentage of heavy elements, which can vary slightly within the same spectral class. At the same time, this relation is not perfect because the main sequence is not a thin line, but it extends by one magnitude in either direction. Then, there is the matter of estimating the term $A(d)$. Usually, it can be found by comparing the theoretical colour of the star (based on its spectral class) with the one observed. While the colours would be the same with no extinction, the interstellar medium primary absorbs shorter wavelengths, therefore making the star appear redder (see Sec. 8.4). An estimate of the extinction can therefore be obtained from the colour indices (B−V, U−B) of the star.

9.3 Standard Candles

A standard candle is an astronomical object whose intrinsic brightness can be accurately calculated from measurements conducted on Earth. Examples of standard candles are Cepheid variables and supernovae.
Cepheid variables are a particular type of variable stars, whose brightness varies in a predictable and regular way due to the instability of their surface. In fact, even a small increase in pressure tends to expand the surface of the star and lower its temperature, thus making it appear fainter (Pr. 9.2). In turn, the initial increase in volume causes a decrease in pressure, which then leads the star to compress, increasing its temperature and, therefore, its brightness. By modelling Cepheid variables as thermodynamic engines, Eddington was the first to explain how this periodic cycle sustains itself. Imagine a combustion engine, where a spark triggers the combustion when the gas is maximally compressed by the piston. In the same way, the upper layers of the star become more opaque during the expansion phase, therefore absorbing more energy and being pushed outwards. Then, during the compression phase, they become more transparent, absorbing less energy and being pushed inwards by gravity. This cycle is called κ *mechanism*, since "κ" is the Greek letter which conventionally denotes the opacity of a medium.
The name of this class of stars comes from δ Cephei, discovered in 1784 by John Goodricke. The relationship between period and luminosity was first discovered in 1912 by Henrietta Swan Leavitt. After studying thousands of variable stars in the Magellanic Clouds, she understood that the period of variation of brightness was only a function of the intrinsic brightness of the Cepheid, given that all of the stars she was observing were approximately at the same distance from Earth. For Cepheid variables, the logarithm of the period of variation of brightness is proportional to the logarithm of the average brightness (Fig. 9.4, overleaf). Owing to this relationship, and to the great precision with which the pulsation period can be measured, Cepheid variables can be used as standard candles in determining the distance of globular clusters and galaxies in which they are contained. The period-luminosity relationship for classic Cepheid variables is:

$$\log \frac{\langle L \rangle}{L_\odot} = 1.15 \log P_d + 2.47 \,, \tag{9.6}$$

where $\langle L \rangle$ denotes the average luminosity and P_d is the period of pulsation, measured in days.
Edwin Hubble was the first to identify a Cepheid variable in the Andromeda galaxy. He estimated its distance at $9 \cdot 10^5$ light years, thus proving it had to be located outside our galaxy. This discovery showed that the Milky Way is just one of countless galaxies in the universe.

Today Cepheid variables are divided into two types: Type I Cepheids (or Classical Cepheids) and Type II Cepheids. While the former belong to the (younger) population I stars, the latter belong to (older) population II stars, usually poor in heavy metals, and hence fainter. Because Type I Cepheids are brighter, they can be used to measure greater distances compared to Type II Cepheids. There is also another class of variable stars, called RR Lyrae. However, these are fainter than both types of Cepheids.

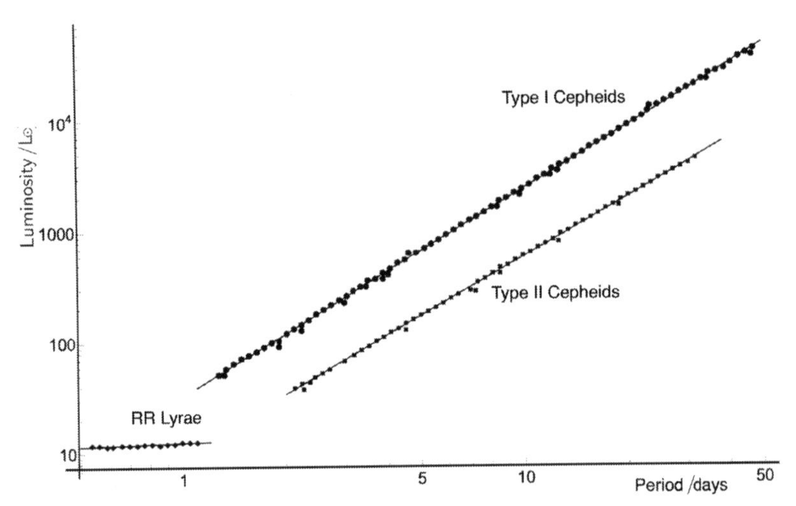

Figure 9.4: Period-luminosity relationship for three types of variables.

9.4 Tully-Fisher Relation

All spiral galaxies rotate around their centre. This rotation gives rise to a blueshifted spectrum in the part of the galaxy that is rotating towards us, and a redshifted spectrum in the part that is rotating away from us. Let v_c be the radial velocity of the centre of the galaxy, $v_{\rm rot}$ the maximum velocity of rotation of the stars around the centre and i the angle between the line of sight and the perpendicular to the plane of the galaxy. To an observer on Earth, the extremes appear to be moving at radial velocities $v_1 = v_c - v_{\rm rot}\sin i$ and $v_2 = v_c + v_{\rm rot}\sin i$ (Fig. 9.5). The displacements in the spectral lines $\Delta\lambda_1 = \lambda_1 v_1/c$ and $\Delta\lambda_2 = \lambda_2 v_2/c$ are:

$$\begin{cases} \dfrac{\Delta\lambda_1}{\lambda} = \dfrac{v_c - v_{\rm rot}\sin i}{c} & (9.7) \\[2ex] \dfrac{\Delta\lambda_2}{\lambda} = \dfrac{v_c + v_{\rm rot}\sin i}{c} & (9.8) \end{cases}$$

Subtracting the two equations:

$$v_{\rm rot} = \frac{c}{\lambda\sin i}(\Delta\lambda_2 - \Delta\lambda_1)\,. \qquad (9.9)$$

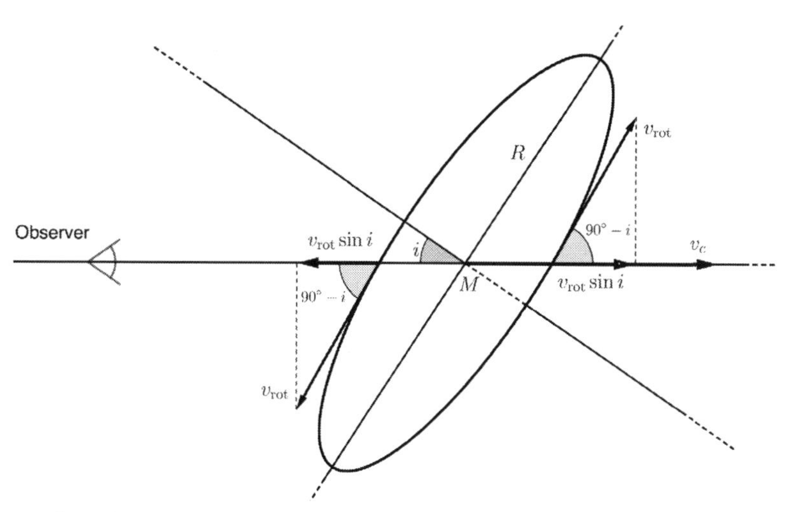

Figure 9.5: Consider a galaxy whose plane forms an angle $90° - i$ with the line of sight, and in which the maximum velocity of stars is v_{rot}. To an observer on Earth, the extremes appear to be moving at radial velocities $v_c - v_{rot} \sin i$ and $v_c + v_{rot} \sin i$, where v_c is the velocity of the centre of the galaxy.

The above equation allows us to find the rotational velocity of the galaxy from quantities measurable on Earth.
The faster the galaxy rotates, the greater $\Delta\lambda_1$ and $\Delta\lambda_2$. A galaxy rotates because of its own gravity, therefore, the greater its mass, the greater its rotational velocity. The rotational velocity is related to the mass (Eq. 10.18) by:

$$v_{rot}^2 = \frac{GM}{R} \Rightarrow M = \frac{v_{rot}^2 R}{G} .$$

The greater the mass, the larger the number of stars, and therefore the brighter the galaxy. Hence, galaxies rotating faster are brighter. Assuming spiral galaxies to be comprised (on average) of Sun-like stars, the luminosity-to-mass ratio must be a constant for all galaxies, hence $C_{LM} = L/M$. Then:

$$L = C_{LM} \frac{v_{rot}^2 R}{G} .$$

Assuming that all galaxies have approximately the same surface luminosity, we substitute $C_{LR} = L/R^2$ for R:

$$L = \frac{C_{LM}^2}{C_{LR}} \frac{v_{rot}^4}{G^2} = C v_{rot}^4 ,$$

where $C = C_{LM}^2/C_{LR}$. Hence, we obtain the Tully-Fisher relation:

$$\boxed{L = C v_{rot}^4} . \tag{9.10}$$

An alternative form relates the magnitude to the rotational velocity. Taking as a reference a galaxy with rotational velocity $v_{\text{rot},0}$ and absolute magnitude M_0, we use Eq. 8.4:

$$M = M_0 - 2.5 \log \frac{L}{L_0}$$

$$= M_0 - 2.5 \log \frac{v_{\text{rot}}^4}{v_{\text{rot},0}^4} = M_0 - 10 \log \frac{v_{\text{rot}}}{v_{\text{rot},0}} .$$

Hence, the Tully-Fisher relation can also be written as:

$$\boxed{M_1 - M_2 = -10 \log \frac{v_{\text{rot},1}}{v_{\text{rot}_2}}} . \tag{9.11}$$

It is therefore possible to compute the absolute magnitude of a galaxy from the absolute magnitude of another galaxy and the ratio of their rotational velocities, which can in turn be determined from Eq. 9.9. However, Eq. 9.11 is not exact, and the constant in front of the logarithm (10, in this case) actually varies according to the type of spiral galaxy under consideration. We can divide spiral galaxies into at least three classes (Sa, Sb, Sc), depending on their surface luminosity C_{LR} and luminosity-to-mass ratio C_{LM}. We then obtain three equations:

$$\begin{cases} M = -9.95 \cdot \log v_{\text{rot}} + 3.15 & \text{for Sa galaxies} \quad (9.12a) \\ M = -10.2 \cdot \log v_{\text{rot}} + 2.71 & \text{for Sb galaxies} \quad (9.12b) \\ M = -11.0 \cdot \log v_{\text{rot}} + 3.31 & \text{for Sc galaxies} . \quad (9.12c) \end{cases}$$

Although not exact, the theoretical model described in this section gives a reasonable estimate for the absolute magnitude of a spiral galaxy.
The spectral line that is often used to measure the Doppler shift is the hydrogen line at 21 cm. Being very narrow, this line allows the measurement of the smallest displacements. At the same time, because of its long wavelength, it is not significantly absorbed by the interstellar medium. This method allows us to measure distances up to 200 Mpc, beyond which the displacement in the spectral lines becomes too tiny to observe.

9.5 Hubble's Law

In 1929, cosmology achieved one of its most important discoveries when Edwin Hubble found a relationship that linked the velocities of galaxies to their distances from Earth, thus showing that the universe is expanding. It was the first experimental verification that confirmed what had already been predicted by Georges Lemaître, in 1927. The linear relationship obtained at that time is

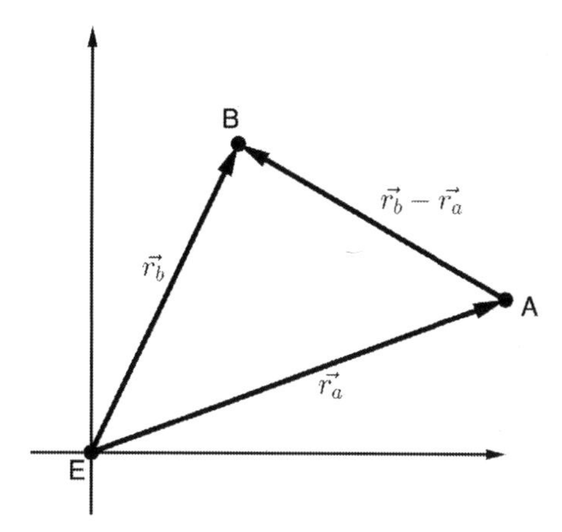

Figure 9.6: An observer on Earth sees A and B receding with velocities $\vec{v_a} = H_0\vec{r_a}$ and $\vec{v_b} = H_0\vec{r_b}$, respectively. Therefore, an observer in A sees B moving away with a velocity of $\vec{v_b} - \vec{v_a} = H_0(\vec{r_b} - \vec{r_a})$ (directed radially outwards from A). Hence, also A can be regarded as the centre of expansion of the universe.

actually valid only at relatively small distances from Earth, where the acceleration of the expansion of the universe is negligible. Hubble's law states that:

$$\boxed{v = H_0\, d}\,, \tag{9.13}$$

where H_0 is *Hubble's constant* and d is the distance. According to the latest measurements, $H_0 = 68 \pm 5.5\,\mathrm{km\,s^{-1}\,Mpc^{-1}}$. In the case of small velocities, the redshift $z = \Delta\lambda/\lambda$ is given by:

$$z = \frac{v}{c} = \frac{H_0\, d}{c}\,.$$

In the more general case, when special relativity is taken into account, we can use Eq. 7.20 to obtain:

$$z = \frac{\sqrt{1 + \frac{H_0\, d}{c}}}{\sqrt{1 - \frac{H_0\, d}{c}}} - 1\,.$$

We now want to show that every point in the universe can be regarded as the centre of this expansion, described by the same Hubble's law. Let us place the Earth at the centre of a Cartesian system and consider two galaxies A and B, with distances $\vec{r_a}$ and $\vec{r_b}$ from Earth (Fig. 9.6). According to Hubble's law, the magnitudes of the recession velocities of A and B are $v_a = H_0\, r_a$ and $v_b = H_0\, r_b$, respectively. Since the recession velocities are directed along the radial direction (Earth-galaxy, in both cases), they can be written in vector form as:

$$\vec{v_a} = H_0\vec{r_a}$$
$$\vec{v_b} = H_0\vec{r_b}\,.$$

The recession velocity of galaxy B, observed from A, is then:

$$\vec{v_b} - \vec{v_a} = H_0 \, \vec{r_b} - H_0 \, \vec{r_a} = H_0(\vec{r_b} - \vec{r_a}) \, .$$

But $\vec{r_b} - \vec{r_a}$ is the distance of B from A, and is directed radially outwards from A. Hence, also A can be regarded as the centre of expansion of the universe, with the same Hubble's law.

9.6 Exercises

1. (NAZ 2015, Th.J, q.4) Compute the distance to Procyon (α CMi) in parsecs and in light years, given its parallax $\pi_p = 0.286''$. Procyon has an apparent magnitude of $m = 0.34$. If its distance were 10 times greater, would it still be visible to the naked eye?

2. A spiral galaxy in the Virgo Cluster rotates around its centre with a maximum velocity of $v_{\text{rot},1} = 175\,\text{km/s}$, and its apparent magnitude is $m_{v,1} = 11.87$. Another spiral galaxy, in the Coma Cluster, has $v_{\text{rot},2} = 245\,\text{km/s}$ and $m_{v,2} = 14.63$. Find the ratio of the distances to the galaxies.

3. If the uncertainty in the measurement of the absolute magnitude of a Cepheid variable is $\Delta M = 0.5$, what is the uncertainty in its distance?

4.* The observed rotational velocity of a spiral galaxy around its centre is $v_{\text{obs}} = 300\,\text{km/s}$. The ratio between its semi-major and semi-minor axes is $a/b = 1.74$, and the apparent magnitude is $m = 18$ (H band). The redshift is $z = 0.20$. Obtain the value of Hubble's constant. You are given the Tully-Fisher relation in the H-band:

$$M_{\text{H}} = -9.5 \log v_{\text{rot}} + 2.08$$

5. (CAO 2018, Th., q.5) A given elliptical galaxy has an apparent magnitude of $m = 18$ and redshift of $z = 0.1$. Estimate the mass of the galaxy. Interstellar absorption is neglected.

6. (MyAO 2018, Th., q.10) Estimate the age of the Universe from Hubble's constant.

7.* (CzAO 2018, Th.AB, q.C) Suppose that the universe contains only matter that interacts only by gravity. Modelling the universe as a sphere of radius R that expands radially due to Hubble's law, assuming it has a constant density of ρ, determine the total energy of a spherical shell with radius r and thickness dr. What is the critical density for which the energy of the universe is zero? (Hint: use Eq. 10.22).

8.* (USAAAO 2020, Th.2, q.3) The Lyman-break galaxy selection technique makes use of the fact that any light from galaxies with wavelength shorter than the Lyman limit (the shortest wavelength in the Lyman series) is essentially totally absorbed by neutral gas surrounding the galaxies. The ionization energy of hydrogen is 13.6 eV. Suppose that we are observing galaxies in the V band, whose effective midpoint is 551 nm and bandwidth is 99 nm.

- At what range of redshifts would we begin to see galaxies "disappear" (break) from images in the V band?
- What range of recessional velocities (km/s) and distances (Mpc) does this correspond to? Assume only Hubble expansion contributes to the radial velocity and redshift.

9.** (IAO 2012, Th.α, β, q.6) Astronomers have discovered a distant galaxy that appears to have the same colour as ϵ Eridani, but which is 1000 times fainter. It appears, however, that this galaxy is only composed of stars similar to the Sun in physical characteristics. Find the number of stars in the galaxy. You know the following about ϵ Eridani: temperature $T = 4900\,\mathrm{K}$, magnitude $m = 3.74$, parallax $\pi_p = 0.311''$ and mass $M = 0.82 M_{\odot}$.

10.* (IOAA 1016, Th., q.6) The star β-Doradus is a Cepheid variable star with a pulsation period of 9.84 days. We make a simplifying assumption that the star is brightest when it is most contracted (radius being R_1) and it is faintest when it is most expanded (radius being R_2). For simplicity, assume that the star maintains its spherical shape and behaves as a perfect black body at every instant during the entire cycle. The bolometric magnitude of the star varies from 3.46 to 4.08. From Doppler measurements, we know that during pulsation the stellar surface expands or contracts at an average radial speed of 12.8km/s. Over the period of pulsation, the peak of (intrinsic) thermal radiation of the star varies from 531.0 nm to 649.1 nm. Find:

- the ratio of radii of the star in its most contracted and most expanded states;
- the radii of the star (in metres) in its most contracted and most expanded states;
- the flux of the star when it is in its most expanded state.

9.7 Problems

1.* The Astronomical Unit

Determining the astronomical unit (au) was one of the biggest problems in celestial mechanics. Although the ratios between the distances of the planets from the Sun had already been obtained, a solution had not yet been found to measure one of the distances involved. By observing the apparent motion of Venus with a telescope, it was possible to measure a maximum elongation of $\epsilon = 46.3°$. Limiting ourselves, for simplicity, to the plane of the orbits, consider two stations P_1 and P_2 on the equator, separated by a longitude difference of $120\,°$. From these stations, we observe Venus (at maximum elongation) and a star at a much greater distance. From station P_1, the angle between the star and Venus is $\phi_1 = 0\,°37'15'' \pm 3''$; from station P_2 it is $\phi_2 = 0\,°36'52'' \pm 3''$. Determine the value of the astronomical unit, estimating the relative error, given that the Earth's radius is $R_\oplus = 6380\,\mathrm{km}$. (Hint: read Ch. 11 first).

2.* Luminosity-radius relationship

By differentiating the Stefan-Boltzmann law, derive a relationship between the infinitesimal variations of brightness, radius and temperature. Assuming that the star compresses adiabatically ($TV^{\gamma-1}$ is constant), find a relationship between the infinitesimal variations of brightness and radius only. Hence, show that its brightness increases when the star compresses (Hint: assume $\gamma = 5/3$).

3. Period-density relationship**

The radial oscillations within a star are the result of sound waves that propagate within it. An estimate of the period of oscillation can be obtained by considering the time that a wave takes to cross the star. Knowing that the speed of the waves is:

$$v = \sqrt{\frac{\gamma P}{\rho}}$$

where γ is the adiabatic expansion coefficient, P the pressure and ρ the density; using for P the expression of hydrostatic equilibrium (Eq. 7.29), compute the period of oscillation as a function of density.

Part III

Celestial Mechanics

10

Gravitation and Kepler's Laws

The primary task of celestial mechanics is to explain and predict the motions of the planets and their satellites. In the past, several empirical models, such as the epicycles and Kepler's laws, were employed to describe these motions, but none were based on a firm theoretical ground. In 1687, Newton came up with a simple explanation: the law of universal gravitation. We begin the chapter with a summary of Newtonian mechanics and the universal law of gravitation, from which Kepler's laws can be derived. Although the exact proof of Kepler's laws is given in Appendix B, in this chapter we show, when possible, some simple derivations.

10.1 Newtonian Mechanics

Newtonian mechanics accurately describes nature for most of the phenomena we are familiar with. It is based on the following four principles, of which the first is attributed to Galileo (which we will call *law zero*), and the other three to Newton:

0. The laws of physics are the same in all *inertial reference systems*, i.e. systems moving at a constant velocity with respect to one another.
1. In an inertial reference system, every object in a state of uniform motion remains in that state of motion unless acted upon by an external force.
2. The resultant force acting on a body is equal to its mass times its acceleration: $\vec{F} = m\vec{a}$.
3. If A exerts a force $\vec{F}$ on B, then B exerts a force $-\vec{F}$ on A (same magnitude, opposite direction).

10.2 Universal Law of Gravitation

In 1687, Newton published the "Philosophiae Naturalis Principia Mathematica", in which, for the first time, he described the law of universal gravitation. According to this law, each point mass m_1 attracts every other point mass m_2

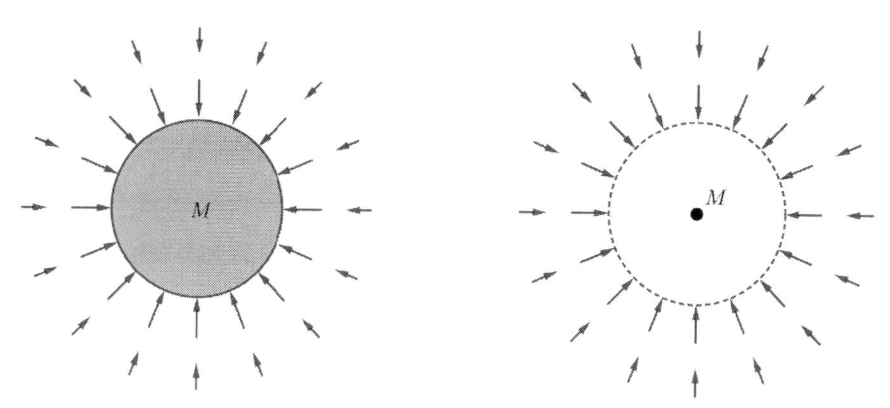

Figure 10.1: Left: gravitational field created by a spherically symmetric body. Right: the same field is produced by a point mass, of the same mass as the body, placed at its centre.

with a force F which is directly proportional to the product of the masses and inversely proportional to the square of their distance r:

$$\boxed{F = \frac{Gm_1m_2}{r^2}}, \tag{10.1}$$

where F is directed along the radius vector connecting the two bodies and is always attractive, and G is the universal gravitational constant, first calculated by Cavendish in 1798, with a value of:

$$G = 6.674 \cdot 10^{-11}\,\mathrm{Nm^2\,kg^{-2}}\,.$$

Two important points, which you can prove in Pr. 10.1, are:

1. the gravitational field created outside a spherically symmetric body is the same as that created by a point mass, of the same mass, placed at its centre (Fig. 10.1);
2. if the system has particular symmetries, we can consider a surface (usually a sphere or a cylinder) such that only the mass inside it contributes to the gravitational field, while the mass outside gives a net zero contribution (Fig. 10.2, overleaf).

From the first point, it follows that the gravitational force created by a spherically symmetric planet is equal to that created by a point mass, of the same mass, placed at its centre, which can be calculated using Eq. 10.1. Therefore, the gravitational force acting on a body of mass m_0 on the surface of a planet of mass M and radius R can be obtained by substituting $m_1 = M$, $m_2 = m_0$ and $r = R$ in Eq. 10.1. According to Newton's second law, the gravitational force is also equal to $F = m_0\,g$, where g is the acceleration of the body. Hence,

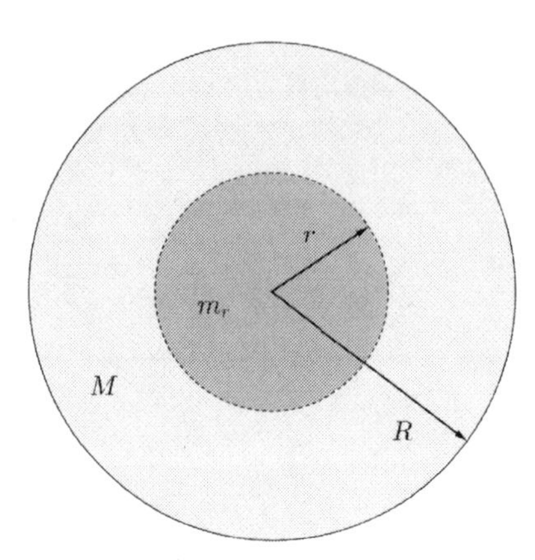

Figure 10.2: At a distance r from the centre of a spherical symmetric body, only the mass m_r inside a sphere of radius r contributes to the gravitational field. Outside, part of the mass attracts the body to the left and part of it to the right. As it turns out, these two contributions cancel each other.

the gravitational acceleration on the surface of the planet is:

$$\boxed{g = \frac{GM}{R^2}} . \tag{10.2}$$

Note that the acceleration does not depend on the mass of the body m_0. Therefore, it is possible to define a vector field, called the *gravitational field*, equal at each point in space to the acceleration that a hypothetical body would have if it were placed at that point. On Earth, $M_\oplus = 5.972 \cdot 10^{24}\,\text{kg}$ and $R_\oplus = 6378\,\text{km}$, hence:

$$g = 9.81\,\text{m/s}^2 ,$$

which is the mean value of the gravitational acceleration on Earth.
The second point can greatly simplify a problem if particular symmetries are present. Let us find, as an example, the gravitational field at a distance r from the centre of a spherically symmetric planet, of mass M and radius R. For $r \geqslant R$, the answer is given by Eq. 10.2, with R replaced by r. However, inside the body, at distance r from the centre, the only contribution to the gravitational field is given by the mass m_r inside a sphere of radius r (Fig. 10.2). Assuming constant density, mass is proportional to volume, which in turn is proportional to the radius cubed, i.e. $m_r = (r/R)^3 M$. Using Eq. 10.2 (with M replaced by m_r and R by r), we finally obtain:

$$g = \frac{GM}{R^3} r \text{ for } \quad r < R .$$

Let us now discuss an extension of the second point. It is possible to divide any closed surface into many infinitesimal elements having surface area $\Delta\vec{A}$, with magnitude equal to the area of that element, direction perpendicular to

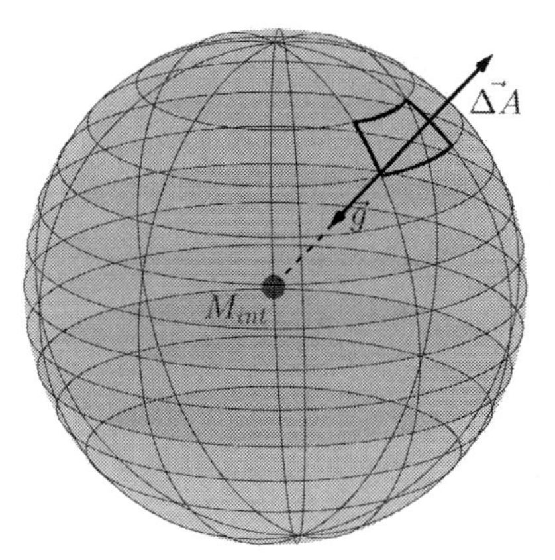

Figure 10.3: It is possible to divide a spherical surface into many infinitesimal elements with surface area $\Delta\vec{A}$, so that $\vec{g}\cdot\Delta\vec{A}$ is constant for every element.

the surface (which, at that point, is effectively planar) and pointing outwards. We then define the *flux* of a vector field $\vec{g}$ on the surface S, as the sum of the scalar product $\vec{g}\cdot\Delta\vec{A}$, for ΔA that tends to zero on the whole surface. In this limit, the sum becomes an integral:

$$\Phi = \lim_{\Delta\vec{A}\to 0}\sum\vec{g}\cdot\Delta\vec{A} = \int_S \vec{g}\cdot d\vec{A}, \tag{10.3}$$

where $\int_S$ denotes the integral over the entire (not necessarily closed) surface S. Now, from the second point, it follows that the flux of the gravitational field on a closed surface S depends only on the mass inside the surface. The form of this dependence can be found by considering the simple case of a point mass placed at the centre of a spherical surface (Fig. 10.3). In this case, the flux is:

$$\Phi = \oint\vec{g}\cdot d\vec{A} = \oint -\frac{GM_{\text{int}}}{R^2}\hat{r}\cdot d\vec{A} = -\frac{GM_{\text{int}}}{R^2}\oint dA = -\frac{GM_{\text{int}}}{R^2}(4\pi R^2) = -4\pi GM\,,$$

where we have taken the constant GM_{int}/R^2 outside the integral, and equated the integral over the closed surface ($\oint dA$) to the area of the sphere ($4\pi R^2$). The negative sign arises because $\hat{r}$ and $\vec{g}$ have opposite directions, i.e. because gravity is attractive. Therefore, in general:

$$\boxed{\Phi = \oint\vec{g}\cdot d\vec{A} = -4\pi GM_{\text{int}}}\,. \tag{10.4}$$

To use the above equation, it is necessary to find a surface with the right symmetry (otherwise the integral cannot be easily solved). If working with a spherical distribution, take a sphere; for a cylindrical distribution (for instance, a cylinder or a plane), take a cylinder (Ex. 10.18).

10.3 Kepler's Laws

Based on the observations made by the Danish astronomer Tycho Brahe, after 16 years of work, Kepler was able to calculate the parameters of the orbit of Mars and, subsequently, those of the Earth. The transition from observations to the derivation of an empirical law required a great deal of work, as the orbit of the planet is observed from the Earth, which, in turn, orbits the Sun. In what Kepler himself called a flash of genius, he realized that every 687 days (the orbital period of Mars), Mars returned to exactly the same position in the Solar System. Kepler's analysis of the planetary motion can be summarized in three laws, which bear his name:

1. All planets move in elliptical orbits, with the Sun at one focus.
2. The radius vector joining any planet to the Sun sweeps out equal areas in equal times.
3. The square of the orbital period of any planet is directly proportional to the cube of the semi-major axis of its orbit.

First Law

Kepler's first law states that each planet moves in an elliptical orbit around the Sun. The Sun is in one focus of the ellipse, with nothing in the other focus. In Appendix B we prove this statement, starting from Newton's law of gravitation. Here, we briefly discuss the properties of the ellipse which are useful in astronomy, while you should refer to Appendix A.2 for a more mathematical treatment.
In astronomy, the points of the ellipse closest and farthest from the focus in which the Sun is located are called *perihelion* and *aphelion*, respectively. For bodies orbiting the Earth, these points are instead called *perigee* and *apogee*. By convention, a is the semi-major axis, b the semi-minor axis, c the semi-focal distance and e the eccentricity, defined as $e = c/a$ (Fig. 10.4). If the Sun is at $\mathrm{F_1}$, the distances of the planet to the Sun in perihelion and aphelion are, respectively:

$$d_p = \overline{\mathrm{AF_1}} = \overline{\mathrm{AO}} - \overline{\mathrm{F_1O}} = a - c\,,$$
$$d_a = \overline{\mathrm{F_1B}} = \overline{\mathrm{OB}} + \overline{\mathrm{F_1O}} = a + c\,.$$

In astronomy we rarely know b, while it is more common to be given some combination of a, e, d_a and d_p. If we know a and e, then we can find d_p and d_a using the formulae:

$$d_p = a - c = a(1-e)\,,$$
$$d_a = a + c = a(1+e)\,.$$

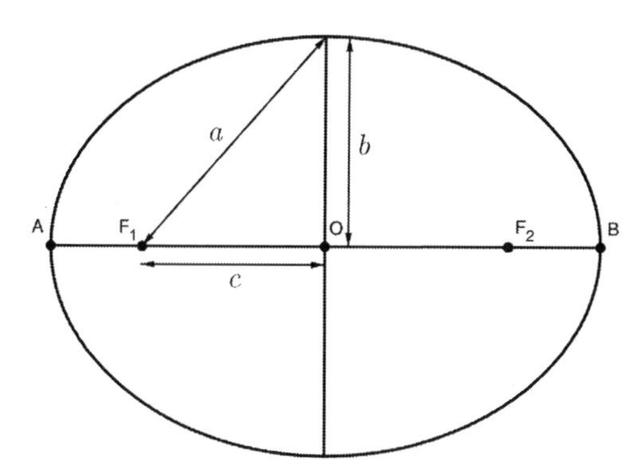

Figure 10.4: a is the semi-major axis, b the semi-minor axis and c the semi-focal distance. The Sun can be in either F_1 or F_2.

If, instead, we know d_a and d_p, we can then find a and e by taking the sum and difference of the above equations:

$$a = \frac{d_a + d_p}{2}, \tag{10.5}$$

$$c = \frac{d_a - d_p}{2}. \tag{10.6}$$

Hence, the eccentricity e can be written as:

$$e = \frac{c}{a} = \frac{d_a - d_p}{d_a + d_p}. \tag{10.7}$$

In the case $e \to 1$, the orbit becomes a line. Therefore, Kepler's laws can also be used to study the motion of a body that falls vertically in the gravitational potential (Ex. 10.10).

The Second Law

With regard to Kepler's second law, it is important to define the *angular momentum* of a body, i.e. the vector product of its position vector and the linear momentum:

$$\boxed{\vec{L} = m\vec{r} \times \vec{v}}. \tag{10.8}$$

Let θ be the angle between the position and velocity vectors. Then, the magnitude of the angular momentum is $L = m\,r\,v\,\sin\theta$ (see Appendix A.1). In a circular orbit, the position vector is always perpendicular to the velocity, therefore $\theta = \pi/2$, hence $L = m\,r\,v$. In an elliptical orbit, the position vector is perpendicular to the velocity only at aphelion and perihelion.
We define the *torque*, with respect to a fixed origin, as the vector product of the position vector (*lever arm*) and the force:

$$\boxed{\vec{\tau} = \vec{r} \times \vec{F}}. \tag{10.9}$$

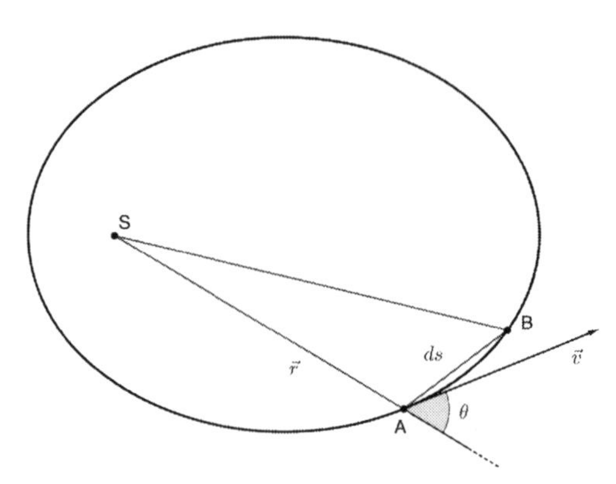

Figure 10.5: It can be proven that the rate at which the radius vector $\vec{r}$ sweeps out area is equal to $L/2m$, where L is the angular momentum. Since L is constant, it follows that the radius vector sweeps out equal areas in equal times, which is Kepler's second law.

As before, the magnitude of torque is $\tau = F\,r\,\sin\theta$. Note that, if a body is only subject to a central force (i.e. directed radially inwards, such as gravity), $\theta = 0$, hence $\tau = 0$.
The principal equation of rotational mechanics relates $\vec{L}$ and $\vec{\tau}$:

$$\boxed{\frac{\mathrm{d}\vec{L}}{\mathrm{d}t} = \vec{\tau}}\,, \tag{10.10}$$

which follows from the definition of the force as the rate of change of momentum. But why consider angular momentum? In an isolated system in which the only force is gravity, the torque acting on a body is zero, as previously shown. Then, from Eq. 10.10, it follows that the rate of change of angular momentum is zero: the angular momentum is constant. In physics, we are especially interested in all the quantities that are constant in time, and angular momentum, like energy, is one of those.
Could it be that the constant rate at which the radius vector sweeps out area (Kepler's second law) is a consequence of the angular momentum being constant? After an infinitesimal time dt, the planet covers a distance of $ds = v\,dt$ along its orbit (Fig. 10.5). Since ds is small compared to r, it follows that $\angle\text{ABS} \approx \theta$, where θ is the angle between the radius vector and the velocity. The area swept by the radius vector joining the planet to the Sun is equal to the area of triangle ABS, which in turn is equal to half its base ds times its height $r\sin(\angle\text{ABS})$:

$$dA = \frac{1}{2} ds\, r\sin(\angle\text{ABS}) = \frac{1}{2} r\,v\sin\theta\,dt = \frac{L}{2m}dt\,.$$

Dividing both sides by dt, we find:

$$\frac{\mathrm{d}A}{\mathrm{d}t} = \frac{L}{2m}\,.$$

Since L is constant, it follows that dA/dt is constant, thus proving Kepler's second law.
Kepler's second law can then be used to compute the time t that it takes a planet to cover an area of A_s. Let T be the period of revolution and $A = \pi a\, b$ the total area of the orbit. We then have the relation:

$$\frac{t}{T} = \frac{A_s}{A} \Rightarrow t = T \cdot \frac{A_s}{A} \,. \tag{10.11}$$

Because of the conservation of angular momentum, we expect the velocity of a planet to be maximum at perihelion and minimum at aphelion. Indeed, since the velocity is perpendicular to the radius vector at those two points:

$$d_a v_a = d_p v_p \,, \tag{10.12}$$

where d_a, d_p and v_a, v_p are the distances and velocities at aphelion and perihelion, respectively. In Ex. 10.9 you will use Eq. 10.12, together with the conservation of energy, to compute v_a and v_p, given only a and e.

The Third Law

Kepler's third law states that, for every body in the Solar System, the ratio of the square of the orbital period (T) and the cube of the semi-major axis (a) is constant:

$$\frac{T^2}{a^3} = k \,. \tag{10.13}$$

Writing the previous equation for a generic planet and for the Earth:

$$\frac{T^2}{a^3} = \frac{T_\oplus^2}{a_\oplus^3} \,,$$

or, equivalently:

$$\left(\frac{T}{T_\oplus}\right)^2 = \left(\frac{a}{a_\oplus}\right)^3 .$$

But $T/T_\oplus$ is just the period T measured in years, whilst $a/a_\oplus$ is the semi-major axis of the planet measured in astronomical units. Hence:

$$\boxed{{T_{\rm yr}}^2 = {a_{\rm au}}^3} \,. \tag{10.14}$$

In the following section, we derive Kepler's third law from Newton's law of gravitation. We will only consider the case of circular orbits, while the general case is treated in Appendix B. We will also assume that the mass of the planet is much smaller than the mass of the Sun. This is a good approximation for the Solar System, but it is not strictly valid for binary systems, in which the orbiting bodies can have similar masses.

Proof of Kepler's Third Law

The derivation of Kepler's third law in the case of circular orbits is a good opportunity to look at an application of Newton's law of gravitation, and to define some recurring quantities.
It is useful to define the instantaneous angular velocity, equal to the angle travelled per unit of time:

$$\omega = \lim_{\Delta t \to 0} \frac{\Delta\theta}{\Delta t} = \frac{\mathrm{d}\theta}{\mathrm{d}t} . \tag{10.15}$$

In uniform circular motion, ω is constant and equal to $\omega = 2\pi/T$, since the body covers an angle of $360°$ or 2π rad every period T.
It is possible to write ω as a function of the instantaneous tangential velocity and the radius vector. After the body covers an infinitesimal angle $d\theta$, the distance ds_t covered in the tangential direction is:

$$ds_t = R\, d\theta .$$

Dividing both sides by dt, and recognizing $v_t = ds_t/dt$, $\omega = d\theta/dt$:

$$\boxed{v_t = \omega R} . \tag{10.16}$$

In general, $v_t = v \sin\theta$, where θ is the angle between the velocity and the radius vector. In the special case of circular motion, $v = v_t$, since the path (and therefore the velocity) is always tangent to the radius vector. The velocity is then equal to the ratio of the circumference and the period of rotation, i.e. $v = 2\pi R/T$, which gives $\omega = 2\pi/T$, as before. Let us now find an expression for the force necessary to keep a body moving in a curved orbit. After a time dt, the body covers an angle of $d\theta$, and its velocity changes by $dv = v_t\, d\theta$ (see Fig. 10.6). Therefore, its acceleration is:

$$a_c = \frac{\mathrm{d}v}{\mathrm{d}t} = \frac{v_t\, d\theta}{dt} = v_t \frac{\mathrm{d}\theta}{\mathrm{d}t} = v_t\, \omega .$$

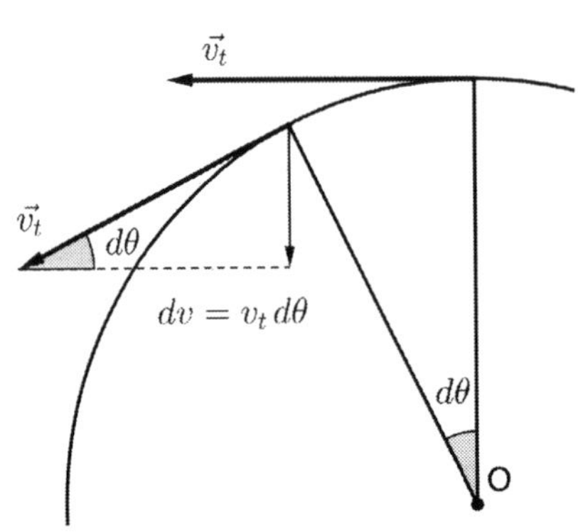

Figure 10.6: After the body covers an angle of $d\theta$, its velocity changes by $dv = v_t\, d\theta$.

Using Eq. 10.16, the previous formula can be written as:

$$\boxed{a_c = \frac{v_t^2}{R} = \omega^2 R} . \tag{10.17}$$

Eq. 10.17 gives the required acceleration as a function of the instantaneous tangential velocity (or angular velocity) and the radius of curvature. This is known as the *centripetal* acceleration, since it is directed radially towards the centre of rotation. The centripetal force is then given by Newton's second law, i.e. $F_c = ma_c$. If a planet is found to rotate in a circular orbit, the necessary centripetal force is provided by the force of gravity:

$$\frac{GmM}{R^2} = ma_c\,.$$

Using Eq. 10.17 to substitute a_c in terms of the velocity:

$$\boxed{v = \sqrt{\frac{GM}{R}}}\,. \tag{10.18}$$

Since the circumference is $2\pi R$, the time (period of revolution) taken to cover it is $T = 2\pi R/v$. Substituting for v, given by Eq. 10.18:

$$\boxed{T^2 = \frac{4\pi^2}{GM}R^3}\,. \tag{10.19}$$

This is the canonical form of Kepler's third law. Eq. 10.19 implies:

$$\frac{T^2}{R^3} = \frac{4\pi^2}{GM} = k\,,$$

which is what we wanted to show. Eq. 10.19 is also valid for elliptical orbits, but R must be replaced with the semi-major axis a.
Let us now derive Kepler's third law without neglecting the gravitational field created by the planet itself. Let m_1,m_2 be the masses of the two bodies and r_1, r_2 the radii of their circular orbits. The bodies rotate around their common centre of mass with the same angular velocity ω, being diametrically opposite and at a constant distance $r_1 + r_2$ from each other at all times. In the Solar System, even if all planets aligned, the centre of mass would still be below the surface of the Sun. For the Earth-Moon system however, this phenomenon is of such importance that it gives rise to two tides every day. In accordance with Newton's law of gravitation, the force that the bodies exert on each other is:

$$F_{g,1} = F_{g,2} = \frac{Gm_1m_2}{(r_1+r_2)^2}\,.$$

For the bodies to remain in a circular orbit, the centripetal force must have the constant magnitude of:

$$F_{c,1} = m_1\omega^2 \cdot r_1\,,$$
$$F_{c,2} = m_2\omega^2 \cdot r_2\,.$$

But the centripetal force is the gravitational force, hence:

$$\frac{Gm_2}{(r_1+r_2)^2} = \omega^2 \cdot r_1 ,$$
$$\frac{Gm_1}{(r_1+r_2)^2} = \omega^2 \cdot r_2 .$$

Adding the above equations:

$$\frac{G(m_1+m_2)}{(r_1+r_2)^2} = \omega^2 \cdot (r_1+r_2)$$
$$\Rightarrow G(m_1+m_2) = \omega^2 \cdot (r_1+r_2)^3 .$$

Let $r_1 + r_2 = a$ be the distance between the bodies. Since $\omega = 2\pi/T$, we find:

$$\boxed{T^2 = \frac{4\pi^2}{G(m_1+m_2)} a^3} . \tag{10.20}$$

In the case of the Solar System, the mass of the Sun is much greater than the mass of any planet, therefore the sum of the two masses at the denominator reduces to the mass of the Sun only, and we recover Kepler's third law.
Eq. 10.20 is also valid for elliptical orbits, but in this case a must be interpreted as the sum of the semi-major axes a_1 and a_2. We require the centre of mass to be fixed, hence $a_1 m_1 = a_2 m_2$, and substituting either a_1 or a_2 in $a = a_1 + a_2$, we get $a = (1 + m_1/m_2)a_1 = (1 + m_2/m_1)a_2$.
Kepler's third law brings with it a great application: by measuring the orbital period of the planets, it is possible to obtain the ratios of their semi-major axes to that of the Earth. However, unless we measure one of the distances involved, it is as if we had a map with no scale. Try solving Pr. 9.1, for an example of how one of these distances can be measured.

10.4 Energy of an Orbiting Body

The total energy of an orbiting body is equal to the sum of its potential and kinetic energy. Imagine two bodies, at a distance r from each other. The gravitational potential energy of this configuration is equal to the work required to bring one of them to a distance r from the other (which remains fixed), starting from infinity. During this process, we must apply a force equal and opposite to that of gravity, so that the speed remains constant, and the change in kinetic energy is zero (we want to isolate the contribution of the potential energy). Therefore, the potential energy is the opposite of the work done by the force of gravity to bring one of the two bodies at a distance r from the other, starting from infinity:

$$\Delta U = -\Delta W . \tag{10.21}$$

Eq. 10.21 is valid for any *conservative* force (i.e. a force for which the work done is independent of the path taken, such as gravity). Work is defined as the scalar product of force and displacement, hence, considering an infinitesimal displacement dx towards the fixed body, the work done by the force of gravity is $dW_g = F_g\,dx$. The total work ΔW_g is given by the sum (integral) of all these infinitesimal contributions dW_g. Since gravity is a conservative force, to evaluate ΔW_g, we can chose a path that is a straight line:

$$\begin{aligned}\Delta W_g &= \int dW_g = -\int_{\infty}^{r} F_g\,dx = \int_r^{\infty} \frac{Gm_1m_2}{x^2}dx = Gm_1m_2 \int_r^{\infty} \frac{dx}{x^2} \\ &= Gm_1m_2\Big[-\frac{1}{x}\Big]_r^{\infty} = Gm_1m_2\Big[\frac{1}{r} - 0\Big] = \frac{Gm_1m_2}{r}\,.\end{aligned}$$

As expected, ΔW_g is positive, since the force of gravity and the displacement both point towards the body. Therefore, the potential energy is:

$$\boxed{U = -\frac{Gm_1m_2}{r}}\,. \tag{10.22}$$

If the body only moves in a small region of space, we can regard the gravitational field $\vec{g}$ as a constant. Therefore:

$$\begin{aligned}\Delta U &= -\Delta W = -\vec{F}\cdot\vec{h} = mgh \\ &\Rightarrow U(h) = mgh + \text{const.}\,,\end{aligned}$$

where h is the projection of the distance covered by the body in the direction of the gravitational field (positive if moving away from it), and $U(h)$ is defined within an arbitrary constant, which depends on the choice of zero-height. In Eq. 10.22, the arbitrary constant is zero, since, by convention, an object at infinity has zero potential.

The kinetic energy is defined as the work done by a force to increase the velocity of a body from zero to v, while at constant potential:

$$\Delta K = \Delta W\,, \tag{10.23}$$

where:

$$\Delta W = \int F\,dx = \int ma\,dx = \int_0^v m\frac{dv'}{dt}\,v'dt = \int_0^v mv'\,dv' = \Big[\frac{1}{2}mv'^2\Big]_0^v = \frac{1}{2}mv^2\,.$$

In the above derivation, we used v' as a variable of integration, to distinguish it from the upper limit v. Hence:

$$\boxed{K = \frac{1}{2}mv^2}\,.$$

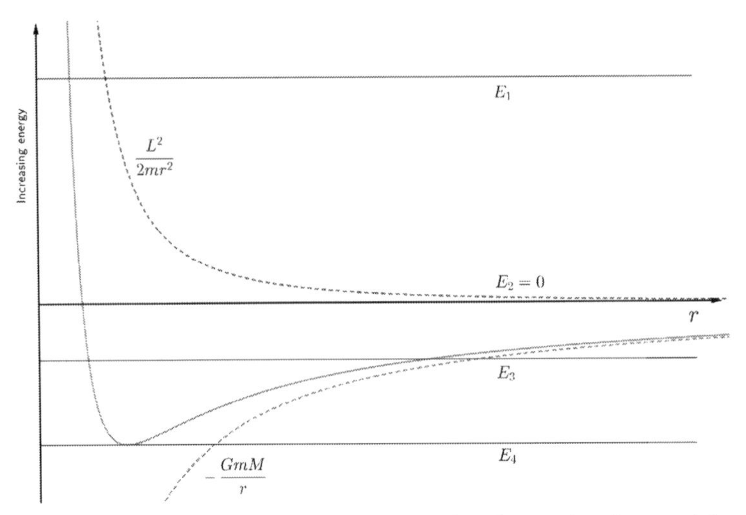

Figure 10.7: The dashed lines represent the gravitational and centrifugal potentials; the solid line their sum, known as the effective potential. The horizontal lines E_1, E_2, E_3, E_4 are the energies of an hyperbolic, parabolic, elliptic and circular orbit, respectively, and their intersections with the effective potential give the turning points of the orbits.

Therefore, the total energy of a body of mass m, orbiting at a distance r from a body of mass M (with $m \ll M$), is:

$$E = U + K = -\frac{GmM}{r} + \frac{1}{2}mv^2 . \tag{10.24}$$

The total energy is conserved if the system is isolated, i.e. if no *dissipative* forces are present. Indeed, a conservative force changes U and K by $\Delta U = -\Delta W$ (Eq. 10.21) and $\Delta K = \Delta W$ (Eq. 10.23). Hence $\Delta U + \Delta K = 0$, i.e. the energy is conserved. Decomposing the velocity into radial and tangential components ($v^2 = {v_r}^2 + {v_t}^2$), and substituting v_t as a function of the angular momentum ($v_t = L/mr$), we finally obtain:

$$E = \left(\frac{L^2}{2mr^2} - \frac{GmM}{r}\right) + \frac{1}{2}m{v_r}^2 . \tag{10.25}$$

In the above equation, the terms in brackets depend only on the variable r, therefore they can be regarded as an effective potential:

$$U_{\text{eff}} = \frac{L^2}{2mr^2} - \frac{GmM}{r} .$$

The effective potential is plotted in Fig. 10.7. The dashed lines represent the gravitational ($-GmM/r$) and centrifugal ($L^2/2mr^2$) potential terms; the solid line their sum, the effective potential. The horizontal lines show the different energies an orbiting body may have. Since the kinetic energy must always be positive, the body can never be found in regions where the effective potential is greater than the total energy. The intersection points between the horizontal lines and the effective potential are therefore known as *turning* points, since the body must invert its (radial) motion at those points. Looking at Fig. 10.7, when $E < 0$ (E_3), the horizontal line always intersects the effective potential at two points, which (from left to right) are the minimum and maximum distances. When E has the smallest possible value (E_4), the two intersection points coincide: this is the case of a circular orbit. When $E = 0$ (E_2) or $E > 0$ (E_1), there is only one turning point and the orbit is unbound: it reaches a minimum distance from the centre, but moves infinitely far away afterwards. Hence, we essentially have four types of orbits:

$$\begin{cases} E > 0\,, & \text{hyperbola}\,; \\ E = 0\,, & \text{parabola}\,; \\ E < 0\,, & \text{ellipse}\,; \\ E \text{ min}\,, & \text{circle}\,. \end{cases}$$

Closed Orbits

As the name suggests, a body in a closed orbit periodically rotates around the centre of attraction. Closed orbits can be either circular or elliptical.
In the case of circular orbits, the potential energy is constant and equal to:

$$U_c = -\frac{GMm}{R}\,.$$

According to Eq. 10.18, the velocity is:

$$v_c = \sqrt{\frac{GM}{R}}\,.$$

Hence the kinetic energy is:

$$K_c = \frac{1}{2}mv_c^2 = \frac{1}{2}m\frac{GM}{R} = \frac{GmM}{2R}\,.$$

The total energy is the sum of the potential and kinetic energies:

$$\begin{aligned} E_c = U_c + K_c &= -\frac{GmM}{R} + \frac{GmM}{2R} \\ \Rightarrow E_c &= -\frac{GmM}{2R}\,. \end{aligned}$$

The equation above is also valid for an elliptical orbit, if R is replaced by the semi-major axis a:

$$\boxed{E = -\frac{GmM}{2a}}\,. \tag{10.26}$$

Hence, it follows that the total energy of circular and elliptical orbits is always negative. As a tends to infinity, the energy goes to zero; for small a, it grows increasingly negative. Eq. 10.26, used in conjunction with the conservation of angular momentum, should allow you to solve (almost all) problems in Celestial Mechanics.

Open Orbits

In the Solar System, the planets are gravitationally bound to the Sun, travelling in elliptical, almost circular orbits. If a body does not rotate periodically around the centre of attraction, we say that its orbit is *unbound*, and it can be either *parabolic* or *hyperbolic*. In this case, the body first approaches the centre, reaches a minimum distance, and then moves infinitely far away. Since the body can reach arbitrarily large distances from the centre of attraction, for some point infinitely far away its potential energy must be zero. If the body is at rest when it is at an infinite distance, then its total energy is $E = 0$, and it can be shown that it travels in a parabolic orbit. Of course, the body is never exactly at infinity, and the velocity is never exactly zero (therefore it is actually able to *cover* the orbit). The situation is reversible, so that the minimum amount of energy that a body needs in order to escape from a system is $E = 0$. If, on the other hand, the velocity of the body at an infinite distance is non-zero, then the total energy is $E > 0$, and it can be shown that the body follows a hyperbolic orbit (Appendix B). An object in a parabolic or hyperbolic orbit must come from outside the Solar System.

10.5 Escape Velocity

The escape velocity is the minimum velocity required to move indefinitely far away from the centre of attraction. If the initial velocity of a body is exactly equal to the escape velocity, as it moves away from the centre, the velocity decreases until, asymptotically, it tends to zero. Therefore, the final kinetic and potential energies are zero, hence the total energy is zero as well. Because energy is conserved:

$$-\frac{GmM}{R} + \frac{1}{2}mv_{\text{esc}}^2 = 0\,,$$

where $m \ll M$ and R is the initial distance between the body and the centre of attraction. Solving for v_{esc}, we find:

$$\boxed{v_{\text{esc}} = \sqrt{\frac{2GM}{R}}}\,. \tag{10.27}$$

If a body has a velocity lower than the escape velocity, it follows a circular or an elliptical orbit, being gravitationally bound to the centre of attraction. If the velocity is exactly equal to the escape velocity, the orbit is parabolic, if it is greater, the orbit is hyperbolic. In general, the escape velocity is greater for planets with a larger mass. This aspect is important in the discussion of planetary atmospheres, covered in Sec. 7.3.

Schwarzschild Radius

The Schwarzschild radius is the size of the event horizon of a black hole, i.e. the distance from the singularity within which the escape velocity from the black hole is greater than the speed of light. At a distance from the singularity equal to the Schwarzschild radius (R_s), the escape velocity is exactly equal to c:

$$v_{\text{esc}} = c = \sqrt{\frac{2GM}{R_s}}$$

$$\Rightarrow \boxed{R_s = \frac{2GM}{c^2}}. \tag{10.28}$$

Although this equation was obtained in the framework of classical physics, it turns out that this is the same result predicted by the theory of general relativity. The above derivation is given to facilitate the memorization of Eq. 10.28.

10.6 Virial Theorem

For an inverse square force law, such as gravity, the Virial theorem relates the average potential and kinetic energies:

$$\boxed{\overline{U} = -2\overline{K}}. \tag{10.29}$$

The total energy, as usual, is the sum of the potential and kinetic energies:

$$\overline{E} = \overline{U} + \overline{K} = \overline{U} - \frac{\overline{U}}{2} = \frac{\overline{U}}{2}. \tag{10.30}$$

For example, for a body in circular orbit, the potential and kinetic energies are constant. Since $\overline{U} = U_c = -GmM/a$, it follows that the energy is also constant and equal to:

$$E_c = \frac{1}{2}U_c \Rightarrow E_c = -\frac{GmM}{2a},$$

which is the result given in Eq. 10.26. In Appendix C, we prove the Virial theorem for a general potential of the from $U = -\alpha r^{-n}$, in which case:

$$\boxed{\overline{K} = -\frac{n}{2}\overline{U}}. \tag{10.31}$$

For the gravitational potential, $n = 1$, hence we recover Eq. 10.29.

10.7 Exercises

1. Compute the ratio between the gravitational acceleration on the surface of the Sun and the Earth.

2. (INT 2015, Th.S, q.3) The commander of the interstellar mission Kout-Iz-Paut wants to attempt a descent on the planet Sweets-3 and it is essential for him to know the gravitational acceleration g_{SW3} on the surface of the planet. The only data he is given is the density ($\rho_{\mathrm{SW3}} = 4\rho_{♃}$) of the planet Sweets-3, and its radius ($R_{\mathrm{SW3}} = R_{♃}/4$). Compare the value of g_{SW3} with the gravitational acceleration on Jupiter $g_{♃}$.

3.* (IOAA 2012, Th., q.5) Compute the ratio between the mean densities of the Earth and Sun, using only the following data:

- mean angular diameter of the Sun ($\theta_{\odot} = 32'$), as seen from Earth;
- gravitational acceleration on the surface of the Earth ($g = 9.80\,\mathrm{m/s^2}$);
- length of the sidereal year ($T_{\mathrm{sid}} = 365.2564^d$);
- the fact that one degree in longitude corresponds to $s = 111\,\mathrm{km}$ on the surface of the Earth.

4. The Hubble Space Telescope rotates around the Earth, in a circular orbit, at an average distance of $H = 560\,\mathrm{km}$ from the surface. Calculate the average velocity v_c and the orbital period T of HST.

5. Determine the distance d of a geostationary satellite from the surface of the Earth, and its velocity v.

6. (NAZ 2015, Th.S, q.2) Consider a satellite that orbits the Earth along the equatorial plane, in the same direction as Earth's rotation. At what minimum height (h_{min}) from the Earth's surface should the satellite be, in order for an observer to see it move in the sky from east to west?

7. CoRoT-7b is the closest exoplanet to its star that has ever been discovered. The semi-major axis of its orbit is just 0.0172 au, while its period of revolution is only 0.8536 days. Estimate the mass of the star CoRoT-7, assuming that the mass of the planet is negligible compared to the mass of the star.

8. In its orbit around the Sun, a comet reaches a maximum distance of 31.5 au, and a minimum of 0.5 au. What is its orbital period? At what rate does the radius vector joining the comet to the Sun sweep out area, in square kilometres per year?

9.* A planet is in an elliptical orbit around the Sun, with semi-major axis a and eccentricity e. Compute its velocities at perihelion and aphelion.

10.* Two asteroids of masses M_1 and M_2 (negligible compared to that of the Earth) are on a collision course with Earth, at a distance of $1.2 \cdot 10^6$ km and 10^6 km, respectively. Assuming their initial velocities are zero, find:

- the time it takes the first asteroid to impact the Earth;
- the time between the two impacts;
- if the impacts happen on the equator, what is the distance (measured along the Earth's surface) between the two craters?

11.* (IOAA 2011, Th., q.8) A spaceship orbits very close to the asteroid Seneca, transmitting data on Earth. Due to the relative motion between the asteroid and the Earth, the time taken by a signal transmitted by the spaceship to arrive on Earth varies between 2 and 39 minutes. Assuming that the Earth moves around the Sun in a circular orbit that does not intersect the orbit of Seneca, calculate:

- the semi-major axis a_s and the eccentricity e_s of Seneca's orbit;
- the period T_s of Seneca's orbit.

12.* (IAO 2015, Th.α, q.5) A satellite moves around the Earth in a slightly elliptical, equatorial, orbit and passes through the perigee point at a height of $d_p = 428.0$ km above sea level, with a speed 0.6 % greater than that of a circular orbit at the same height. Estimate the time taken for the satellite to reach a height of $d = 498.0$ km.

13.* (BAAO, 2019, Th., q.1, adapted) Calculate the time the Parker Solar Probe (PSP) spends at a distance from the Sun less than the length of the latus rectum of its orbit (i.e. the point where the angle between the radius vector and the semi-major axis, known as the true anomaly, is 90°). First, solve this exercise using Kepler's second law and Eq. A.7 (Appendix A.2); then using the equations given in Pr. 10.12.

14.** (IOAA 2014, Th., q.2) In a gravitational catastrophe, the mass of the Sun suddenly halves. Find the new orbital period if, at that time, the Earth was at aphelion or perihelion.

15. What should the radius of a planet be, if it has the same density as the Earth and if a human can escape from it with a simple jump?

16. Estimate the number of stars in a globular cluster of diameter 40 pc, if the escape velocity from its outer edge is 6 km/s. Assume that, on average, the cluster is composed of Sun-like stars.

17. If the Sun collapsed to form a non-rotating black hole, what would the radius of its event horizon be? Calculate the average density.

18.* A planet is shaped as a long cylinder. Its mean density, radius and rotational period are the same as on Earth.

- Compute the first cosmic velocity $v_{c,1}$ (i.e. the orbital velocity on the surface of the planet).
- What is the height of a geostationary satellite, measured from the surface?
- What can you say regarding the second cosmic velocity (i.e. the escape velocity)?

19.* A homogeneous sphere of radius a contains a cavity of radius $a/4$, whose centre is $3a/8$ from the surface. The diameter passing through the centre of the sphere and the cavity meets the surface at points A and B. Calculate the ratio of the gravitational acceleration at these two points.

20.** A projectile is launched from the surface of the Earth with the first cosmic velocity, at an angle $\theta = 30°$ with respect to the local horizon. Neglecting friction with air, find:

- the semi-major axis a of the orbit;
- the maximum height of the projectile with respect to the Earth's surface, in units of $R_\oplus$;
- the range of the projectile, i.e. the distance, along the Earth's surface, between the launch and landing points;
- the time of flight.

21.** A probe is launched from the north pole of the Earth with the first cosmic velocity, in such a way that it lands at the equator. Neglecting friction with air, find:

- the semi-major axis a of the orbit;
- the maximum height of the probe, with respect to the Earth's surface, in units of $R_\oplus$;
- the time of flight.

22.** In the absence of atmosphere on Earth, what are the maximum and minimum velocities of Sun-orbiting asteroids that impact our planet?

23.*** Find the maximum time a comet, following a parabolic trajectory around the Sun, can spend within the orbit of the Earth. Assume that the Earth's orbit is circular and in the same plane as that of the comet.

24.*** What should the shape of a planet be (of fixed volume and density), such that it produces the maximum possible gravitational acceleration at a certain point in space? How much greater is the gravitational acceleration in this case, compared to that of a spherical planet with the same mass and density?

25.*** Two comets with identical masses and speeds are found to be approaching the Sun along parabolic trajectories that lie in the same plane. The comets collide at their common perihelion P, and break into many pieces that then travel in all directions, but with identical initial speeds. What shape is the envelope of the subsequent trajectories of the pieces of debris?

10.8 Problems

1.** **Planets as point masses**
For a spherical shell, show that:

- at its exterior, the gravitational field is the same as that produced by a point mass, of the same mass, placed at its centre;
- at its interior, the gravitational field is zero.

Hence, prove the two points stated in Sec. 10.2.

2.* **Using Gauss' theorem**
Compute the gravitational acceleration at a distance r from the centre of the Earth, assuming its density is constant.

3.* **Deep tunnels**
On a future lunar base, one wonders what the fastest way is to transport a load from a place on the surface of the Moon to its antipode. Two solutions have been proposed:

- introducing the load into a circular orbit;
- dropping the load into a straight tunnel connecting the two places.

Which method enables a faster transport? Assume the density of the Moon is constant.

4.** **Potential energy of a spherical body**
Compute the gravitational potential energy of a star with mass M and radius R. You can imagine assembling the star in spherical shells, brought from infinity.

5.** **Escape velocity**
Obtain an expression for the escape velocity in the case that the masses of the two bodies are comparable.

6.** **Roche limit**
Obtain an expression for the Roche limit, defined as the minimum distance from the centre of a celestial body such that a secondary body (held together only by the force of gravity) does not disintegrate due to tidal forces. Your answer must be a function of the density of the two bodies and the radius of the central body. Give a numerical value for the Roche limit in the case of the Sun-Earth and Earth-Moon systems.

7.** **Jeans limit**
Obtain an equation for the Jeans limit, i.e. the critical mass above which a nebula collapses to form a star. Write the result as a function of the temperature T, the molar mass μ of the gas that constitutes the nebula and its density ρ.

8.** **Hill sphere and Lagrangian points L_1, L_2**
Calculate the distance to the Sun of the Lagrangian points L_1 and L_2, i.e. those points on the line joining the Earth and the Sun, placed in between the Sun and the Earth and beyond the Earth, respectively, having the same orbital period as the Earth (see Fig. 15.15, pag. 252). Hence, find the radius of the Hill sphere, defined as the distance within which a secondary body can retain satellites while orbiting around a central body. Give a numerical value for the radius of Earth's Hill sphere.

9.*** **Position of Lagrangian point L_3**
Find the position of the Lagrangian point L_3, i.e. that point on the line joining the Earth and the Sun, on the opposite side of the Earth relative to the Sun, having the same orbital period as the Earth.

10.*** **The three-body problem and Lagrangian points L_4, L_5**
A system consists of three bodies with masses m_1, m_2 and m_3 placed at distances $r_{1,2}$, $r_{1,3}$ and $r_{2,3}$. Find a relationship between their distances so that the triangle formed by the three masses always maintains a constant shape. Hence, calculate the orbital period of the system. Explain how this problem is related to the calculation of the Lagrangian points L_4 and L_5, positioned (on Earth's orbit) 60° before and after the Earth, and having the same orbital period.

11.** **The fate of abandoned satellites**
At the end of their useful lives, satellites lose energy in the upper layers of the atmosphere, before finally burning up when they reach the denser, lower, layers. It can be shown that satellites originally moving along circular trajectories will continue to travel in approximately circular orbits, with their orbital radii slowly decreasing. Consider a satellite with mass $m = 500\,\text{kg}$, in an approximately circular orbit, when it is abandoned.

The friction force on the satellite is approximately $F_a = c\rho v^2$, where $c = 0.23\,\text{m}^2$, ρ is the density of air at the height of the satellite and v is its instantaneous velocity.

- Does the satellite decelerate or accelerate as a result of the air drag? Explain your answer from the point of view of dynamics.
- Find a (simple) equation that relates the drag force to the tangential acceleration of the satellite.
- What is the density of air at an altitude of 200 km, if the orbital radius of the satellite decreases by 100 m during a single revolution?

12.*** Kepler's equation

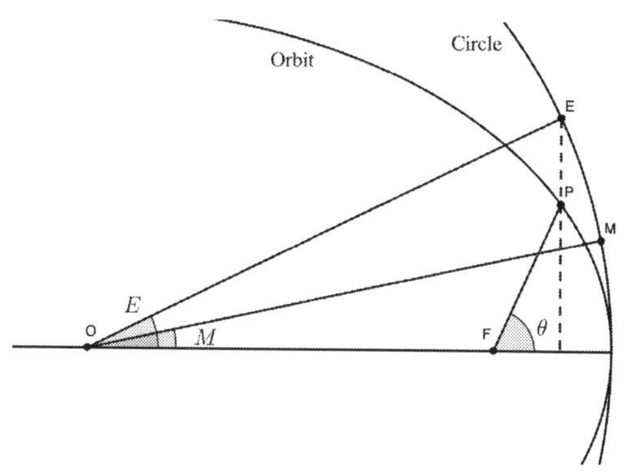

Figure 10.8: True (θ), eccentric (E) and mean (M) anomalies.

Let O be the centre of an elliptical orbit, F the focus around which the body orbits and P the position of the body (see Fig. 10.8).
The *true anomaly* (θ) is defined as the angle between the direction of perihelion and the position of the body, as seen from F. Prove that the distance r of P from F is related to the true anomaly by:

$$\boxed{r(\theta) = \frac{a(1-e^2)}{1+e\cos\theta}}. \tag{10.32}$$

Let E be the intersection between the perpendicular to the major axis passing through P and the circle. The *eccentric anomaly* (E) is defined as the angle between the line OE and the direction of perihelion. Prove that the distance r is related to the eccentric anomaly by:

$$\boxed{r = a(1-e\cos E)}. \tag{10.33}$$

The *mean anomaly* (M) is defined as the angle corresponding to the fraction of the period of an elliptical orbit that has elapsed since the body passed through perihelion. Hence, $M = \omega(t - t_0)$, where $\omega = \sqrt{GM/a^3}$ is the mean angular velocity of revolution and t_0 is the time at perihelion. Prove that the mean anomaly is related to the eccentric anomaly by:

$$\boxed{M = E - e\sin E}\,, \tag{10.34}$$

this is known as Kepler's equation (Hint: use Eqs. 10.32, 10.33 and 5.9).

13.** Relaxation time**

Consider a system of stars, such as a galaxy or a cluster. In these systems, collisions between stars are very rare, since the average distance between them is much greater than their size. A physical model can approximate the system with a continuous distribution of matter, thus neglecting its granularity. However, the real force acting on each star is slightly different from the force obtained using the physical model. We want to estimate the time scale for which the deviation from the continuous distribution model becomes significant. Systems for which this deviation is small are called "collision-free", and the time scale, called *relaxation time*, is usually larger than the age of the universe.

You can use the following model: a star of mass m and velocity v approaches a stationary star (of the same mass m) with impact parameter b. If the distance between the two stars is large, their interaction is small and the star moves approximately on a straight line at a distance b, undergoing a small deviation in the direction perpendicular to its velocity. Find the variation of the square of the velocity in the perpendicular direction as a function of the impact parameter b. Then, integrating over all the values of b, estimate the relaxation time, i.e. the time necessary for the variation in the mean square speed to be equal to the square of the velocity of the star: $\overline{dv^2_{\perp\text{tot}}} \approx v^2$.

11

Motion of the Planets

The Sun and the Moon always rise in the east and set in the west. However, over the course of a year, they describe great circles on the celestial sphere in the opposite direction to the apparent daily rotation of the sky, with a variable but almost constant velocity. The trajectory described by the planets on the celestial sphere is, instead, more complex: it contains knots and loops, that is, points in which the motion reverses, from direct (same direction as the Sun, from west to east) to retrograde, or vice-versa.

11.1 Apparent Motion of the Planets

According to their apparent motion, we distinguish *inferior* planets (such as Mercury and Venus) from *superior* planets (all others, excluding Earth).
Inferior planets always appear to be moving close to the Sun, their angular separation never exceeding the *maximum elongation* ($\epsilon_{\rm max}$). Considering the triangle $\mathrm{SE_wT}$ in Fig. 11.1, we see that:

$$\boxed{\sin \epsilon_{\rm max} = d/d_\oplus}, \tag{11.1}$$

where d and $d_\oplus$ are the Sun-planet and Sun-Earth distances, respectively. Because of the eccentricity of the orbits, $\epsilon_{\rm max}$ varies according to the position of the Earth and the planet, taking the values $\epsilon_{\rm max} = 18^\circ - 28^\circ$ for Mercury and $\epsilon_{\rm max} = 45^\circ - 48^\circ$ for Venus.
At the point of maximum eastern elongation $\mathrm{E_e}$, the planet is visible in the evening, shortly after sunset. In fact, since the sky rotates from east to west, the Sun, further west, sets earlier than the planet, which is therefore visible for a brief period of time during twilight. Subsequently, the planet moves closer to the Sun until, at the moment of *inferior conjunction* C_i, it is aligned with the Sun and the Earth. If the alignment is perfect, the planet appears completely dark, eclipsing the portion of the Sun it passes through. If, instead, the planet does not transit on the solar disk, it is still possible to see a small illuminated portion of it. The planet then moves away from the Sun, with a retrograde motion towards the west, and is now visible just before dawn. Its apparent

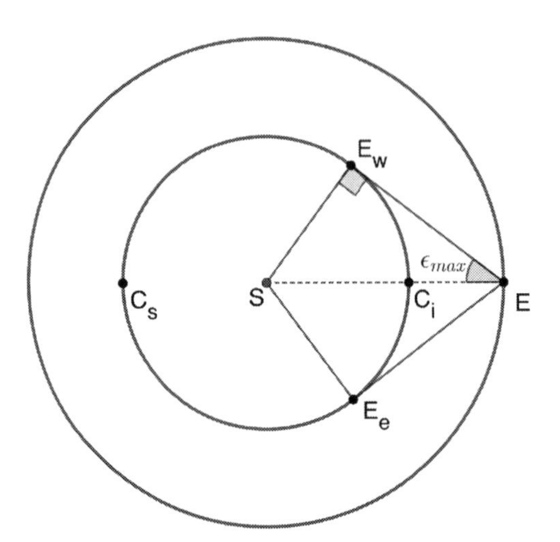

Figure 11.1: Positions occupied by an inferior planet. The angular distance between the Sun and the planet never exceed the maximum elongation $\epsilon_{\max}$. The planet appears to be stationary, as viewed from Earth (E), at the points of maximum eastern and western elongation (E_e and E_w, respectively), while it appears to have the greatest velocity in superior and inferior conjunction (C_s and C_i, respectively).

velocity decreases, until it appears stationary at the point of maximum (western) elongation E_W. Here, the planet reverses its motion, now approaching the Sun at an increasing velocity. The moment when the planet is on the opposite side of the Earth, relative to the Sun, is called *upper conjunction* C_S. If the alignment is perfect, the planet disappears behind the solar disk. The planet continues its motion in the direct direction, moving away from the Sun. Then, at the point of maximum (eastern) elongation, it reverses its motion again, and the cycle starts over.

The motion of the superior planets on the celestial sphere is somewhat different, since their elongation can take any value. Now, the moment when the planet is on the opposite side of the Earth relative to the Sun is called *conjunction*, denoted by C in Fig. 11.2. When in conjunction, the planet is difficult to observe, since it rises and sets very close to the Sun. It then proceeds in the direct direction, moving with a decreasing velocity away from the Sun, until it becomes stationary at the point of *eastern quadrature* Q_e. Now the motion is retrograde, and the planet moves away at an increasing velocity. The velocity of the planet is maximum when it is aligned with the Sun and the Earth, on the same side of the Earth. This is the point of *opposition* O. Thereafter, the planet moves at a decreasing velocity until it stops, at the point of *western quadrature* Q_W. It then moves with direct motion again, until reaching the point of conjunction, where the cycle repeats itself.

Explaining the change in direction of the motion of a planet is relatively simple if we assume that the Earth orbits around the Sun. In the past, however, it was believed that the Earth sat motionless at the centre of the universe (*geo-*

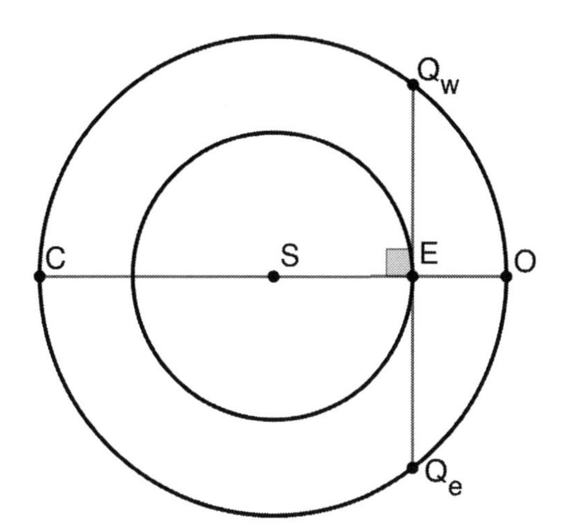

Figure 11.2: Positions occupied by a superior planet. Unlike inferior planets, whose angular distance from the Sun cannot exceed the *maximum elongation*, superior planets can have any elongation. The planet appears stationary, as viewed from Earth (E), at the points of eastern and western quadrature (Q_e and Q_w, respectively), while its apparent velocity reaches a local maximum in conjunction and opposition (C and O, respectively).

centric model). The Ptolemaic system (the best known geocentric model) tried to explain direct and retrograde motions by assuming that planets rotate on circles called *epicycles*, which in turn move on greater circles, called *deferentials.* Moreover, to explain the difference in the length of the seasons, the centre of the Sun's deferential was assumed to be at some distance from the Earth, on a point called the *eccentric.* Often, the simplest explanation (i.e. the *heliocentric* model) is the right one.

11.2 Synodic Period

The synodic period is the interval of time between two successive conjunctions or oppositions of a planet. After a sidereal period, the planet has completed an orbit around the Sun, returning to the same position within the Solar System. However, during that time, the Earth has moved around the Sun, therefore the position of the planet as seen from Earth will, in general, be different.
All planets orbit the Sun anti-clockwise when viewed from the north pole of the Earth. Hence, to calculate the relative angular velocity, we take the difference between the angular velocities of the Earth (ω_e) and the planet (ω_p). For inferior planets $\omega_p > \omega_e$, therefore:

$$\omega_r = \omega_p - \omega_e \,,$$

while, for superior planets $\omega_p < \omega_e$, thus:

$$\omega_r = \omega_e - \omega_p \,.$$

Since $\omega = 2\pi/T_{\text{syn}}$, $\omega_p = 2\pi/T_{\text{p}}$ and $\omega_e = 2\pi/T_{\text{yr}}$, we find, for inferior and superior planets, respectively:

$$\frac{1}{T_{\text{syn}}} = \frac{1}{T_p} - \frac{1}{T_{\text{yr}}},$$
$$\frac{1}{T_{\text{syn}}} = \frac{1}{T_{\text{yr}}} - \frac{1}{T_p}.$$

For inferior planets:

$$\boxed{T_{\text{syn}} = \frac{T_p T_{\text{yr}}}{T_{\text{yr}} - T_p}}. \tag{11.2}$$

For superior planets:

$$\boxed{T_{\text{syn}} = \frac{T_p T_{\text{yr}}}{T_p - T_{\text{yr}}}}. \tag{11.3}$$

These formulas can be applied to any pair of bodies in orbit around a common centre, if we substitute for T_p and T_{yr} the periods of their orbits. If they travel in opposite directions, the relative angular velocity is the sum of the angular velocities, hence the minus sign at the denominator in Eqs. 11.2 and 11.3 becomes a plus sign, and we are left with one equation only.
We can express T_{syn} directly in years by dividing both sides of the previous equations by T_{yr}. For inferior planets then:

$$\frac{T_{\text{syn}}}{T_{\text{yr}}} = \frac{T_p/T_{\text{yr}}}{1 - T_p/T_{\text{yr}}} \Rightarrow T_{\text{syn, yr}} = \frac{T_{p,\text{yr}}}{1 - T_{p,\text{yr}}}.$$

While, for superior planets:

$$\frac{T_{\text{syn}}}{T_{\text{yr}}} = \frac{T_p/T_{\text{yr}}}{T_p/T_{\text{yr}} - 1} \Rightarrow T_{\text{syn, yr}} = \frac{T_{p,\text{yr}}}{T_{p,\text{yr}} - 1}.$$

11.3 Exercises

1. (INT 2014, Th.S, q.1) Two consecutive oppositions of an external planet have been observed. The time interval between the two events is 398.85 days. Which planet is it?

2. A planet has a synodic period of 584 days. What is its distance from the Sun and what planet is it?

3. (INT 2013, Th. S, q.3) Two space stations are in circular polar orbit around the Earth, travelling in the same direction. Stations A and B are located at heights $h = 200$ km and $H = 400$ km. Determine the time between two consecutive alignments of the two stations.

4. The elongation of a planet is 150°. Is it an inferior or superior planet?

5. In the evening, which planets rise in the east?

6. The synodic period of a planet is equal to its sidereal period. What is its sidereal period?

7. Calculate the time it takes Mars to move from opposition to quadrature. You may assume the orbits to be circular.

8.* The largest and smallest maximum elongations of Mercury are $e_{\max,1} = 28°$ and $e_{\max,2} = 18°$, respectively. Find the eccentricity of its orbit.

9.* (CAO 2019, Th., q.3) An observer on earth measured the apparent magnitude of an asteroid at every opposition. The asteroid's period is 3.9 years. Estimate the eccentricity of its orbit, if the difference between the highest and lowest apparent magnitudes is $\Delta m = 2.5$. Assume the Earth's orbit is circular.

10.* (INAO 2020, Th., q.1) On the evening of the autumnal equinox, Siddhant noticed that Mars was exactly along the north-south meridian at the moment when the sun was setting. If the orbital radius of Mars is 1.52 au, at what time will Mars rise on the next autumnal equinox?

11.4 Problems

1.* **Great opposition**
The orbit of Mars is more eccentric than Earth's orbit. Of the various oppositions, some occur when Mars is at perihelion. These (great) oppositions are very favourable for observing the planet. How often do they occur?

12

Orbital Manoeuvres

In this chapter we present a simplified approach for placing into orbit and changing the orbits of satellites. Most of the time the limiting factor is the available fuel, which is related, through the *rocket equation*, to a quantity called Δv.

12.1 The Rocket Equation

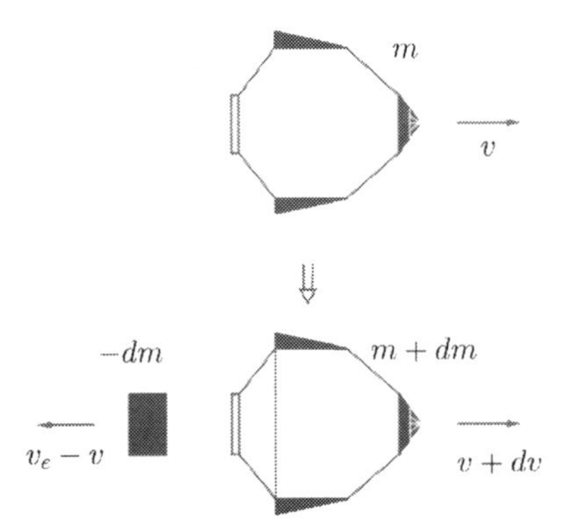

Figure 12.1: After the exhaustion of mass $-dm$ at velocity $v_e - v$, the mass of the rocket will be $m + dm$ and its velocity $v + dv$.

Let us assume that, in the reference system of the rocket, the exhaustion speed of the fuel is constant and equal to v_e. This velocity depends on the amount of energy generated by chemical reactions in the fuel. For an inertial observer, the instantaneous velocity of the rocket is v (Fig. 12.1). The rocket then expels a mass of fuel $-dm$ with velocity $v - v_e$, and consequently its velocity and mass increase by dv and dm, respectively (note that we will be integrating from the initial to the final mass, hence the increment dm is negative). Requiring momentum to be conserved:

$$mv = (m + dm)(v + dv) - (-dm)(v_e - v)\,.$$

Simplifying:

$$dv = -v_e \frac{dm}{m}\,.$$

Integrating both sides:

$$\int_{v_i}^{v_f} dv = -v_e \int_{m_i}^{m_f} \frac{dm}{m}\,.$$

Hence:

$$\boxed{\Delta v = v_f - v_i = v_e \ln \frac{m_i}{m_f}}\,. \tag{12.1}$$

An important result, which we will use to justify the Oberth effect, is that Δv does not depend on the initial speed of the rocket, but only on the exhaustion speed and the amount of fuel burned.

12.2 Oberth Effect

The Oberth manoeuvre consists in firing the rocket when the velocity is maximum, for example at perihelion. In some cases, it may be convenient to bring the rocket closer to a celestial body to take advantage of this effect.
As shown in Sec. 12.1, Δv depends only on the amount of fuel burned. For sufficiently small Δv, the change in the energy of the rocket is:

$$\Delta K = K_f - K_i \approx mv\Delta v\,.$$

Therefore, for a fixed Δv, the energy gained is directly proportional to the velocity of the spaceship. Note that m remains fixed, since we are only interested in the kinetic energy of the rocket (and not in that carried by the fuel).

12.3 Hohmann Transfer Orbit

Hohmann's manoeuvre describes the most efficient way of transferring a satellite between two circular orbits, around the same centre of force and coplanar to each other. It consists of two stages (Fig. 12.2):

- at point P the probe is given an impulse to bring it into an intermediate elliptical orbit, tangent to the final circular orbit;
- at point A the probe is given a second impulse, to introduce it into the final circular orbit.

Let R_1 and R_2 be the radii of the initial and final circular orbits, respectively. We want to find the two Δv-s that the probe must be given in P and A. After the first impulse, the probe travels on an ellipse with semi-major axis $2a = R_1 + R_2$. The energy of this orbit is given by Eq. 10.26:

$$E = -\frac{GmM}{R_1 + R_2}\,.$$

The kinetic energy of the probe in P is then:

$$K = E - U = -\frac{GmM}{R_1 + R_2} + \frac{GmM}{R_1} = \frac{GmM}{R_1}\left(1 - \frac{1}{1 + R_2/R_1}\right),$$

which corresponds to a velocity of:

$$v_\mathrm{P} = \sqrt{\frac{2}{m}K} = \sqrt{\frac{2GM}{R_1}\left(1 - \frac{1}{1 + R_2/R_1}\right)} = v_0\sqrt{\frac{2R_2/R_1}{1 + R_2/R_1}},$$

where v_0 is the velocity of the circular orbit with radius R_1. The Δv in P is:

$$\boxed{\Delta v_\mathrm{P} = v_0\left(\sqrt{\frac{2R_2/R_1}{1 + R_2/R_1}} - 1\right)}\,. \tag{12.2}$$

The velocity of the probe in A can be obtained applying conservation of angular momentum at perihelion and aphelion (Eq. 10.12):

$$m v_\mathrm{P} R_1 = m v_\mathrm{A} R_2 \Rightarrow v_\mathrm{A} = \frac{R_1}{R_2} v_\mathrm{P}\,.$$

The final velocity is that of a circular orbit with radius R_2, i.e. $v_f = \sqrt{GM/R_2}$. Hence, the Δv that must be applied in A is:

$$\Delta v_\mathrm{A} = v_f - v_\mathrm{A} = v_f\left(1 - \sqrt{\frac{2}{1 + R_2/R_1}}\right).$$

Since $v_f = \sqrt{R_1/R_2}\, v_0$, the last equation can be rewritten in terms of v_0:

$$\boxed{\Delta v_\mathrm{A} = v_0\sqrt{\frac{R_1}{R_2}}\left(1 - \sqrt{\frac{2}{1 + R_2/R_1}}\right)}\,. \tag{12.3}$$

Assuming that both impulses are instantaneous and using Kepler's third law, we find the time required to enter the new circular orbit:

$$\boxed{T^2 = \frac{\pi^2}{8GM_\oplus}(R_1 + R_2)^3}\,. \tag{12.4}$$

The Hohmann transfer orbit requires the starting and arrival points to be in specific positions, therefore the probe can be launched only in certain windows of time. For example, the required alignment for space missions between Earth and Mars occurs every 26 months. An even more favourable time is during great oppositions (Pr. 11.1), when the Earth-Mars distance is minimum. Often, Hohmann's manoeuvre is not the fastest way to achieve a transfer, but it does allow the rocket to carry the largest payload. Due to its reversible nature, this manoeuvre can also be used to bring a satellite from a higher to a lower orbit. In this case, the impulses must be directed opposite to the velocity.

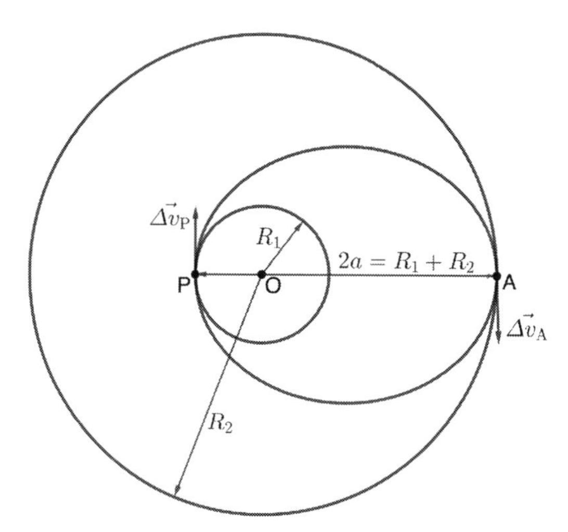

Figure 12.2: Hohmann transfer between two circular orbits of radii R_1 and R_2

12.4 Exercises

1. Calculate the velocity increments Δv_P and Δv_A that a probe must be given, to be transferred from a circular orbit of radius $R_1 = 1\,\text{au}$ around the Sun, to another circular orbit of radius $R_2 = 5.2\,\text{au}$ (Jupiter's orbit) and coplanar with the first. Determine the transfer time T.

2.* The sum of the positive speed increments necessary to transfer a probe around the Sun from a circular orbit of radius R_1 to another circular orbit of radius R_2 and coplanar to the first is 84.73 km/s. Determine the radii of the two orbits if the transfer time is $8^h3^m10^s$. You will need to solve an equation numerically.

3.* A probe is intended to leave Earth's gravitational field but the fuel in the main engine is a little less than necessary. The probe is also equipped with an auxiliary engine, capable of operating for a short period of time. When is it convenient to operate the auxiliary engine: at take-off, or when the probe is almost stationary, or is it not important?

4.** A construction project for a space lift requires it to remain fixed relative to some point on Earth and suspended in equilibrium, without falling towards the Earth or moving away into space. What should the length of the lift be? What is the maximum tension in the rope?

5.* (ARAO 2018, Th.X, q.2) A probe was launched to another planet in the Solar System along an energetically optimal trajectory. After flying close to the planet, the probe immediately set off on its way back to Earth. Throughout the mission, the probe (with no engines) made one revolution around the Sun and returned to our planet, at exactly the same starting position. For which planet closest to the Sun is this possible?

6.* (BAAO 2018, Th., q.4, adapted) The most energetic chemical reaction we can use in a rocket is hydrogen-oxygen, which gives $v_e = 4.46$ km/s, and engineering limits us to a rocket design with a maximum of 96 % of launch mass being fuel. Assuming that the density of rocky planets is:

$$\rho = 2.32 + 3.18\frac{R_p}{R_\oplus}\ \mathrm{g/cm^3},$$

derive the maximum R_p above which any alien civilization would be unable to escape their planet's gravity using simple propulsion systems.

12.5 Problems

1.* **Circular orbits**
What is the minimum velocity required to place a satellite into a circular orbit around the Earth? How much faster should the satellite be, if it is to be placed in a polar orbit? Where should the satellites be launched from, in the two cases? What is the minimum speed required to escape the Earth's gravitational field?

2.* **Changing the plane of the orbit**
Find the minimum Δv required to change the inclination of the circular orbit of a satellite, moving with velocity v_0, by an angle i. Estimate the additional mass of fuel required, at launch, to change the inclination of the circular orbit of a 1000 kg satellite by 1°. Assume the satellite orbits the Earth 200 km above the surface, and that $v_e = 4.5$ km/s.

3.** **Third cosmic velocity**
What is the minimum Δv required for a probe, launched from Earth, to escape the Solar System? The gravitational field of the Earth is not negligible compared to that of the Sun. What requires more energy: leaving the Solar System or reaching the Sun?

4.** **Rocket in a gravitational field**
Find the rocket equation in a constant gravitational field of magnitude g, if the (constant) rate at which fuel is expelled is r.

5.** **Reaching the centre**
For what n does the potential $U = \alpha r^{-n}$ allow a body to reach the centre of attraction ($r \to 0$), if the angular momentum is non-zero?

6.*** **Gravitational slingshot**
What is the best planet to use as a gravitational slingshot? Taking advantage of this planet, what is the minimum Δv required to escape the Solar System, if launching from Earth?

13

Binary Stars

Often, stars appear very close to each other, and one might wonder whether their proximity is only apparent, or if they really do form a binary system.
If the linear distance between the stars is large (the limit depends on the mass of the components, but is rarely greater than a few light years), they do not orbit each other and we call them *optical* binaries. In fact, just under half of all stars are in isolated systems, thus you are more likely to observe a system of two or more stars, rather than a single star.
Based on the method of their discovery, binary systems fall into different categories. Two stars form a *visual* binary if their angular distance is greater than the minimum resolution of a telescope on Earth. Even if the system cannot be directly resolved, other methods can still be used to identify the binary.
In a binary system, the stars orbit around their common centre of mass. This motion gives rise, as a result of the Doppler effect, to an observable shift in the emission or absorption lines in their spectra. Binary systems that are discovered on the basis of the Doppler shift are known as *spectroscopic* binaries. By analysing the spectra of both stars, we can obtain the minimum mass of each component, as well as their mutual distance. However, if one of the stars is much brighter than its companion, the spectrum of the system can be completely dominated by that of the brightest star. In this case, the system is called an *astrometric* binary. Unless the mass of the brightest star can be estimated from its spectral type, it is not possible to compute the mass of its companion. If the orbital plane of the system coincides with the plane of observation, the stars periodically eclipse each other and are called *photometric* binaries. From the variation in brightness of the system, it is possible to infer the period of rotation and the relative sizes of the stars. We can then analyse their spectrum to obtain more information.
Binary systems can also be classified according to the distance between the components. If the two stars are very far apart, we call them *distant* binaries, with a typical period of revolution of around tens or thousands of years. If, instead, the stars are at a distance on the order of one astronomical unit, they are called *close* binaries. In this case, tidal forces may be strong enough

to deform the stars significantly, giving rise to an anisotropic emission of energy from their surface, which can be observed as a periodic variation in their brightnesses. Finally, if the stars of a close binary system come into contact, we call them *contact* binaries. In the following sections, we will study in more detail each of the cases listed above.

13.1 Distance Between Two Stars

Let S_1 and S_2 be two stars, at distances d_1, d_2 from Earth, with declinations δ_1, δ_2 and right ascensions α_1, α_2, respectively. Considering a spherical polar coordinate system centred on Earth, their positions can be written as:

$$\begin{aligned} S_1 &= (d_1 \cos\delta_1 \cos\alpha_1,\ d_1 \cos\delta_1 \sin\alpha_1,\ d_1 \sin\delta_1)\,, \\ S_2 &= (d_2 \cos\delta_2 \cos\alpha_2,\ d_2 \cos\delta_2 \sin\alpha_2,\ d_2 \sin\delta_2)\,. \end{aligned}$$

Then, the distance between S_1 and S_2 is:

$$\begin{aligned} d^2 &= (\Delta x)^2 + (\Delta y)^2 + (\Delta z)^2 \\ &= d_1^2 + d_2^2 - 2d_1 d_2 \cos\delta_1 \cos\delta_2 \cos\alpha_1 \cos\alpha_2 - 2d_1 d_2 \cos\delta_1 \cos\delta_2 \sin\alpha_1 \sin\alpha_2 \\ &\quad - 2d_1 d_2 \sin\delta_1 \sin\delta_2 \end{aligned}$$

Applying the cosine addition formula (Eq. A.16), we find:

$$\boxed{d^2 = d_1^2 + d_2^2 - 2d_1 d_2[\cos\delta_1 \cos\delta_2 \cos(\alpha_2 - \alpha_1) + \sin\delta_1 \sin\delta_2]}\,. \tag{13.1}$$

When analysing possible candidates for binary systems, we usually take into consideration stars with very similar right ascensions and declinations. Let $\delta_1 = \delta$, $\delta_2 = \delta + \Delta\delta$ and $\alpha_2 - \alpha_1 = \Delta\alpha$, with $\Delta\alpha, \Delta\delta \ll 1$. In this approximation, Eq. 13.1 becomes (see Ex. 13.1):

$$d^2 = d_1^2 + d_2^2 - 2d_1 d_2\Big[1 - \frac{\Delta\delta^2}{2} - \frac{\Delta\alpha^2}{2}\cos^2\delta\Big]\,. \tag{13.2}$$

If $\Delta\delta = \Delta\alpha = 0$, the stars are on the same line of sight, hence Eq. 13.1 reduces to $d = |d_1 - d_2|$, as expected. If d is less than a few light years, there is a good chance that the stars do form a binary system. It is then possible to observe the relative motion of the visual binary to obtain the shape of the orbits.

13.2 Visual Binaries

The motion of visual binaries can be studied using the generalized version of Kepler's law (Eq. 10.20):

$$T^2 = \frac{4\pi^2}{G(m_1 + m_2)} a^3\,,$$

where m_1 and m_2 are the masses of the two stars and $a = a_1 + a_2$ is the sum of the semi-major axes of their orbits. The stars orbit around the common centre of mass, placed in one of the two foci of their elliptical orbits (Fig. 13.1). The semi-major axes obey the relationship:

$$m_1 a_1 = m_2 a_2 \Rightarrow \frac{a_1}{a_2} = \frac{m_2}{m_1}.$$

By observing the system, it is possible to obtain the period T, while a_1 and a_2 can be estimated only if we know the distance of the system from Earth. In this case, it is possible to determine the mass of both components:

$$m_1 = \frac{4\pi^2}{GT^2}(a_1 + a_2)^2 a_2 \,,$$
$$m_2 = \frac{4\pi^2}{GT^2}(a_1 + a_2)^2 a_1 \,.$$

13.3 Astrometric Binaries

The stars that form this type of binary system have a significant difference in brightness, therefore the spectrum of the faintest star is completely dominated by that of the brightest. If the mass of the brightest star can be estimated from its spectral class, then the mass of the faintest can also be determined. If not, there is no way to determine the mass of the system. Sirius was the first star to be classified as an astrometric binary, following the observation of its periodic motion, which indicated the presence of a smaller companion — the first example of a *white dwarf*. This was later directly observed thanks to more powerful telescopes.

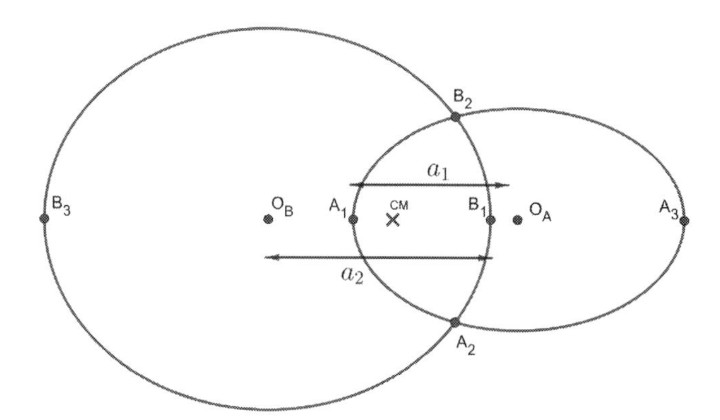

Figure 13.1: Orbits of two stars in a visual binary.

13.4 Spectroscopic Binaries

Spectroscopic variables appear as a single star even to the most powerful telescopes, but their spectra reveal a periodic displacement of the absorption or emission lines. Unlike astrometric binaries, in this case it is possible to observe the spectra of both stars.
As seen in Sec. 7.6, for small velocities, the displacement of the spectral lines is directly proportional to the radial velocity, according to the equation $\Delta\lambda = \lambda(v/c)$. Therefore, the displacement is maximum when the stars are directly approaching or moving away from the observer. Let us assume that the perpendicular to the orbital plane of the binary system forms an angle i with the direction of observation. Let v be the velocity obtained from the shift $\Delta\lambda$ of the spectral lines. The velocity measured in this way is actually the projection along the line of sight of the velocities of the two stars (see Sec. 9.4, for a similar reasoning). Therefore, the observed velocity (v) is related to the orbital velocity (v_0) by:

$$v = v_0 \sin i\,. \tag{13.3}$$

From the periodic shift of the spectral lines, we can obtain the orbital period T of the system. Let us assume that the orbits of the stars are circular, with radii a_1 and a_2, respectively. The maximum orbital velocity of the first star is:

$$v_{0,1} = \frac{2\pi a_1}{T}\,.$$

The observed velocity is given by Eq. 13.3:

$$v_1 = \frac{2\pi a_1 \sin i}{T}\,. \tag{13.4}$$

In the case of spectroscopic binaries, we cannot directly observe the two components, therefore the empirical determination of a_1 and a_2 is precluded. Let $a = a_1 + a_2$ be the distance between the two bodies. Considering the centre of mass frame, we find:

$$m_1 a_1 = m_2 a_2 \Rightarrow a_2 = \frac{m_1}{m_2} a_1\,. \tag{13.5}$$

We want to express a_1 as a function of m_1, m_2 and a only:

$$a = a_1 + a_2 = a_1\left(1 + \frac{m_1}{m_2}\right) \Rightarrow a_1 = a\frac{m_2}{m_1 + m_2}\,.$$

From Kepler's third law (Eq. 10.20):

$$T^2 = \frac{4\pi^2}{G(m_1 + m_2)} a^3$$

$$\Rightarrow a = \left[\frac{GT^2(m_1 + m_2)}{4\pi^2}\right]^{1/3}\,.$$

Hence:

$$a_1 = \frac{m_2}{m_1 + m_2}\left[\frac{GT^2(m_1+m_2)}{4\pi^2}\right]^{1/3} = m_2\left[\frac{GT^2}{4\pi^2(m_1+m_2)^2}\right]^{1/3}.$$

Substituting a_1 in Eq. 13.4

$$v_1 = m_2\left[\frac{2\pi G}{(m_1+m_2)^2 T}\right]^{1/3} \sin i\,. \tag{13.6}$$

Hence, we find:

$$\frac{{m_2}^3 \sin^3 i}{(m_1+m_2)^2} = \frac{{v_1}^3 T}{2\pi G}\,. \tag{13.7}$$

If the spectrum of the system is dominated by that of the brightest star (astrometric binaries), only T and v_1 can be empirically determined, therefore it is not possible to compute the total mass or the mass of the individual components. If, however, it is possible to record the spectral lines of both stars, then v_1, v_2 and T can be obtained, and the mass of both stars can be determined, as long as we know i. Indeed, since the stars are always diametrically opposite, their angular velocities are the same:

$$\omega = v_1/a_1 = v_2/a_2 \Rightarrow \frac{v_1}{v_2} = \frac{a_1}{a_2}\,. \tag{13.8}$$

Using Eq. 13.5, we can write $a_1/a_2 = m_2/m_1$, hence:

$$m_1 = m_2\frac{v_2}{v_1} \tag{13.9}$$

Substituting m_1 in Eq. 13.7, we find:

$$\frac{m_2 \sin^3 i}{(v_1+v_2)^2} = \frac{v_1 T}{2\pi G}\,. \tag{13.10}$$

If we know the inclination i, it is straightforward to find m_2 and, hence, m_1. If the two stars eclipse each other (eclipsing binaries), it is possible to determine the inclination, which is close to $i \approx 90°$. In all other cases, we can only state a lower limit for the mass of the two stars. Isolating m_1 and m_2:

$$m_1 = \frac{T}{2\pi G \sin^3 i} v_2 (v_1+v_2)^2\,,$$
$$m_2 = \frac{T}{2\pi G \sin^3 i} v_1 (v_1+v_2)^2\,.$$

The minimum values for m_1 and m_2 are obtained by setting $i = 90°$:

$$m_{1,\text{min}} = \frac{T}{2\pi G} v_2 (v_1+v_2)^2\,,$$
$$m_{2,\text{min}} = \frac{T}{2\pi G} v_1 (v_1+v_2)^2\,.$$

This reasoning is only valid for circular orbits, whereas for increasing values of eccentricity the behaviour of the system deviates from the circular case. In fact, most of the spectroscopic binaries have nearly circular orbits, since tidal forces tend to reduce the eccentricity on a time scale which is short compared to the lifetime of the system.

13.5 Photometric Binaries

In photometric binaries, the presence of the secondary star can be inferred by analysing the light curve of the system. The brightness of the binary system can change over time for several reasons. If the stars periodically eclipse each other, we call them *eclipsing* binaries. It is possible to observe a variation in brightness even if the two stars do not eclipse each other. In these systems, called *ellipsoidal* binaries, at least one component has been significantly deformed by the gravity of the other, so that, at different moments in time, the projection of the surface area of the star in the direction of the observer is different. This effect, together with the fact that the temperature, and therefore the energy radiated per unit area, is lower for parts of the surface farther from the centre, produces a periodic variation in the amount of energy that reaches the observer. Ellipsoidal binaries are often close or contact binaries.
Based on the characteristics of its light curve, it is possible to determine the mechanism underlying the change in brightness of a system. Eclipsing binaries are also called Algol stars (β Persei), named after the first observed system of this type. Ellipsoidal binaries are instead divided into β Lyrae, which are mainly close binaries, and WUMa, mainly contact binaries.
Eclipsing binaries usually consist of stars with very different brightnesses (for example, a giant and a dwarf star). For the stars to eclipse each other, the angle i between the perpendicular to the plane of the system and the line of sight must be close to $90°$. The light curve is roughly constant during most of the period, when the stars are far from each other and both are visible from Earth. The light curve (Fig. 13.2) is characterized by two minima of very different depths: the deeper corresponds to the large and cold star totally eclipsing the small and hot star; the shallower corresponds to the small and hot star partially eclipsing the large and cold one. In fact, according to the Stefan-Boltzmann law, the hotter star emits more energy per unit area compared to the colder star, whilst the area of the eclipsed region, equal to the cross-section of the smaller star, is the same in both cases. The shape of the minimum is different depending on whether the eclipses are partial or total. In a partial eclipse, the light curve varies gently; in a total eclipse, the minimum is flat and constant throughout the time when one of the two stars is completely eclipsed.
Consider an eclipsing binary system, at a distance d from Earth, consisting of stars having temperatures T_1, T_2 and radii R_1, R_2, with $R_1 > R_2$. When both

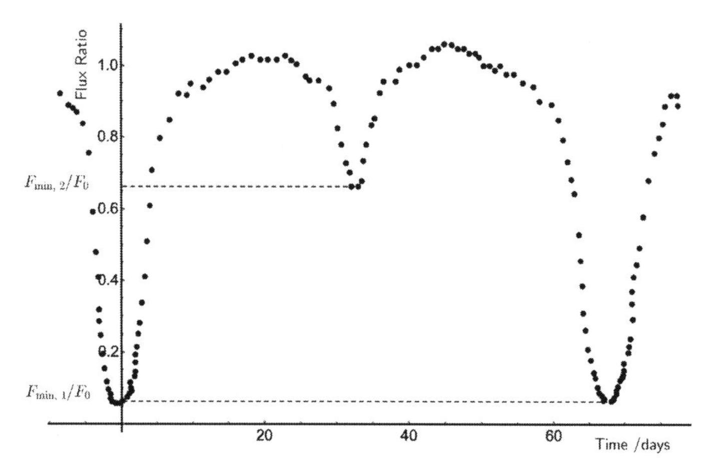

Figure 13.2: Periodic variation in the brightness of an eclipsing binary.

stars are visible, the total flux reaching the Earth is:

$$F_0 = F_1 + F_2 = \frac{\sigma}{4d^2}({R_1}^2{T_1}^4 + {R_2}^2{T_2}^4)\,.$$

The first minimum corresponds to the first star covering the second completely. If $T_2 > T_1$, this is the deeper minimum, otherwise, if $T_2 < T_1$, this is the shallower minimum. The flux is:

$$F_{\text{min},1} = \frac{\sigma}{4d^2}({R_1}^2{T_1}^4)\,.$$

The second minimum corresponds to the second star partially eclipsing the first. The area of the eclipsed portion is $\pi {R_2}^2$, hence light from the first star only reaches the observer from a portion of area $\pi({R_1}^2 - {R_2}^2)$:

$$F_{\text{min},2} = \frac{\sigma}{4d^2}\left[({R_1}^2 - {R_2}^2){T_1}^4 + {R_2}^2{T_2}^4\right].$$

The ratios of the minima to the average luminosities are thus:

$$\boxed{\frac{F_{\text{min},1}}{F_0} = \frac{{R_1}^2{T_1}^4}{{R_1}^2{T_1}^4 + {R_2}^2{T_2}^4}}\,, \tag{13.11}$$

$$\boxed{\frac{F_{\text{min},2}}{F_0} = \frac{({R_1}^2 - {R_2}^2){T_1}^4 + {R_2}^2{T_2}^4}{{R_1}^2{T_1}^4 + {R_2}^2{T_2}^4}}\,. \tag{13.12}$$

13.6 Exercises

1. Prove Eq. 13.2 from Eq. 13.1.

2. Mizar and Alcor have parallaxes of $0.03936''$, $0.03991''$, and their coordinates are $\alpha_m = 13^h23^m56^s, \delta_m = +54°55'31''$ and $\alpha_a = 13^h25^m14^s, \delta_a = 54°59'17''$. What is the distance between them in parsecs?

3. An eclipsing binary consists of two stars of radii R_1 and R_2, with $R_1 = 4R_2$ and $T_1 = 3T_2$. Compute the ratio between the maximum and minimum fluxes. Hence, find the maximum magnitude difference.

4. By what percentage does the luminosity of a Sun-like star decrease during the transit of a gas giant (i.e. a planet similar to Jupiter) and an Earth-like planet? What about a red dwarf?

5.* In an eclipsing binary, the difference between the two lowest magnitudes of the system is $\Delta_1 = 1.5$, whilst the difference between the maximum and minimum magnitudes is $\Delta_2 = 2.3$. The system is made up of a large, cold star, and a small, warm star. Find the ratios of their radii and temperatures.

6.* Two stars with masses m and $2m$ orbit around their common centre of mass. If the orbits are circular, they do not intersect. What is the smallest value of the eccentricity for which they intersect?

7.* The presence of a planet orbiting a star can be revealed by a small periodic shift in its absorption or emission lines. What is the order of magnitude of the frequency shift that an optimally positioned observer, outside the Solar System, should be able to measure, if the presence of Jupiter is to be inferred in this way? Assume that the frequency of observation is $f \approx 6 \cdot 10^{14}$ Hz.

8.** (ARAO 2018, Th.XI, q.6) Suppose that, in our galaxy, there is a special class of binaries, whose components are identical to the Sun, and rotate in circular orbits, separated by 1 au. The concentration of such systems in space is constant (in particular, does not depend on the distance from the disk of the Galaxy) and is equal to $0.001\,\text{pc}^{-3}$. There are observatories at your disposal in the northern and southern hemispheres on Earth. Each of them has a *photometer* that can observe stars up to a magnitude of 15, with an accuracy of 0.001 magnitudes, a *spectrograph* with a resolution $\lambda/\Delta\lambda = 10^5$ and limiting magnitude 12, and an *astrograph* with an angular resolution of $0.1''$ and limiting magnitude 20. How many such pairs will be discovered as optical, eclipsing and spectroscopic binaries, respectively? Neglect the interstellar absorption of light.

Part IV

Solutions

14

Exercise Solutions

1. Celestial Coordinate Systems

1.1

Assuming the Earth to be perfectly spherical, the length (l) of the arc that subtends an angle $\alpha = \phi_2 - \phi_1 = 10^\circ$ is:

$$\frac{l}{10^\circ} = \frac{2\pi R_\oplus}{360^\circ} \Rightarrow l = \frac{\pi}{18} R_\oplus \approx 1,113\,\text{km}\,.$$

1.2

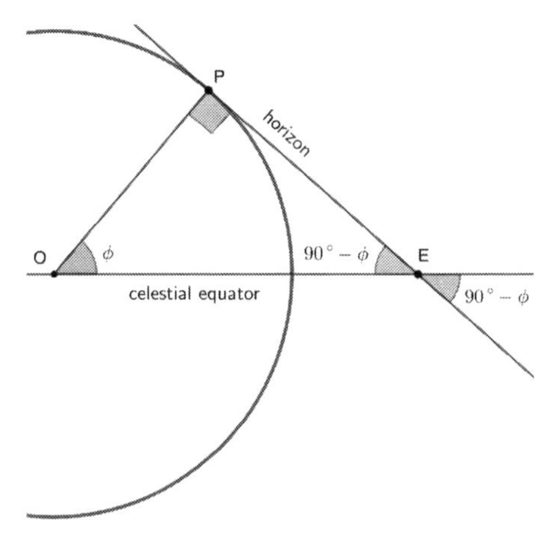

Figure 14.1: Height of the celestial equator on the (astronomical) horizon.

As you can see in Fig. 14.1, the height of the equator on the (astronomical) horizon is:

$$h_e = 90^\circ - \phi\,.$$

1.3

Stars move on minor circles parallel to the celestial equator (i.e. their declination is constant), hence the altitude of a star is constant only if the horizon and the equator coincide. This only happens at the poles, therefore the condition is $\phi = \pm 90^\circ$.

1.4

The vernal equinox (♈) is one of the two points of intersection between the celestial equator and the ecliptic. The declination is defined as the angle to the celestial equator, hence the declination of ♈ is zero. By definition, the right ascension is the angle to the vernal equinox, therefore the right ascension of ♈ is also zero. To sum up, $\delta_\gamma = \alpha_\gamma = 0$.

1.5

At the north pole, the celestial equator coincides with the horizon. Since the ecliptic is inclined by an angle of $\epsilon = 23°27'$ (*ecliptic obliquity*) to the celestial equator, this is also the angle between the ecliptic and the horizon, at the north pole.

1.6

The north ecliptic pole is inclined by an angle of $\epsilon = 23\,°27'$ with respect to the axis of the Earth. Hence, the observer must be located at a distance ϵ from the north pole, i.e. at a latitude of $\phi = 90\,° - \epsilon = 66\,°33'$ (Fig. 14.2).

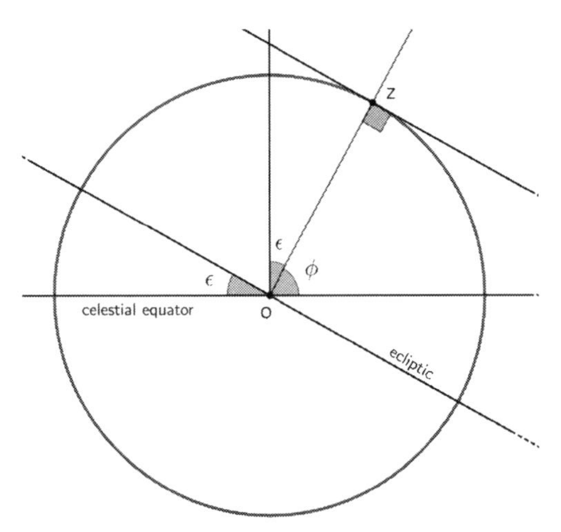

Figure 14.2: The north ecliptic pole appears at the zenith for an observer at a latitude of $\phi = 90\,° - \epsilon = 66\,°33'$.

1.7

The best viewing conditions are obtained for objects at the zenith, since the distance which light travels through the atmosphere is the smallest. The planets orbit the Sun very close to the ecliptic, hence we require the ecliptic to appear at the zenith. This happens at latitudes in the interval $-\epsilon \leqslant \phi \leqslant \epsilon$, where $\epsilon = 23\,°27'$, i.e. for places in between the tropics.

1.8

Both the ecliptic longitude and right ascension are taken anti-clockwise from the vernal equinox, however, the former is measured along the ecliptic, the latter along the celestial equator. Therefore, when $\lambda = \alpha = \pm 90\,°$, the two coordinates have the same value (Fig. 14.3, overleaf). This is the great circle that passes through the celestial and ecliptic poles (P, EP) and the summer and winter solstices (S_s, S_w), and is called the *solstitial colure*. On the other hand, the great circle passing through the poles and the equinoxes is the *equinoctial colure*.

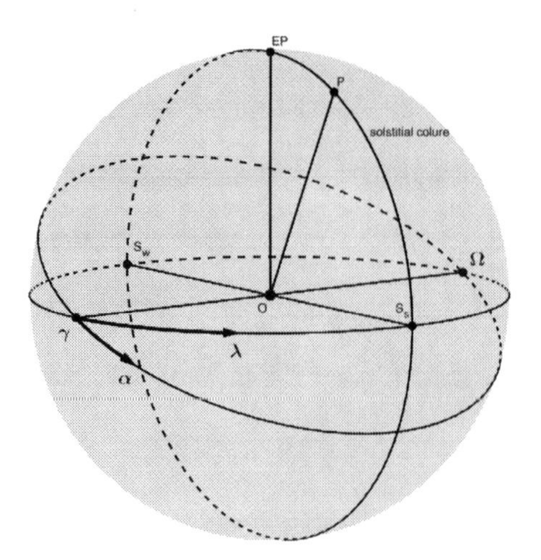

Figure 14.3: The solstitial colure is the great circle passing through the celestial and ecliptic poles (P, EP), and the summer and winter solstices (S_s, S_w).

1.9

Using Eq. 1.2, to compute H from α and ST:

$$\begin{aligned} H &= \mathrm{TS} - \alpha = 4^h 20^m 13^s - 5^h 14^m 28^s \\ &= -(4^h 74^m 28^s - 4^h 20^m 13^s) = -(54^m 15^s) \end{aligned}$$

1.10

By symmetry, the problem is equivalent to finding the distance of the horizon for an observer on top of the skyscraper. Looking at Fig. 1.2, we are interested in finding the length of the segment PT:

$$\begin{aligned} d &= \sqrt{(R+h)^2 - R^2} \\ &= \sqrt{2Rh + h^2} \\ &\approx \sqrt{2Rh} \approx 103\,\mathrm{km}\,, \end{aligned}$$

which is approximately the distance from Dubai to the Iranian sea coast.

1.11

The orbital period of Uranus is 84 years. Its orbit is inclined by only 0.7° relative to the ecliptic, so it is reasonable to assume that its orbital plane coincides with that of the Earth. In 238 years, Uranus completes 238/84 =2.833 orbits, hence its longitude must have been $43° - 2.833 \cdot 360° = -977°$. Normalizing to 360°, we get $\lambda = 103°$. Therefore, Uranus was in the constellation of Gemini.

2. Transformation of Coordinates

2.1

For the star to appear at the zenith, its altitude during upper culmination must be $h_u = 90°$. Then, from Eq. 2.5:

$$h_u = 90° + \delta - \phi \Rightarrow \delta = \phi\,.$$

Hence, the latitude must be equal to the declination of the star.

2.2

Setting $h_l = 0°$ in Eq. 2.6:

$$h_l = \delta + \phi - 90°$$
$$\Rightarrow \phi = 90° - \delta = 90° - (16°42'58'') = 89°59'60'' - (16°42'58'') = 73°17'2''\,.$$

2.3

For an observer in the northern hemisphere, the altitude of the Sun is maximum when its declination is maximum. This happens during the summer solstice, when the declination is equal to the ecliptic obliquity ($\delta_{\text{max}} = 23°27'$). Conversely, the altitude of the Sun is minimum during winter solstice, when $\delta_{\text{min}} = -(23°27')$. Using Eq. 2.5, we find the maximum and minimum altitudes of the Sun, measured from the southern horizon:

$$h_{\text{max}} = 90° + 23°27' - \phi = 113°27' - \phi\,,$$
$$h_{\text{min}} = 90° - (23°27') - \phi = 66°33' - \phi\,.$$

Whenever $h_{\text{max}} > 90°$, i.e. for $\phi < 23°27'$, the maximum altitude of the Sun is $90°$. In this case, the Sun passes through the zenith twice a year, before and after the summer solstice. For an observer in the southern hemisphere, we replace ϕ, δ with $-\phi, -\delta$, obtaining:

$$h_{\text{max}} = 90° + 23°27' + \phi = 113°27' + \phi\,,$$
$$h_{\text{min}} = 90° - (23°27') + \phi = 66°33' + \phi\,.$$

The altitude is maximum when the declination of the Sun is minimum, i.e. during winter solstice. If $h_{\text{max}} > 90°$, i.e. for $\phi > -(23°27')$, the maximum altitude of the Sun is $90°$. In this case, the Sun passes through the zenith twice a year, before and after winter solstice.

2.4

From Eq. 2.7, we find the declination of the star:

$$\delta = \frac{h_l + h_u}{2} = \frac{34°34'23'' + 53°54'45''}{2} = 43.5°44'34''\,.$$

The latitude can be obtained from the altitude of upper culmination, using Eq. 2.5:

$$\phi = 90° + \delta - h_u = 90° + 43.5°44'34'' - (53°54'45'') = 80°19'49''\,.$$

Alternatively, we can use Eq. 2.8:

$$\phi = 90° - \frac{h_u - h_l}{2} = 90° - \frac{53°54'45'' - (34°34'23'')}{2}$$
$$= 90° - \frac{19°20'22''}{2} = 90° - (9°40'11'') = 80°19'49''\,.$$

2.5

In the northern hemisphere, the Sun is circumpolar when $\delta > 90° - \phi$, hence $\phi > 90° - \delta$. Since the maximum declination of the Sun is equal to the ecliptic obliquity $\epsilon = 27°27'$, the Sun is circumpolar at least once a year for $\phi > 90° - 23°27' = 66°33'$. The solution is similar for the southern hemisphere, thus in general $|\phi| > 66°33'$.

2.6

Since the height of the observer is equal to the length of their shadow, it follows that the altitude of the Sun is $h = 45^\circ$. If the Sun is in upper culmination at this moment, its altitude can be related to ϕ and δ through:

$$h_{\mathrm{s}} = 90^\circ - \phi + \delta\,,$$
$$h_{\mathrm{n}} = 90^\circ + \phi - \delta\,,$$

if the Sun culminates to the south or to the north of the observer, respectively. Using $\delta = \epsilon = 23^\circ 27'$ since it is summer solstice, we finally find:

$$\phi = \epsilon \pm (90^\circ - h)\,.$$

Hence, the latitude can be either 68.4° or -21.6°.

2.7

If the Sun were to set, its minimum altitude would be zero, and the length of the shade projected by the vertical stick would be infinite. Since l_{max} is finite, we deduce that the Sun is circumpolar at the place of observation. Therefore, it is indeed possible to find its maximum and minimum altitudes:

$$h_{\mathrm{max}} = 90^\circ + \delta_\odot - \phi\,,$$
$$h_{\mathrm{min}} = \delta_\odot + \phi - 90^\circ\,.$$

The length of the shade is related to the length of the stick and the altitude of the Sun:

$$\frac{l}{l_{\mathrm{min}}} = \tan(h_{\mathrm{max}}) \Rightarrow 90^\circ + \delta_\odot - \phi = \arctan\left(\frac{l}{l_{\mathrm{min}}}\right),$$
$$\frac{l}{l_{\mathrm{max}}} = \tan(h_{\mathrm{min}}) \Rightarrow \delta_\odot + \phi - 90^\circ = \arctan\left(\frac{l}{l_{\mathrm{max}}}\right).$$

Summing and subtracting the two equations, we obtain, respectively:

$$\delta_\odot = \frac{1}{2}\left[\arctan\left(\frac{l}{l_{\mathrm{min}}}\right) + \arctan\left(\frac{l}{l_{\mathrm{max}}}\right)\right] = 20^\circ\,,$$
$$\phi = 90^\circ - \frac{1}{2}\left[\arctan\left(\frac{l}{l_{\mathrm{min}}}\right) - \arctan\left(\frac{l}{l_{\mathrm{max}}}\right)\right] = 80^\circ\,.$$

As we can verify, the Sun is indeed circumpolar ($\delta_\odot > 90^\circ - \phi$).

2.8

For an observer at latitude ϕ, all stars rotate around the celestial north pole, which is inclined by an angle ϕ to the northern horizon. The stars describe circles on the celestial sphere with radii $|90^\circ - \delta|$ and inclinations $90^\circ - \phi$ relative to the horizon. Hence, a star with declination $\delta = \pm 90^\circ$ appears stationary wherever it is observed from, and its azimuth is always constant. At the equator ($\phi = 0^\circ$), all stars travel in circles perpendicular to the horizon. Then, a star with $A_z = 6^h$ has constant azimuth until its passage on the meridian, when its azimuth changes to $A_z = -6^h$. To sum up, the azimuth is constant for at least half a day if: $\delta = \pm 90^\circ$ and the observer is located anywhere on Earth, or $A_z = \pm 6^h$, and the observer is at the equator.
It is also possible to solve the problem using Eq. 2.3b to isolate $\cos h$ and substituting in Eq. 2.3c. If we then require all terms containing the hour angle H to be zero, we find a system of four equations, whose solutions give the desired result.

2.9

This case is simple enough that we do not need to use the equations to convert from ecliptic to equatorial coordinates. As we saw in Ex. 1.8, the point $\alpha = 6^h$ is at $\lambda = 90°$. At this point, the ecliptic is above the celestial equator by an angle of $\epsilon = 23°27'$, while the star is $\delta = 10°$ above the celestial equator. Hence, the ecliptic latitude of the star is $\beta = \delta - \epsilon = 10° - (23°27') = -(13°27')$.

2.10

In the northern hemisphere:

- S: $H = 0^h$, $\delta = -(90° - \phi)$;
- W: $H = 6^h$, $\delta = 0°$;
- N: $H = 12^h$, $\delta = 90° - \phi$;
- E: $H = 18^h$, $\delta = 0°$;
- Z: $H = 0^h$, $\delta = \phi$.

In the southern hemisphere:

- S: $H = 12^h$, $\delta = -(90° + \phi)$;
- W: $H = 6^h$, $\delta = 0°$;
- N: $H = 0^h$, $\delta = 90° + \phi$;
- E: $H = 18^h$, $\delta = 0°$;
- Z: $H = 0^h$, $\delta = \phi$.

2.11

A star that rises exactly in the east and sets exactly in the west stays above the horizon for 12^h. As the points of rising and setting shift north, the star is visible for longer, because more of its path will be above the horizon. Therefore, stars that are visible for more than 12^h rise in the north-east direction.

2.12

Neglecting refraction and parallax, assuming the Moon has zero declination, Gianna will observe the Moon at the meridian once the Earth has rotated by $90°$, that is, after 1/4 of a sidereal day. Therefore, Gianna's longitude is $\lambda = -(12°30') + 90° = 77°30'$ E. The only capital city with this longitude is Washington.

2.13

On the 21st of April, on average 30 days have passed since the vernal equinox. Applying Eq. 2.13:

$$\delta_\odot = \arcsin\left[\sin\epsilon\sin\left(\frac{2\pi}{365.25}30\right)\right] = 11°19' \approx 11° .$$

The Sun is circumpolar at:

$$\phi \geqslant 90° - \delta_\odot = 90° - 11° = 79° .$$

2.14

A star can pass through the zenith only if its declination is equal to the latitude, hence $\delta = \phi$. In this case, the passage happens during upper culmination, when $H_c = 0$. When the sidereal time is 9^h2^m, the hour angle of the star is $H = 0^h + (9^h2^m - 10^m) = 8^h52^m$. Setting $\delta = \phi$ in Eq. 2.4c:

$$\sin h = \cos H\cos^2\phi + \sin^2\phi \Rightarrow 1 - \sin h = \cos^2\phi(1 - \cos H) .$$

Isolating ϕ and substituting $h = 78°12'$:

$$\phi = \arccos\left(\sqrt{\frac{1 - \sin h}{1 - \cos H}}\right) = 83°34^m .$$

2.15

During the equinoxes, the path of the Sun is perpendicular to the horizon, hence its velocity relative to the horizon is maximum. By examining this case, we obtain the minimum velocity that guarantees a successful observation at any other time of the year. The apparent angular velocity of the Sun is:

$$\omega = \frac{360^\circ}{86164.1^s} = 4.178 \cdot 10^{-3\,\circ}/\mathrm{s}\,.$$

At the height of $h = 6\,\mathrm{m}$, the dip of the horizon is:

$$\theta = \arccos\left(\frac{R_\oplus}{R_\oplus + h}\right) = 0.0786^\circ\,.$$

The Sun covers this angle in a time $\Delta t = \theta/\omega = 18.81\,\mathrm{s}$. Hence, the velocity must be greater than $v_c = 6\,\mathrm{m}/\Delta t = 0.32\,\mathrm{m/s}$. If the velocity is exactly v_c, during the equinoxes, the sunset is prolonged by approximately 20 s.

3. Perturbation of Coordinates

3.1

The Barnard star has an angular velocity of $u_b = 10.3''/\mathrm{yr}$. The angular diameter of the Moon is approximately $\theta_☾ = 30'$, i.e. $30 \cdot 60'' = 1800''$. Hence, the time taken is:

$$\Delta t = \frac{\theta_☾}{u_b} = \frac{1800}{10.3} \approx 174.8\,\mathrm{years}\,.$$

3.2

Aldebaran's distance from Earth is $d = 1/\pi''_p = 20.83\,\mathrm{pc}$, hence its tangential velocity is:

$$v_t = u \cdot d = 4.167\,\mathrm{pc} \cdot \mathrm{arcsec/yr}\,.$$

By definition, $\mathrm{pc} \cdot \mathrm{arcsec} = 1\,\mathrm{au}$, therefore:

$$v_t = 4.167\,\mathrm{au/yr} \approx 20\,\mathrm{km/s}\,.$$

The redshift parameter is $z = \Delta\lambda/\lambda = 1.79 \cdot 10^{-4}$. Since $z \ll 1$, we use Eq. 7.21:

$$v_r = c \cdot \Delta\lambda/\lambda \approx 54\,\mathrm{km/s}\,.$$

The total velocity is then:

$$v_t = \sqrt{v_r^2 + v_t^2} \approx 58\,\mathrm{km/s}\,.$$

3.3

Let $u_{\mathrm{pr}} = d\lambda/dt = 50''/\mathrm{yr}$. From Eqs. 3.3:

$$\frac{\mathrm{d}\delta_{\mathrm{pr}}}{\mathrm{d}t} = u_{\mathrm{pr}} \sin\epsilon \cos\alpha\,,$$

$$\frac{\mathrm{d}\alpha_{\mathrm{pr}}}{\mathrm{d}t} = u_{\mathrm{pr}}(\sin\alpha \sin\epsilon \tan\delta + \cos\epsilon)\,.$$

The total rates of change of the coordinates, $d\alpha_{\mathrm{tot}}/dt$ and $d\delta_{\mathrm{tot}}/dt$, are equal to the sum of the perturbations due to the proper motion of the star and to the precession of Earth's axis:

$$\begin{cases} \dfrac{\mathrm{d}\alpha_{\mathrm{tot}}}{\mathrm{d}t} = u_\alpha + u_{\mathrm{pr}}(\sin\alpha \sin\epsilon \tan\delta + \cos\epsilon) \\[2ex] \dfrac{\mathrm{d}\delta_{\mathrm{tot}}}{\mathrm{d}t} = u_\delta + u_{\mathrm{pr}} \sin\epsilon \cos\alpha\,. \end{cases} \tag{14.1}$$

In principle, it is possible to solve this system of coupled differential equations. First, using both equations, solve for α and $d\alpha_{\text{tot}}/dt$ as a function of δ and $d\delta_{\text{tot}}/dt$ only. Then, take the derivative of the second equation with respect to time, and substitute your previous expressions for α and $d\alpha_{\text{tot}}/dt$. You will be left with a second order differential equation, with the only variable being δ and its time derivatives. Although not pretty, it can be solved. In this case however, the period under consideration is short enough so that we may regard $d\delta_{\text{tot}}/dt$ and $d\alpha_{\text{tot}}/dt$ as being constant. The approximate solution is thus:

$$\alpha_{J2000.0} = \alpha_{J1900.0} + \frac{\mathrm{d}\alpha_{\text{tot}}}{\mathrm{d}t}\Delta t \approx 6^h 46^m 26^s ,$$

$$\delta_{J2000.0} = \delta_{J1900.0} + \frac{\mathrm{d}\delta_{\text{tot}}}{\mathrm{d}t}\Delta t \approx 16^\circ 38' 42'' .$$

3.4

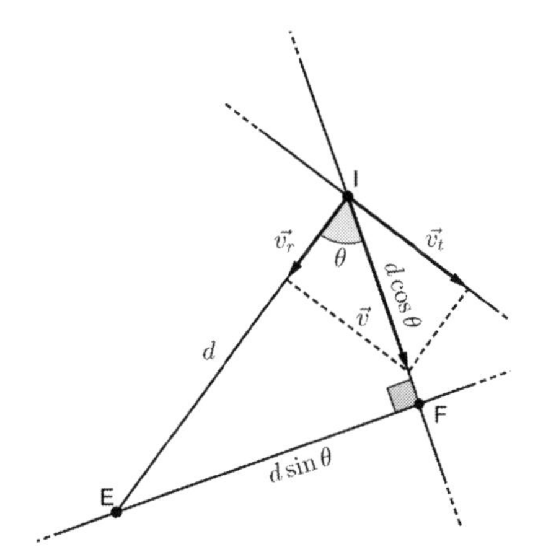

Figure 14.4: I and F are the initial and final positions of Sirius, E is the observer on Earth.

Sirius' distance is $d = 1/\pi''_p = 2.667\,\text{pc}$. The tangential velocity is given by Eqs. 3.8 and 3.7:

$$v_t = u \cdot d = d\sqrt{u_\alpha^2 \cos^2\delta + u_\delta^2} = 1.603 \cdot 10^{-6}\,\text{pc/yr} .$$

The radial velocity is $v_r = 8.182 \cdot 10^{-6}$ pc/yr. The total velocity is then:

$$v = \sqrt{v_r^2 + v_t^2} = 8.337 \cdot 10^{-6}\,\text{pc/yr} .$$

Looking at Fig. 14.4, the distance Sirius has to travel to be the closest to the Sun is:

$$\Delta d = d\cos\theta = d \cdot \frac{v_r}{v} .$$

The time taken is therefore:

$$\Delta t = \frac{\Delta d}{v} = d \cdot \frac{v_r}{v^2} = 2.667 \cdot \frac{8.182}{8.337^2} \cdot 10^6\,\text{years} = 3.14 \cdot 10^6\,\text{years} .$$

At the point of closest approach, the radial velocity is zero. Indeed, if it were negative, the star would be getting closer to the Sun, but this is absurd since we already chose the point of closest approach. If it were positive, because the velocity was initially negative, some time earlier the velocity must have been zero, but then the body was closer at that time. Hence, the velocity must be entirely tangential, i.e. $v_{r,f} = 0$, $v_{t,f} = v$. The final distance is $d_f = d\sin\theta = dv_t/v$, thus $\pi''_{p,f} = 1/d_f = v/(v_t \cdot d) = 8.337/(1.603 \cdot 2.667) = 1.95''$.

3.5

As seen in Sec. 3.5, we can write Snell's law between the layer at $h = 0$ and that at h:

$$\frac{n(h)}{n(h=0)} = \frac{\sin \alpha_i}{\sin \alpha_r}.$$

Since the ray circles the Earth at a height h, the refraction angle is $\alpha_r = 90^\circ$. Instead, the incident angle is $\sin \alpha_i = R/(R+h)$. Substituting $n(h)$ in the previous equation:

$$\frac{n_0}{1+\epsilon h} \cdot \frac{1}{n_0} = \sin \alpha_i = \frac{R}{R+h}.$$

Taking the reciprocal of both sides:

$$1 + \epsilon h = 1 + \frac{h}{R} \Rightarrow R = 1/\epsilon.$$

3.6

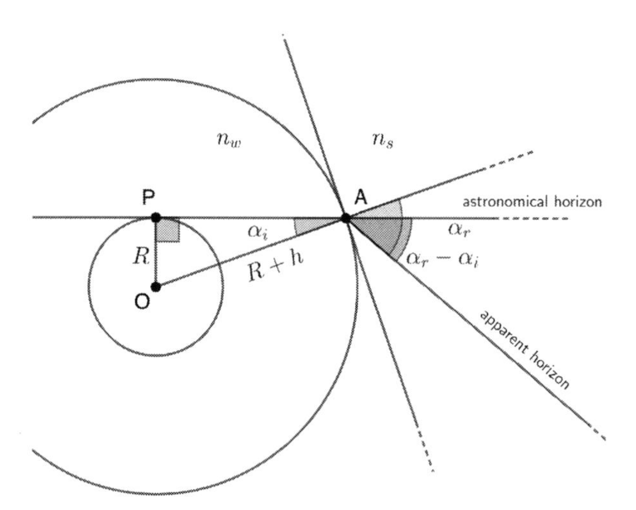

Figure 14.5: Refraction of light on Hydroplanet.

As can be seen in Fig. 14.5, for an object seen on the horizon, Snell's law gives:

$$\frac{\sin \alpha_r}{\sin \alpha_i} = \frac{n_w}{n_s} = \frac{4}{3},$$

where $n_w = 1.33 \approx 4/3$ and $n_s = 1$ are the refractive indices of water and space, respectively. If the ray joining P and A is totally reflected, then it is not possible to see any object on the horizon. This happens when $\alpha_r > 90^\circ$, hence the stars are visible on the horizon only if:

$$\frac{\sin \alpha_i}{\sin 90^\circ} \leqslant \frac{3}{4} \Rightarrow \sin \alpha_i \leqslant \frac{3}{4}.$$

Considering triangle OPA in Fig. 14.5, we see that $\sin \alpha_i = R/(R+h)$, hence the condition on the height is:

$$1 + \frac{h}{R} \geqslant \frac{4}{3} \Rightarrow h \geqslant \frac{R}{3}.$$

From now on we will consider the case $h = R/3$, for which $\alpha_r = 90^\circ$. Now, the outgoing ray (PA) coincides with the apparent horizon, which forms an angle of $\alpha_r - \alpha_i$ with the astronomical horizon. As long as the star is above the apparent horizon, which covers an angle of $180 + 2(\alpha_r - \alpha_i) = 360 - 2\alpha_i$, it is day on Hydroplanet. Denoting with T the period of rotation, the length of the day is:

$$t = \frac{360 - 2\arcsin(3/4)}{360} \cdot T = 0.73 \cdot T = 17^h 31^m 12^s .$$

The value of atmospheric refraction at the horizon is:

$$R = \alpha_r - \alpha_i = 90^\circ - \arcsin\left(\frac{3}{4}\right) = 41.4^\circ ,$$

much larger than $34'$, on Earth.

4. Observation and Instruments

4.1

From Eq. 4.10, we obtain the angular resolution:

$$R \approx 1.22 \frac{\lambda_v}{D_v} = 1.22 \frac{500 \cdot 10^{-9}}{10} = 6.1 \cdot 10^{-8} \,\text{rad} .$$

If a radio telescope observes at $\lambda_r = 20\,\text{cm}$, with the same resolution, its diameter is:

$$D_r \approx 1.22 \frac{\lambda_r}{R} = \frac{\lambda_r}{\lambda_v} D_v = \frac{0.2}{500 \cdot 10^{-9}} \cdot 10\,\text{m} = 4 \cdot 10^6 \,\text{m} .$$

This shows that a single radio telescope cannot achieve the same resolution of an optical telescope, but other strategies, such as interferometry, need to be used.

4.2

Using Eq. 4.2:

$$\alpha_{\text{rad}} = \frac{h_\text{O}}{d} \Rightarrow h_\text{O} = d \cdot \alpha_{\text{rad}} = 55 \cdot 10^6 \,\text{ly} \cdot \frac{400}{206264.8} = 1.06 \cdot 10^5 \,\text{ly} .$$

From Eq. 4.8, we then have:

$$h_\text{I} = \alpha_{\text{rad}} \cdot f_\text{o} = \frac{400}{206264.8} \,\text{m} = 1.94 \,\text{mm} .$$

4.3

The maximum and minimum distances of Mars from the Sun are, respectively:

$$d_a = a(1+e) = 1.5237 \cdot (1 + 0.0934)\,\text{au} = 2.492 \cdot 10^{11} \,\text{m} ,$$
$$d_p = a(1-e) = 1.5237 \cdot (1 - 0.0934)\,\text{au} = 2.067 \cdot 10^{11} \,\text{m} .$$

The radius of the Sun is $R_\odot = 695475\,\text{km}$, hence its maximum and minimum angular diameters are, respectively:

$$\alpha_{\max} = 2R_\odot / d_p = 6.74 \cdot 10^{-3} \,\text{rad} ,$$
$$\alpha_{\min} = 2R_\odot / d_a = 5.59 \cdot 10^{-3} \,\text{rad} .$$

The distance to the Moon is $d_☾ = 384.4 \cdot 10^6$ m. If a solar eclipse is just visible, $\alpha_{\min}$ is equal to the angular size of the satellite, as seen from Mars. Therefore, the diameter is:

$$d = \alpha_{\min} \cdot d_☾ \approx 2150 \,\text{km} .$$

The angular sizes of the Sun observed from Mars and from Earth are in the same ratio as the Mars-Sun and Earth-Sun distances. When Mars is in aphelion, this ratio is equal to $1.5237(1+0.0934) = 1.666$, hence the satellite can be 1.666 times smaller than the Moon, i.e it can have a minimum diameter of $2R_{☾}/1.666 \approx 2085\,\text{km}$; which is approximately the same result we obtained earlier. This result is actually more accurate, since it makes no reference to the Earth-Moon distance and the radius of the Sun. The reason for the discrepancy is that total solar eclipses on Earth only occur when the Moon is close to aphelion, or the Earth is close to perihelion (hence $d_{☾}$ is actually a bit smaller than 2150 km).

4.4

The ratio of the angular diameters is equal to the ratio of the distances from which Mars is observed. During quadrature, the Earth-Mars distance is $d_Q = \sqrt{d_♂^2 - d_⊕^2}$, during opposition $d_O = d_♂ - d_⊕$. Hence, the ratio of the angular diameters is:

$$\frac{d_♂ - d_⊕}{\sqrt{d_♂^2 - d_⊕^2}} = 0.46\,.$$

4.5

Using 4.9, we have:

$$\frac{h_1}{h_0} = \frac{f_o}{d} \Rightarrow f_o = \frac{h_1}{h_0}d = (1254 \pm 10)\,\text{mm}\,.$$

4.6

As explained in Sec. 3.6, the angle $\Delta\alpha$ on the celestial equator corresponds to the angle $\Delta\alpha\cos\delta$ on the parallel of declination δ, since the radius (hence, the arc length) of the parallel is smaller by a factor of $\cos\alpha$. In this case the computation is easy, because both pairs of stars have one common coordinate. Otherwise, we would need to use the spherical law of cosines (Appendix A.4). The angular distances are:

$$\Delta_{1,2} = \delta_1 - \delta_2 = 40' = 2400''\,,$$

$$\Delta_{1,3} = \frac{360^\circ}{24^h}(\alpha_1 - \alpha_2)\cos\delta_1 = 1.224^\circ = 4405''\,.$$

The focal length of the telescope is $f_o = 10 \cdot 200\,\text{mm} = 200\,\text{cm}$. The CCD is a square, with length $L = 4096 \cdot l_{\text{pix}} = 2.62\,\text{m}$ and diagonal $D = \sqrt{2}L = 3.71\,\text{cm}$. To capture the widest possible angle, we use its diagonal:

$$\frac{D}{f_o} = \tan\Delta_{\text{max}} \Rightarrow \Delta_{\text{max}} = \arctan\frac{D}{f_o} = 1.06^\circ \approx 3816''\,.$$

Hence, it is possible to capture Star 1 and 2 together, but not Star 1 and 3.

4.7

The field of view is:

$$\theta_{\text{FOV}} = \frac{360}{86164.1}(2.5\cdot 60)\cdot\cos(46^\circ 0' 14.4'') = 26'\,,$$

where we have used the fact that the angular velocity of a star with declination δ is $\omega_{\text{sid}}\cos\delta$ (see Sec. 3.6). The angular diameter of the Moon is approximately $30'$, hence we cannot see the full Moon using this set up.

4.8

To find an approximation to the FOV of the telescope, we divide the FOV of the eyepiece by the magnification:

$$\mathrm{FOV}_{\mathrm{telescope}} = \frac{\mathrm{FOV}_{\mathrm{eyepiece}}}{\omega} = \frac{\mathrm{FOV}_{\mathrm{eyepiece}}}{f_{\mathrm{telescope}}/f_{\mathrm{eyepiece}}} = \frac{45^\circ}{130 \cdot 5/25} = 1.7^\circ\,.$$

The faintest star that is observable with the naked eye has an apparent magnitude of $m_{\mathrm{eye}} = 6$, and the diameter of the pupil at night is approximately $d = 8\,\mathrm{mm}$. On the other hand, the telescope has a diameter of $D = 130\,\mathrm{mm}$, hence a star viewed with the telescope can be D^2/d^2 fainter, and barely visible. According to Eq. 8.4:

$$m_{\mathrm{eye}} - m_{\mathrm{telescope}} = -2.5\log\frac{D^2}{d^2} = -5\log\frac{D}{d}$$

$$\Rightarrow m_{\mathrm{telescope}} = m_{\mathrm{eye}} + 5\log\frac{D}{d} = 6 + 5\log\frac{130}{8} \approx 12\,.$$

To calculate the angular resolution, we use Eq. 4.10:

$$\theta = \frac{1.22\lambda}{D} \approx 5 \cdot 10^{-6}\,\mathrm{rad}\,,$$

where $\lambda = 550\,\mathrm{nm}$ has been taken as the average wavelength for visible light. Using Eq. 10.14, taking into account that the total mass of the system is $M_{\mathrm{tot}} = (18.9 + 16.2)M_\odot = 35.1M_\odot$, we compute the separation between the stars in astronomical units:

$$\frac{T_{\mathrm{yr}}^2}{a_{\mathrm{au}}^3} = \frac{1}{M_{\mathrm{tot,\ in}\ M_\odot}} \Rightarrow a = (T_{\mathrm{yr}}^2 \cdot M_{\mathrm{tot,\ in}\ M_\odot})^{1/3} \approx 1.45\,\mathrm{au}\,.$$

Therefore, the angular separation is:

$$\alpha = \frac{a}{d} = \frac{1.45}{2.29 \cdot 10^3\,\mathrm{pc} \cdot 206265\mathrm{au/pc}} = 3.1 \cdot 10^{-9}\,\mathrm{rad}\,. \qquad (14.2)$$

The angular separation between the stars is smaller than the angular resolution, therefore the astronomer will observe both stars as a single point in his telescope.

4.9

The curvature on either side of the system of lenses is in the same direction, hence R_{A} and R_{B} are both positive. Taking into account the different indices of refraction in Eq. 4.5:

$$\frac{1}{f} = \frac{(n_{\mathrm{A}} - 1)}{R_{\mathrm{A}}} - \frac{(n_{\mathrm{B}} - 1)}{R_{\mathrm{B}}}\,.$$

Considering red and blue light, respectively:

$$\frac{1}{f_r} = \frac{n_{\mathrm{A},r} - 1}{R_{\mathrm{A}}} - \frac{n_{\mathrm{B},r} - 1}{R_{\mathrm{B}}}\,,$$

$$\frac{1}{f_b} = \frac{n_{\mathrm{A},b} - 1}{R_{\mathrm{A}}} - \frac{n_{\mathrm{B},b} - 1}{R_{\mathrm{B}}}\,.$$

Since $f_r = f_b = f$, equating both sides:

$$\frac{n_{\mathrm{A},b} - n_{\mathrm{A},r}}{R_{\mathrm{A}}} = \frac{n_{\mathrm{B},b} - n_{\mathrm{B},r}}{R_{\mathrm{B}}} \Rightarrow R_{\mathrm{B}} = \frac{n_{\mathrm{B},b} - n_{\mathrm{B},r}}{n_{\mathrm{A},b} - n_{\mathrm{A},r}} R_{\mathrm{A}}\,.$$

Inserting the numerical values, we get $R_{\mathrm{B}} = 2R_{\mathrm{A}}$. Using this condition in the equation for f_r:

$$\frac{1}{f} = \frac{1}{2R_{\mathrm{A}}}\Big[2(n_{\mathrm{A},r} - 1) - (n_{\mathrm{B},r} - 1)\Big] = \frac{2n_{\mathrm{A},r} - n_{\mathrm{B},r} - 1}{2R_{\mathrm{A}}}\,.$$

Hence:

$$R_{\mathrm{A}} = \frac{f/2}{2n_{\mathrm{A},r} - n_{\mathrm{B},r} - 1} = 1\,\mathrm{m}\,.$$

Therefore, $R_{\mathrm{B}} = 2R_{\mathrm{A}} = 2\,\mathrm{m}$.

5. Time Systems

5.1

The difference in longitude is equal to the difference in sidereal time. Indeed, the (fixed) stars return to the same position with respect to the meridian every complete rotation of the Earth, that is, every sidereal period. Hence, the time difference between two locations with a difference in longitude of $\Delta\lambda$ is:

$$\Delta T_{\text{sid}} = \frac{\Delta\lambda}{360^\circ} \cdot 86400^s .$$

Using Eq. 5.2, this can be converted in solar time:

$$\Delta T_{\text{sol}} = \frac{86164.1}{86400} \cdot \Delta T_{\text{sid}} = \frac{\Delta\lambda}{360^\circ} \cdot 86164.1^s .$$

5.2

The difference in longitude between Rome and Greenwich is $\Delta\lambda = 12^\circ 28' 54.48''$, which corresponds to a difference in sidereal time of $49^m 55.63^s$, or $(86164.1/86400) \cdot (49^m 55.63^s) = 49^m 47.45^s$ of solar time. Since Rome is to the east of Greenwich, the time in Rome is greater by $49^m 47.45^s$. When the time in Greenwich is $12:00$ GMT, in Rome the local time is $12^h 49^m 47.45^s$. Since Rome belongs to the time zone GMT+1, when in Greenwich the time is $12:00$ GMT, in Rome the meridian time is $13:00$ GMT+1. The difference between GMT+1 and the local time in Rome is therefore $10^m 12.55^s$.

5.3

The difference in local time is $\Delta t = 11^h 14^m 15^s - 10^h 17^m 14^s = 57^m 1^s$. Local time is measured in solar time, but the difference in longitude is equal to the difference in sidereal time. By converting in sidereal time, using Eq. 5.3, we find $\Delta\lambda = \Delta t \cdot (86400/86164.1) \approx 3431^s = 57^m 11^s$. The longitude of Naples is therefore $\lambda = 14^\circ 17' 39''$.

5.4

The 24 principal meridians corresponding to the time zones are separated in longitude by 15°, starting from Greenwich at $\lambda = 0^\circ$. The meridian closest to the place under consideration is the one at 135° W. The difference in longitude between the place and the meridian is thus $\Delta\lambda = 6^\circ 45' = 27^m$. This corresponds to a difference in solar time of $(86164.1/86400) \cdot 27^m = 26^m 55.5^s$. Hence, the local time is $T_m = 8^h 46^m 22.0^s + 26^m 55.5^s = 9^h 13^m 17.5^s$. From the equation of time, we can then determine the true solar time:

$$\text{E.T.} = T_t - T_m \Rightarrow T_t = \text{E.T.} + T_m = -(14^m 13^s) + 9^h 13^m 17.5^s = 8^h 59^m 4.5^s .$$

Hence, the hour angle of the Sun is:

$$H_t = T_t - 12^h = -(3^h 0^m 55.5^s) .$$

5.5

Solar and sidereal time are equal during the autumnal equinox, when ♈ culminates at midnight.

5.6

Each solar day the stars move west by $3^m 56^s$. Hence, they rise $3^m 56^s$ earlier every day, or $1^h 58^m$ earlier after one month. Therefore, a month later, the time of rising will be $20:58$.

5.7

A star transits the meridian when its hour angle is zero. At this moment, the sidereal time is equal to the right ascension of the star (Eq. 1.2), i.e. ST$= \alpha = 18^h$. Therefore, we need to wait $18^h - (9^h50^m12^s) = 8^h9^m48^s$ of sidereal time, which corresponds to $(86164.1/86400) \cdot (8^h9^m48^s) = 8^h8^m27.8^s$ of solar time. Hence, the star transits the meridian at $0^h + 8^h8^m27.8^s = 8^h8^m27.8^s$ GMT.

5.8

We want the star to be in upper culmination, so that the path which light travels through the atmosphere is the shortest possible. At the moment of upper culmination, the star is on the meridian, hence the sidereal time is equal to the right ascension: ST $= \alpha = 10^h$. Since we want to observe the star at midnight, we require the sidereal time at 12:00 on that day to be $10^h - (86400/86164.1) \cdot 12^h = -2.03^h = 21.97^h$ (normalized to 24^h). Starting from the 21st of March, when ST$= 0$, the sidereal time increases by about 2^h each month. Therefore, ST $= 21.97^h$ corresponds to the end of February.

5.9

When the time in Rome is 12:00 GMT+1, in Greenwich the time is 11:00 GMT. The time elapsed between the vernal equinox and 11:00 GMT on the 13th of April is $24^d3^h25^m = 24.1424^d$. Using Eq. 5.4, we find the sidereal time in Greenwich:

$$\text{ST} = T_m + 11^h53^m + 24.1424 \cdot (3^m56^s) = T_m + 1^h34^m58^s\,.$$

In Rome, the local time is $11^h + (12°28'54.48'')/15° = 11^h49^m56^s$. Hence, the sidereal time in Rome, at $12:00$ GMT+1 on the 13th of April 2000, was:

$$\text{ST} = 11^h49^m56^s + 1^h34^m58^s = 13^h24^m54^s\,.$$

5.10

Neglecting atmospheric refraction, the hour angle of rising and setting can be computed from Eq. 2.9:

$$\cos H = -\tan\delta\tan\phi \Rightarrow H = \pm 108.47° = \pm 7^h14^m\,.$$

Stars rise to the east of the meridian, hence the hour angle of rising must be negative. Therefore, the hour angles of rising and setting are $H_r = -7^h14^m$ and $H_s = 7^h14^m$, respectively. The sidereal times of the two events are then:

$$\text{ST}_r = H_r + \alpha = -(7^h14^m) + 14^h16^m = 7^h2^m\,,$$

$$\text{ST}_s = H_s + \alpha = 14^h16^m + 7^h14^m = 21^h30^m\,.$$

Note that the answer is independent of the date, indeed a star rises and sets at the same sidereal time every day. Between the 10th of January and the vernal equinox, there are around 70 days, therefore the sidereal time on the 10th of January is (Eq. 5.4):

$$\text{ST} = T_m + 11^h53^m - 70 \cdot (3^m56^s) = T_m + 7^h18^m\,,$$

Isolating the mean solar time, we find:

$$T_{m,r} = \text{ST}_r - 7^h18^m = 7^h2^m - (7^h18^m) = 23^h44^m\,,$$

$$T_{m,s} = \text{ST}_s - 7^h18^m = 21^h30^m - (7^h18^m) = 14^h12^m\,.$$

The closest main meridian to Boston is the one at 75° W, hence the meridian time is smaller by $4°/15° = 16^m$ than the local time in Boston (actually $86164/86400 \cdot 16^m = 15^m57^s$). Therefore, the times of rising and setting are $23:30$ and $14:00$ GMT+5 respectively, with a precision of 10^m.

5.11

If the Earth rotated in the opposite direction, the velocity of the Sun relative to the celestial sphere would be $\omega_r = \omega + \Omega$, where ω and Ω are the angular velocities of rotation and revolution, respectively. Therefore:

$$\frac{1}{t_{\rm sol}} = \frac{1}{t_{\rm sid}} + \frac{1}{t_{\rm rev}} \Rightarrow t_{\rm sid} = \frac{t_{\rm sol}\, t_{\rm rev}}{t_{\rm rev} - t_{\rm sol}} = \frac{365.2564}{365.2564 - 1} = 1.002745^d = 1^d 3^m 57.2^s\,.$$

Hence, the sidereal day would be $3^m 57^s$ longer than a solar day.

5.12

The length of a year is 365 days. Every fourth year is a leap year, the years divisible by 100 are not leap years, except the ones that are also divisible by 400. Considering an interval of time of 400 years, starting from 0 until 399, we see that year 0 is a leap year, but years $100, 200, 300$ are not. Hence, compared to the Julian calendar, there are three less leap years every 400 years. Therefore, the average length of a year in the Gregorian calendar is:

$$\overline{T} = 365 + \frac{1}{4} - \frac{3}{400} = 365.2425^d\,.$$

The length of the tropical year is 365.2422^d, so the difference between the true and mean year is $3 \cdot 10^{-4}$ days, every year. Hence, after 100, 500 and 1000 years, the error in the Gregorian calendar will be, respectively:

$$\begin{aligned}\Delta t_{100} &= 100 \cdot 3 \cdot 10^{-4} = 0.03\,\text{days}\,,\\ \Delta t_{500} &= 500 \cdot 3 \cdot 10^{-4} = 0.15\,\text{days}\,,\\ \Delta t_{1000} &= 1000 \cdot 3 \cdot 10^{-4} = 0.3\,\text{days}\,.\end{aligned}$$

5.13

The orbital period of Venus is $t_{\rm sid} = 0.61519$ years. From Eq. 11.2, the synodic period is:

$$T_{\rm syn} = \frac{t_{\rm sid}}{1 - t_{\rm sid}} \approx 1.59869\,\text{years}$$

Between 2012 and 2117 there are 105 years. Of these, $\lfloor (2117 - 2012)/4 \rfloor - 1 = 25$ are leap years, since 2100 is not a leap year. Let x be the number of days between the $1^{\rm st}$ of January 2117 and the day when Venus transits the Sun that same year. Since the last transit happened on the $6^{\rm th}$ of June 2012, i.e. 157 days after the $1^{\rm st}$ of January, the difference in days between the two eclipses is $365 \cdot 105 + 25 + (x - 157) = 38193 + x$. The period of a planet is measured in *sidereal years*, therefore the synodic period is $1.59869 \cdot 365.2564 = 583.9317^d$. The transit happens again when the Earth and the Sun return to the same relative position, hence the number of synodic years between the two transits must be an integer. Then, $(38193 + x)/583.9317 = n$, with n integer. Since $0 \leqslant x \leqslant 365$, the only possible value is $n = 66$, and we obtain $x = 66 \cdot 583.9317 - 38193 = 346.49$. Hence, the next transit will be in 2117, 346 days after the $1^{\rm st}$ of January, i.e. on the $12^{\rm th}$ of December. Furthermore, the eclipse will occur approximately 12^h later than the previous one.

6. The Moon

6.1

The angular distance of the full moon from the Sun (also called elongation) is 180°.
Bodies that orbit the Sun at a distance smaller than the Earth-Sun distance, cannot have an

elongation greater than 90°. For example, the maximum elongations of Mercury and Venus are 28° and 48°, respectively (see Ch. 11). Therefore, Mercury and Venus can never be eclipsed by the full moon. On the other hand, superior planets (all the others, except the Earth) can have any elongation, and can therefore be eclipsed by the full moon.

6.2

During summer, the declination of the Sun is close to the maximum value of $\epsilon = 23°27'$; during winter, it is close to the minimum of $-\epsilon$. The full moon is on the opposite side of the Earth relative to the Sun. Hence, if the ecliptic is $\delta_\odot$ *above* the celestial equator in the direction of the Sun, it is $\delta_\odot$ *below* the equator in the direction of the full moon. Since the orbit of the Moon is inclined by an angle $i = 5°9'$ to the ecliptic, it follows that the declination of the Moon during summer and winter is close to $-\epsilon \pm i$ and $\epsilon \pm i$, respectively. Hence, for an observer in the northern hemisphere, the full moon is higher in winter than in summer. Similarly, the new moon, as well as the Sun, are higher in summer than winter.

6.3

Since the orbit of the Moon is inclined by $i = 5°9'$ to the ecliptic, and the ecliptic is inclined by 23°27′ to the celestial equator, it follows that the declination of the Moon can take any value between $\delta_{\max} = 28°35'$ and $\delta_{\min} = -28°35'$. The condition for a body with declination δ to appear at the zenith is $h_u = 90° = 90° - \phi + \delta$, i.e. $\phi = \delta$. Hence, the Moon can be seen at the zenith from places with latitude:

$$-(28°35') < \phi < 28°35' \Rightarrow |\phi| < 28°35' .$$

6.4

The Moon is circumpolar if $\phi > 90° - \delta$. Since the maximum declination of the Moon is $\delta_{\max} = 28°35'$ (see Ex. 6.3), we find $\phi > 61°25'$. By symmetry, in the southern hemisphere, $\phi < -(61°25')$. Hence, in general $|\phi| > 61°25'$.

6.5

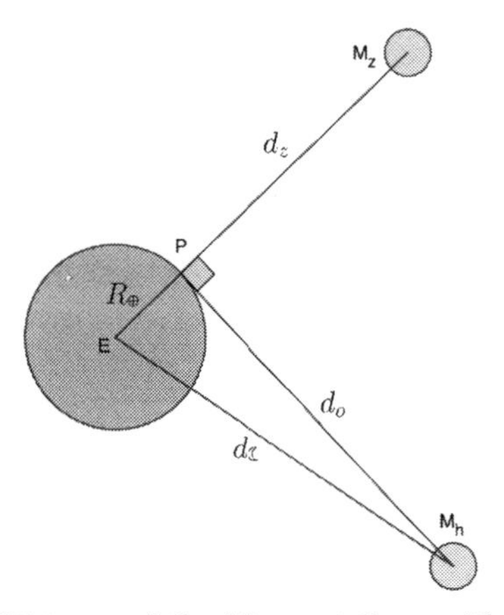

Figure 14.6: Distance of the Moon at the zenith and horizon.

Looking at Fig. 14.6, when the Moon is at the zenith, its distance from Earth is:

$$d_z = d_☾ - R_\oplus = 377629\,\text{km} .$$

When it is on the horizon:

$$d_h = \sqrt{d_{☾}^2 - R_{\oplus}^2} = 383947\,\text{km}\,.$$

Hence, the Moon is closer when at the zenith.

6.6

Using Eq. 10.18, we find the orbital velocity of the Moon:

$$v = \sqrt{\frac{GM_{☾}}{d_{☾}}} = 1017\,\text{m/s}\,.$$

Since $2R_{☾} = 3474.2\,\text{km}$, the time taken to cover a distance equal to its diameter is:

$$\Delta t = \frac{2R_{☾}}{v} = 3416\,\text{s} \approx 1^h\,.$$

6.7

As explained in Sec. "Lunar Phases", the Moon is visible in the morning either if it rises at 6 : 00 and sets at 18 : 00 (new moon), or if it rises at 0 : 00 and sets at 12 : 00 (last quarter). We are told that only half of the Moon is visible, hence there must be a last quarter moon. In this case, the side to the east is illuminated.

6.8

For a planet to be eclipsed by the full moon, it must orbit the Sun close to the plane of the ecliptic, at a distance to the Sun greater than the Sun-Earth distance (see Ex. 6.1). Then, the eclipse occurs when both the full moon and the planet are close to the node line. Since the node line precesses with a retrograde motion of period 18.61 years, the sidereal period of the planet must be exactly 18.61 years, if it is not to be eclipsed by the full moon. Then, from Kepler's third law (Eq. 10.14):

$${T_{\text{pr, yr}}}^2 = a_{\text{au}}^3 \Rightarrow a = 18.61^{2/3}\,\text{au} = 7.022\,\text{au}\,.$$

6.9

The length of the tropical year is $365^d5^h48^m46^s$, that of the lunar year is $354^d8^h48^m$. Hence, the lunar year is shorter than the tropical year by $10^d21^h46^s$. Since the synodic period of the Moon is $29^d12^h44^m33^s = 29.5309^d$, after $10^d21^h46^s = 10.8755^d$, the Moon forms an angle of $10.8755/29.5309 \cdot 360° = 132.6°$ with the Earth and the Sun. Therefore, the Moon will be in between the first quarter (90°) and full moon (180°).

6.10

From Eq. 6.1, we see that the radius of the umbra projected by the Earth at the position of the Moon is:

$$r = R_{\oplus} - \frac{d_{☾}}{d_{\oplus}}(R_{\odot} - R_{\oplus})\,.$$

Hence, the maximum distance of the Sun from the node line can be:

$$\Delta\lambda = \frac{r + R_{☾}}{d_{☾}}\frac{1}{\sin 5.145°} \Rightarrow \Delta\lambda = \frac{1}{\sin 5.145°}\left(\frac{R_{\oplus} + R_{☾}}{d_{☾}} - \frac{R_{\odot} - R_{\oplus}}{d_{\oplus}}\right).$$

Now, $\Delta\lambda_{\text{max}}$ is obtained when $d_{☾}$ is minimum and $d_{\oplus}$ is maximum, whilst $\Delta\lambda_{\text{min}}$ is obtained when $d_{☾}$ is maximum and $d_{\oplus}$ is minimum. Let $e_{\oplus}$, $e_{☾}$ be the eccentricities of the orbits of the Earth and Moon, respectively. Then, $d_{\oplus,\max} = (1+e_\oplus)d_\oplus$, $d_{\oplus,\min} = (1-e_\oplus)d_\oplus$, while $d_{☾,\max} = (1+e_☾)d_\oplus$, $d_{☾,\max} = (1-e_☾)d_☾$. Hence:

$$\Delta\lambda_{\max} = \frac{1}{\sin 5.145^\circ}\left[\frac{R_\oplus + R_☾}{(1-e_☾)d_☾} - \frac{R_\odot - R_\oplus}{(1+e_\oplus)d_\oplus}\right] \approx 11.38^\circ\,,$$

$$\Delta\lambda_{\min} = \frac{1}{\sin 5.145^\circ}\left[\frac{R_\oplus + R_☾}{(1+e_☾)d_☾} - \frac{R_\odot - R_\oplus}{(1-e_\oplus)d_\oplus}\right] \approx 9.79^\circ\,.$$

6.11

We can solve the exercise both from scratch, using a similar approach to that described in Sec. "Solar Eclipse", or by using Eq. 6.1 with the substitution $d_☾ \to -d_☾$, since, during a solar eclipse, the Moon is on the opposite side of the Earth compared to a lunar eclipse. Adopting the second strategy:

$$r = R_\oplus + \frac{d_☾}{d_\oplus}(R_\odot - R_\oplus) \Rightarrow \Delta\lambda = \frac{1}{\sin 5.145^\circ}\left(\frac{R_\oplus + R_☾}{d_☾} + \frac{R_\odot - R_\oplus}{d_\oplus}\right).$$

Following a similar reasoning to Ex. 6.10:

$$\Delta\lambda_{\max} = \frac{1}{\sin 5.145^\circ}\left[\frac{R_\oplus + R_☾}{(1-e_☾)d_☾} + \frac{R_\odot - R_\oplus}{(1-e_\oplus)d_\oplus}\right] \approx 17.3^\circ\,,$$

$$\Delta\lambda_{\min} = \frac{1}{\sin 5.145^\circ}\left[\frac{R_\oplus + R_☾}{(1+e_☾)d_☾} + \frac{R_\odot - R_\oplus}{(1+e_\oplus)d_\oplus}\right] \approx 15.7^\circ\,.$$

6.12

The distance from the Moon to the surface of the Earth is $d_☾ - R_\oplus$, whereas the Sun-Moon distance during a solar eclipse is $d_\oplus - d_☾$. To compute the radius of the umbra on the surface of the Earth, it is possible to use Eq. 6.1, substituting $R_\oplus \to R_☾$, $d_☾ \to d_☾ - R_\oplus$ and $d_\oplus \to d_\oplus - d_☾$. The radius of the umbra is maximum when the Earth is at aphelion and the Moon is at perigee:

$$r_u = R_☾ - \frac{d_☾(1-e_☾) - R_\oplus}{d_\oplus(1+e_\oplus) - d_☾(1-e_☾)}(R_\odot - R_☾) \approx 116\,\text{km}\,.$$

Therefore, the maximum length of the umbra is about 230 km.
If we take the distances $d_\oplus, d_☾$ equal to their average values (i.e. the semi-major axes), we obtain a negative value of r. This means that, for a total solar eclipse to occur, the Earth must be close to aphelion or the Moon close to perigee (otherwise the eclipse is anular).
Let us now derive the radius of the penumbra on Earth. Looking at Fig. 6.2, it suffices to replace $R_\odot$ in the previous formula with $-R_\odot$ (or, equivalently, $R_☾ \to -R_☾$, but then we also need to take the opposite sign overall). We then get, on average:

$$r_p = R_☾ + \frac{d_☾ - R_\oplus}{d_\oplus - d_☾}(R_\odot + R_☾)\,.$$

Neglecting $R_\oplus$ and $R_☾$ in the two factors at numerator, and $d_☾$ at denominator, we obtain:

$$r_p \approx R_☾ + \frac{d_☾}{d_\oplus}R_\odot = R_☾\left(1 + \frac{R_\odot/d_\oplus}{R_☾/d_☾}\right),$$

but $R_\odot/d_\oplus$ and $R_☾/d_☾$ are the angular diameters of the Sun and the Moon, therefore the fraction is approximately unity (actually 1.03, on average). Hence, it follows that the radius of the penumbra on Earth is approximately twice the radius of the Moon: $r_p \approx 2R_☾$.
Depending on the relative positions of the Earth and the Moon, the radius of the penumbra can vary, deviating from the average by at most a few hundred kilometres.

7. Radiation Mechanisms

7.1

Applying Wien's law (Eq. 7.10):

$$\lambda_{\max} = \frac{k}{T} = \frac{2.897 \cdot 10^{-3}\,\mathrm{m\,K}}{4000\,\mathrm{K}} \approx 724\,\mathrm{nm}\,.$$

which corresponds to the colour red.

7.2

Applying Stefan-Boltzmann law (Eq.7.13) to the star under consideration and the Sun:

$$L = 4\pi\sigma R^2 T^4\,,$$

$$L_\odot = 4\pi\sigma {R_\odot}^2 {T_\odot}^4\,.$$

Dividing the two equations and using $T_\odot = 5778\,\mathrm{K}$:

$$\frac{L}{L_\odot} = \left(\frac{R}{R_\odot}\right)^2 \left(\frac{T}{T_\odot}\right)^4 = 2.5^2 \cdot \left(\frac{6800}{5778}\right)^4 \approx 12\,.$$

7.3

Let r be the radius of a dust particle and a its distance to the Sun. Light is incident on an area of πr^2, hence the radiation force is given by Eq. 7.4:

$$F_r = \frac{F}{c} \cdot \pi r^2 = \frac{L_\odot}{4\pi a^2 c}\pi r^2 = \frac{L_\odot r^2}{4a^2 c}\,.$$

The gravitational force, written as a function of the density and radius of the dust particle is:

$$F_g = m\frac{GM_\odot}{a^2} = \frac{4}{3}\pi r^3 \rho \cdot \frac{GM_\odot}{a^2}\,.$$

Equating the forces:

$$\frac{L_\odot r^2}{4a^2 c} = \frac{4\pi}{3}\rho r^3 \frac{GM_\odot}{a^2}\,.$$

Isolating r, we find:

$$r = \frac{3L_\odot}{16\pi\rho G M_\odot c} \approx 0.6\,\mu\mathrm{m}\,.$$

Both the gravitational force and the radiation pressure grow as a^{-2}, hence the radius of equilibrium does not depend on the distance a. The gravitational force is proportional to the mass of the particle, which grows as r^3, while the radiation force is proportional to the cross section area, which grows as r^2. For this reason, there is only one value of r for which these two forces are equal.

7.4

The electronvolt eV is equal to the kinetic energy gained by an electron accelerating from rest through an electric potential difference of one volt. Therefore:

$$1\mathrm{eV} = e \cdot (1\,\mathrm{J/C}) \Rightarrow 1\mathrm{eV} = 1.6 \cdot 10^{-19}\,\mathrm{J}\,,$$

since the electron charge is $e = -1.6 \cdot 10^{-19}\,\mathrm{C}$. Every reaction releases an energy of:

$$E_\mathrm{r} = 26.8 \cdot 10^6 \cdot 1.6 \cdot 10^{-19}\,\mathrm{J} = 4.3 \cdot 10^{-12}\,\mathrm{J}\,,$$

and produces two neutrinos, hence $N_\mathrm{n} = 2N_\mathrm{r}$. The solar constant is approximately $k = 1367\,\mathrm{W/m^2}$, therefore:

$$N_\mathrm{n} = \frac{2k}{E_\mathrm{r}} \approx 6.4 \cdot 10^{14}\,\mathrm{neutrinos/(s \cdot m^2)}\,.$$

7.5

The action of the laser beam is different in the "near zone", where the entire beam is incident on the base of the spaceship, and in the "far zone", where the ship will only intercept part of the beam. The distance at which the size of the beam is equal to the base diameter is:

$$d_0 = \frac{D_0}{\Theta},$$

where D_0 is the base diameter and Θ is the divergence of the laser beam, expressed in radians. As long as the distance of the spaceship is less than d_0, the laser beam gives rise to a constant force of:

$$F_0 = \frac{2P}{c},$$

where P is the power emitted by the laser, and the factor of 2 appears because the base of the spaceship is perfectly reflecting. After travelling a distance d_0, the spaceship will have gained a kinetic energy of:

$$K_0 = F_0 d_0 = \frac{2PD_0}{c\Theta}.$$

When the spaceship moves at distances greater than d_0, only part of the laser beam will be incident on its base. Hence, the total power reflected by the spaceship will be:

$$P(d) = P\left(\frac{d_0}{d}\right)^2.$$

As a result, the force decreases with distance according to:

$$F(d) = \frac{2P}{c}\left(\frac{d_0}{d}\right)^2.$$

We can find the total kinetic energy gained by the spaceship in the "far zone" either by integrating the force from d_0 to infinity, or by recognizing that this force is similar to the gravitational force, as it is proportional to the inverse of the square of the distance, but has opposite direction. Hence, we can associate a "radiation potential" of:

$$U = \frac{2Pd_0^2}{cd}.$$

Since, at infinity, $U \to 0$, while initially $U_0 = 2Pd_0/c$, from the conservation of energy it follows that the kinetic energy must increase by U_0 (equal to K_0). Hence, at an infinite distance from the laser, the total kinetic energy of the spaceship will be:

$$K_f = 2K_0 = \frac{4PD_0}{c\Theta},$$

and its velocity:

$$v = \sqrt{\frac{2K_f}{m}} = \sqrt{\frac{4PD_0}{mc\Theta}} \approx 1\,\text{km/s}.$$

7.6

In Eq. 7.8, we can substitute the mass m (measured in kg) and k_B, with μ (measured in kg/mol) and R, respectively. Indeed $m = \mu/N_a$, where N_a is Avogadro constant, while the Boltzmann constant is numerically equal to $k_B = R/N_a$. Therefore, the root mean square velocity can be written as:

$$v_{\text{rms}} = \sqrt{\frac{3RT}{\mu}} \tag{14.3}$$

Substituting $\mu_{\mathrm{H_2}} = 2.016\,\mathrm{g/mol}$, $\mu_{\mathrm{O_2}} = 32.000\,\mathrm{g/mol}$ e $\mu_{\mathrm{N_2}} = 28.013\,\mathrm{g/mol}$:

$$v_{\mathrm{rms,H_2}} = \sqrt{\frac{3RT}{\mu_{\mathrm{H_2}}}} = 1884.4\,\mathrm{m/s}\,,$$

$$v_{\mathrm{rms,O_2}} = \sqrt{\frac{3RT}{\mu_{\mathrm{O_2}}}} = 473.0\,\mathrm{m/s}\,,$$

$$v_{\mathrm{rms,N_2}} = \sqrt{\frac{3RT}{\mu_{\mathrm{N_2}}}} = 505.5\,\mathrm{m/s}\,.$$

At sea level, the escape velocity from the Earth is $v_{\mathrm{esc}} = \sqrt{2GM/R} = 11182\,\mathrm{m/s}$. Hence, we need to numerically evaluate the integral:

$$f = 4\pi\left(\frac{\mu}{2\pi RT}\right)^{3/2} \int_{v_f}^{\infty} v^2 e^{-\mu v^2/2RT}\,dv\,.$$

Which gives:

$$f_{\mathrm{H_2}} = 1.72 \cdot 10^{-24}\,,$$
$$f_{\mathrm{O_2}} = 7.28 \cdot 10^{-356}\,,$$
$$f_{\mathrm{N_2}} = 1.86 \cdot 10^{-311}\,.$$

The fractions of O_2 and N_2 with velocities greater than the escape velocity are extremely small. In practice, the probability that a molecule close to the surface will escape is insignificant. This is due to the escape velocity being greater at the surface, and to the fact that the mean free path of a molecule is very small when the gas density is high. Thus, molecules are more likely to leave from the uppermost layers of the atmosphere. The *critical layer* is defined as the height at which a molecule, moving upward, has a probability 1/e of hitting another molecule, before escaping. The region of the atmosphere above the critical layer is called the *exosphere*. Earth's exosphere starts at a height of 500 km, where the kinetic temperature of the gas is 1500 – 2000 K.

7.7

Similarly to the previous exercise, we need to evaluate the integral:

$$f = 4\pi\left(\frac{\mu}{2\pi RT}\right)^{3/2} \int_{v_f}^{\infty} v^2 e^{-\mu v^2/2RT}\,dv\,.$$

Using the fact that $v_f = 3v_{\mathrm{rms}} = 3\sqrt{3RT/\mu}$:

$$f = 4\pi\left(\frac{3\sqrt{3}}{2\pi v_f}\right)^3 \int_{v_f}^{\infty} v^2 e^{-\frac{27}{2v_f^2}v^2}\,dv = 1 - \mathrm{erf}\left(\frac{33}{\sqrt{2}}\right) + \frac{18}{\sqrt{6\pi}}e^{-27/2} = 5.8874 \cdot 10^{-6}\,.$$

In the case $v_f = 5\sqrt{3RT/\mu}$, we have:

$$f = 4\pi\left(\frac{5\sqrt{3}}{2\pi v_f}\right)^3 \int_{v_f}^{\infty} v^2 e^{-\frac{75}{2v_f^2}v^2}\,dv = 1 - \mathrm{erf}\left(\frac{53}{\sqrt{2}}\right) + \frac{30}{\sqrt{6\pi}}e^{-75/2} = 3.6233 \cdot 10^{-16}\,.$$

7.8

Taking the derivative of Eq. 7.6 with respect to v, we find:

$$4\pi\left(\frac{\mu}{2\pi RT}\right)^{3/2}\left[2-v^2\frac{\mu}{RT}\right]ve^{-\frac{27}{2v_f^2}v^2}=0\,.$$

It follows that:

$$2-v^2\frac{\mu}{RT}=0\Rightarrow v=\sqrt{\frac{2RT}{\mu}}\,.$$

7.9

Let us consider an element of atmosphere in the shape of a parallelepiped, with infinitesimal height dh, base surface A and mass $dm=\rho(h)A\,dh$. The pressure on the upper face is $P+dP$, and that on the lower face is P. The pressure difference gives rise to a force (directed upwards) that keeps the parallelepiped afloat, against the gravitational force exerted by the Earth:

$$F_g=F_p\Rightarrow g\rho A\,dh=A\,dP\,,$$

where g can be taken as a constant. Indeed, we are assuming the temperature is constant as well, hence our model will be restricted to relatively small heights. From the ideal gas law:

$$PV=nRT\Rightarrow\rho=\frac{P\mu}{RT}\,.$$

Substituting the density in the first equation:

$$dP=-g\frac{P\mu}{RT}\,dh\,.$$

Separating variables and integrating:

$$\int_{P_0}^{P_h}\frac{dP}{P}=-g\frac{\mu}{RT}\int_0^h dh$$

$$\ln\frac{P_h}{P_0}=-g\frac{\mu}{RT}h\,.$$

It follows that:

$$P_h=P_0\exp\left\{-g\frac{\mu}{RT}h\right\}\,.$$

Near the surface of the Earth, the pressure and density of air decrease exponentially. This model is only valid for heights up to a few kilometres, since the temperature actually decreases by about 10K/km. A better approximation is the model of the adiabatic atmosphere.
Consider an element of air that, due to a small turbulence, is displaced upwards slightly. Air pressure decreases with height, so the cube expands. The air is a bad heat conductor, therefore the exchange of energy between the element and the surrounding air is small and the process can be regarded as adiabatic. If the temperature of the air around the element is now higher that the temperature of the element itself, the element will be denser than the surrounding air and, consequently, will be pushed down by gravity, returning to its original position. In the adiabatic model, therefore, the atmosphere is in a continuous dynamic balance. In the limiting case, we can assume that the temperature of the layers above the element is the same as the temperature the element would acquire if it rose adiabatically. In this case, the element will just find a new equilibrium position slightly upwards. In reality, this model is valid only in the absence of convective motion and if the air is perfectly dry. It can be shown that convective motion takes place when the temperature gradient satisfies $-dT/dz>g/c_p\mu\approx 0.2857\,\mathrm{K/km}$, i.e. if the temperature decreases by more than 1 degree every 3.5 km, approximately. For air, the temperature gradient is more than 30 times higher. If the air is humid, the water inside the cube condenses as it rises. Condensation releases heat, which heats up the cube, making it less dense than the surrounding air and therefore able to rise even higher. This is the phenomenon that gives rise to clouds.

7.10

Consider a circular annulus of radius r, thickness dr and mass $dm = \rho(r) \cdot 2\pi r\, dr$. Let P be the pressure on the inner rim and $P + dP$ that on the outer rim. The force caused by the pressure difference must equal the centripetal force due to the rotation of the spacecraft. The density of air can be found from the ideal gas law:

$$PV = nRT \Rightarrow \rho = \frac{P\mu}{RT}$$

where μ is the molar mass of the gas. Hence:

$$dm = 2\pi \frac{P\mu}{RT} r\, dr\,.$$

The pressure differential is due to the centrifugal force:

$$dP = dm\frac{\Omega^2 r}{A} = \left(2\pi\frac{P\mu}{RT} r\, dr\right)\cdot\left(\frac{\Omega^2 r}{2\pi r}\right) \Rightarrow P_c = \frac{P\mu\Omega^2}{RT} r\, dr\,.$$

Integrating:

$$\int_{P_c}^{P_r} \frac{dP}{P} = \frac{\mu\Omega^2}{RT}\int_0^r r'\, dr'$$

$$\ln\frac{P_r}{P_c} = \frac{\mu\Omega^2}{2RT} r^2$$

$$\Rightarrow P_r = P_c \exp\left\{\frac{\mu\Omega^2}{2RT} r^2\right\}.$$

Since the station provides an acceleration equal to g at the outer edge, we have $\Omega^2 R_0 = g$. The pressure P_0 on the outer edge is:

$$P_0 = P_c \exp\left\{\frac{\mu g}{2RT} R_0\right\}.$$

7.11

Space debris can sometimes move at high relative velocities with respect to satellites, therefore constituting a serious danger to orbiting bodies. Let P be the pressure inside the spaceship and assume the pressure outside to be zero. We will also assume that the air near the hole obeys Bernoulli's law, whereby its velocity (v) is related to its density (ρ) and the pressure difference (P) by:

$$P = \frac{1}{2}\rho v^2 \Rightarrow v = \sqrt{\frac{2P}{\rho}}\,.$$

The rate at which air leaves the spaceship through the hole of area A is:

$$dV = A \cdot v = A\sqrt{\frac{2P}{\rho}}\, dt\,.$$

The air density can be written as a function of the pressure using the ideal gas law:

$$\rho = \frac{m_{\text{tot}}}{V} = \frac{n\mu}{V} = \frac{P\mu}{RT}\,,$$

where μ is the molar mass of the gas. Differentiating the ideal gas law, we obtain the change in pressure following the expulsion of an infinitesimal volume of air, assuming nRT to be instantaneously constant:

$$PV = \text{const} \Rightarrow PdV + VdP = 0 \Rightarrow dV = -\frac{V}{P}dP\,.$$

Substituting ρ and dV in the initial equation, we obtain:

$$\frac{dP}{P} = -\frac{A}{V}\sqrt{\frac{2RT}{\mu}}\,dt\,.$$

Integrating:

$$\int_{P_0}^{P(t)} \frac{dP}{P} = -\frac{A}{V}\sqrt{\frac{2RT}{\mu}}\int_0^t dt\,.$$

Hence:

$$\ln\frac{P(t)}{P_0} = -\frac{A}{V}\sqrt{\frac{2RT}{\mu}}\,t \Rightarrow P(t) = P_0 \exp\left\{-\frac{A}{V}\sqrt{\frac{2RT}{\mu}}\,t\right\} \approx P_0 e^{-0.000405\cdot t}\,.$$

Hence, the pressure halves after a time:

$$t = \frac{\ln 2}{0.000405} \approx 1710\,\mathrm{s} = 28^m 30^s\,.$$

7.12

Using the formula suggested:

$$\begin{aligned}\langle v^2\rangle &= \int_0^\infty v^2 f(v)\,dv \\ &= 4\pi\left(\frac{m}{2\pi k_B T}\right)^{3/2}\int_0^\infty v^4 e^{-mv^2/2k_BT}\,dv\,.\end{aligned}$$

The gaussian integral is:

$$I_0 = \int_0^\infty e^{-\alpha x^2}\,dx = \frac{1}{2}\sqrt{\frac{\pi}{\alpha}}\,.$$

If, instead, we have:

$$I_1 = \int_0^\infty x e^{-\alpha x^2}\,dx = \left[-\frac{1}{2\alpha}e^{-\alpha x^2}\right]_0^\infty = \frac{1}{2\alpha}\,.$$

Adding another x in the integrand:

$$I_2 = \int_0^\infty x^2 e^{-\alpha x^2}\,dx = \left[-\frac{1}{2\alpha}x e^{-\alpha x^2}\right]_0^\infty + \frac{1}{2\alpha}\int_0^\infty e^{-\alpha x^2}\,dx = \frac{1}{4}\sqrt{\pi}\alpha^{-3/2}\,,$$

were we have used integration by parts on $v' = xe^{-\alpha x^2}\,dx$ and $u = x$. It is easy to see that:

$$I_4 = \frac{3}{8}\sqrt{\pi}\alpha^{-5/2}\,.$$

Hence:

$$\begin{aligned}\langle v^2\rangle &= 4\pi\left(\frac{m}{2\pi k_B T}\right)^{3/2}\cdot\frac{3}{8}\sqrt{\pi}\left(\frac{2k_BT}{m}\right)^{5/2} \\ &= 4\frac{3}{8}\pi\sqrt{\pi}\pi^{-3/2}\frac{2k_BT}{m} \\ &= \frac{3k_BT}{m}\,.\end{aligned}$$

Therefore, the root mean square speed is:

$$\sqrt{\langle v^2\rangle} = \sqrt{\frac{3k_BT}{m}}\,,$$

as we wanted to show. Another interesting result is that the root mean square speed for molecules hitting the walls of a container is $\sqrt{4k_BT/m}$, which is larger than the result found above. Indeed, faster molecules hit the walls more often, hence the mean speed is larger. In this case, the integral we need to calculate is $\int_0^\infty v^3 f(v)\,dv$, where the additional v comes from the number of collisions with the wall being proportional to the particle's velocity. Try it!

8. Flux and Magnitude

8.1

According to the Stefan-Boltzmann law (Eq. 7.13):

$$L_\odot = 4\pi\sigma {R_\odot}^2 {T_e}^4 = 3.84 \cdot 10^{23}\,\mathrm{W}\,.$$

The flux arriving on Earth is:

$$F = \frac{L_\odot}{\pi d_\oplus^2} = 1367\,\mathrm{W}\,.$$

This is the solar constant $k = 1367\,\mathrm{W}$, equal to the power incident on Earth per square meter. The Earth only intercepts photons on an effective cross-sectional area of $\pi R_\oplus^2$. Hence, the fraction of the Sun's luminosity that arrives on Earth is:

$$f = \frac{\pi R_\oplus^2}{4\pi d_\oplus^2} \approx 4.5 \cdot 10^{-10}\,.$$

A hypothetical sphere that completely encompasses a star is called a *Dyson sphere*. Some have hypothesized that, once the energy requirements of an intelligent civilization exceed what can be provided by their home planet's resources alone, a Dyson sphere would be built, in order to gather most of the energy generated by fission inside their star. To get a feeling of the amount of energy produced by the Sun, imagine we were able to gather all of its energy for just one second: this would be enough to meet our energy demands for the next half-million years (assuming energy consumption stays the same).

8.2

According to the Stefan-Boltzmann law (Eq. 7.13), the power emitted by the Earth is:

$$P_{\mathrm{out}} = 4\pi R_\oplus \sigma T_\oplus^4\,,$$

where $T_\oplus = 300\,\mathrm{K}$. Since the Earth is in thermal equilibrium, the Sun provides a thermal power of P_{out}. Hence, the power incident on the surface of the Earth, per unit area, is:

$$F = \frac{P_{\mathrm{out}}}{4\pi R_\oplus^2} = \sigma T_\oplus^4 \approx 460\,\mathrm{W/m}^2\,. \tag{14.4}$$

The ratio between the flux from the Sun and the flux due to radioactive decay is therefore $460/(5 \cdot 10^{-2}) \approx 10^4$. If the only energy source were radioactive decay, the temperature of the Earth would be approximately $\sqrt[4]{10^4} = 10$ times smaller, i.e. only $30\,\mathrm{K}$.

8.3

The solar constant is equal to $k = 1367\,\mathrm{W/m}^2$. Approximately 40 % of the solar radiation is reflected back into space or absorbed by clouds, whilst 15 % is absorbed by the atmosphere. Hence, the energy arriving on the surface is 45 % of the solar constant, which corresponds to a flux of $k_{\mathrm{eff}} = 0.45 \cdot k = 615\,\mathrm{W/m}^2$. The distance at which a light bulb of power $L = 100\,\mathrm{W}$ would have to be placed to appear equally as bright as the Sun is:

$$k_{\mathrm{eff}} = \frac{L}{4\pi d^2} \Rightarrow d = \sqrt{\frac{L}{4\pi k_{\mathrm{eff}}}} = 0.114\,\mathrm{m}\,.$$

8.4

Using Eq. 8.5:

$$\frac{F_a}{F_b} = 10^{-0.4(m_a - m_b)} = 11070\,.$$

8.5

The ratio of the fluxes is found using Eq. 8.5:

$$\frac{F_b}{F_t} = 10^{-0.4(m_b - m_t)} = 10^{-3}\,.$$

8.6

We know the apparent magnitude of the Sun, as seen from Earth, and the ratio of the planet-Sun to Earth-Sun distances (i.e. the distance of the planet written in au). The apparent magnitude of the Sun, as viewed from each planet, is found using Eq. 8.4:

$$m_p = m_e - 2.5\log\frac{{d_e}^2}{{d_p}^2} \Rightarrow m_p = m_e + 5\log d_{p,\mathrm{au}}\,.$$

Substituting the numerical values, we find:

$$\begin{aligned} m_{\mathrm{mer}} &= -26.8 + 5\log 0.3871 = -28.9\,, \\ m_{\mathrm{ven}} &= -26.8 + 5\log 0.723 = -27.5\,, \\ m_{\mathrm{mar}} &= -26.8 + 5\log 1.524 = -25.9\,, \\ m_{\mathrm{jup}} &= -26.8 + 5\log 5.203 = -23.2\,, \\ m_{\mathrm{sat}} &= -26.8 + 5\log 9.539 = -21.9\,. \end{aligned}$$

8.7

Let F_0 be the initial flux and $F_1 = 1.1\,F_0$ the flux after the increment. Using Eq. 8.4:

$$\Delta m = m_1 - m_0 = -2.5\log\frac{F_1}{F_0} = -0.103\,,$$

for both the Sun and Jupiter. Therefore, their apparent magnitudes will be $m_s = -26.84$ and $m_{♃} = -2.3$, respectively. On Earth, the increase in brightness of the Sun is observed after light has travelled a distance equal to the Earth-Sun distance. Hence:

$$\Delta t_{\odot} = \frac{d_{\oplus}}{c} = 499\,\mathrm{s} = 8^m 19^s\,.$$

After reaching the Earth, light must travel to Jupiter and back. The time this takes is:

$$\Delta t_{♃} = \frac{2(d_{♃} - d_{\oplus})}{c} = 4195s = 1^h 9^m 55^s\,.$$

8.8

The intrinsic luminosity of the stars is the same, hence the ratio of their fluxes depends only on their distances from Earth:

$$\frac{F_2}{F_1} = \frac{L/(4\pi {d_2}^2)}{L/(4\pi {d_1}^2)} = \left(\frac{d_1}{d_2}\right)^2.$$

Using Eq. 8.4, we find the difference in magnitudes from the flux ratio:

$$\Delta m = m_2 - m_1 = -2.5\log\left(\frac{d_1}{d_2}\right)^2 = -5\log\frac{d_1}{d_2}\,.$$

Isolating d_1/d_2:

$$\frac{d_1}{d_2} = 10^{-0.2\Delta m} \approx 0.2\,.$$

8.9

From Eq. 8.7, writing the distance d in parsecs:

$$M = m - 5\log d + 5 = -1.45 - 5\log\left(\frac{8.6}{3.262}\right) + 5 = 1.44\,.$$

Knowing that $M_\odot = 4.83$, we use Eq. 8.5 to find the ratio of the luminosities:

$$\frac{L}{L_\odot} = 10^{-0.4(M - M_\odot)} = 22.7\,.$$

8.10

Using Eq. 8.7, isolating the distance d:

$$M - m = -5\log d + 5 \Rightarrow d = 10^{\frac{m-M}{5}+1} = 11.22\,\text{pc}\,.$$

The parallax is given by Eq. 9.3:

$$\pi_p'' = \frac{1}{d} = \frac{1}{11.22} = 0.0891''\,.$$

8.11

Writing down Eq. 8.7 for Sirius and the Sun, respectively:

$$M_s - m_s = -5\log d_s + 5\,,$$
$$M_\odot - m_\odot = -5\log d_\odot + 5\,.$$

The stars must appear equally as bright, therefore their apparent magnitudes must be the same, i.e. $m_s = m_\odot$. Subtracting the two above equations and using the logarithm property $\log d_s - \log d_\odot = \log d_s/d_\odot$, we find:

$$M_s - M_\odot = -5\log\frac{d_s}{d_\odot} \Rightarrow \frac{d_s}{d_\odot} = 10^{-0.2(M_s - M_\odot)} = 4.85\,.$$

Hence, Sirius would be equally as bright as the Sun if its distance were 4.85 au.

8.12

The maximum diameter of the pupil at night is around $2r = 8\,\text{mm}$. Since the apparent magnitude of the Sun is -26.74, and its flux on Earth is $k = 1367\,\text{W/m}^2$, the flux of a star with magnitude $m = 6$ would be:

$$F = k \cdot 10^{-0.4(m - m_\odot)} = 1.1 \cdot 10^{-10}\,\text{W/m}^2\,.$$

The energy of a photon with wavelength λ is $E = hf = hc/\lambda$. Assuming that the star only emits photons with wavelength λ, the rate of incident photons per square meter (effectively, the flux of photons) is:

$$F_f = \frac{F}{E} = \frac{F\lambda}{hc}\,.$$

Hence, the rate of incident photons on our pupils, with total area $A = 2\pi r^2$, is:

$$F_f \cdot A = \frac{F\lambda}{hc} \cdot 2\pi r^2 \approx 3 \cdot 10^4\,\text{photons/s}\,.$$

8.13

The distance of the Moon to the Sun is approximately equal to the Sun-Earth distance $d_\oplus$, therefore the flux incident on the Moon is equal to the solar constant $k = 1367\,\mathrm{W/m^2}$. Hence, the total incident energy on the Moon is $L_{\rm in} = k\pi R_☾^2$, and the reflected energy is $L_{\rm rif} = AL_{\rm in} = \pi k A R_☾^2$. Assuming that the Moon reflects light isotropically (like a perfect sphere), the flux arriving on Earth is:

$$F_☾ = \frac{L_{\rm rif}}{4\pi d_☾^2} = \frac{kA}{4}\left(\frac{R_☾}{d_☾}\right)^2 = \frac{1}{4}kA\alpha_{\rm rad}^2 \,.$$

where $\alpha_{\rm rad}$ is the angular radius of the Moon as seen from Earth. The ratio of the fluxes of the Moon and Sun are:

$$\frac{F_☾}{k} = 10^{-0.4(m_☾ - m_\odot)} \,.$$

Hence, substituting in the first equation:

$$\frac{1}{4}A\alpha_{\rm rad}^2 = 10^{-0.4(m_☾ - m_\odot)} \Rightarrow A = \frac{4}{\alpha_{\rm rad}^2}10^{-0.4(m_☾ - m_\odot)} \approx 0.53\,.$$

In fact, the Moon does not behave like a perfect sphere, but reflects more strongly in the same direction as the incident light. Therefore, the coefficient 4 in the above formula should be replaced by 1, approximately, in which case $A = 0.13$. We have therefore found an upper bound on the albedo of the Moon, whose measured value is $A \approx 0.12$.

8.14

The absolute magnitude of UY Scuti, if we only consider the light reaching us, can be obtained from Eq. 8.8:

$$M_{\rm app} = m + 5 + 5\log\pi_p'' = -3.25\,.$$

Since only 0.5% of the light passes through the envelope surrounding the star, the true value of the absolute magnitude is:

$$M = M_{\rm app} + 2.5\log 0.005 = -9\,.$$

Hence, in the visible range, the luminosity of UY Scuti is:

$$\frac{L}{L_\odot} = 10^{-0.4(M - M_\odot)} = 3.3\cdot 10^5\,.$$

On the other hand, according to the Stefan-Boltzmann law:

$$\frac{L}{L_\odot} = \left(\frac{R}{R_\odot}\right)^2\left(\frac{T}{T_\odot}\right)^4 = \left(\frac{V}{V_\odot}\right)^{2/3}\left(\frac{T}{T_\odot}\right)^4 .$$

Hence, the temperature of UY Scuti is:

$$T = T_\odot\left(\frac{V_\odot}{V}\right)^{1/6}\left(\frac{L}{L_\odot}\right)^{1/4} = 3350\,\mathrm{K}\,.$$

Since $L_\odot = 3.84\cdot 10^{26}\,\mathrm{W}$, the energy radiated by UY Scuti is:

$$L = 3.3\cdot 10^5 L_\odot = 1.26\cdot 10^{32}\,\mathrm{W}\,.$$

Due to radiation, UY Scuti looses mass at a rate of:

$$\frac{\mathrm{d}m_r}{\mathrm{d}t} = \frac{L}{c^2} = 1.4\cdot 10^{15}\,\mathrm{kg/s}\,.$$

Hence, the total rate at which UY Scuti looses mass is:

$$\frac{dm}{dt} = \frac{dm_r/dt}{0.004} = 3.5 \cdot 10^{18}\,\text{kg/s}\,.$$

Assuming this rate stays constant, the total lifetime of UY Scuti is:

$$\Delta\tau = \frac{8M_\odot}{dm/dt} = 1.5 \cdot 10^5\,\text{years}\,,$$

which is a very short time on the cosmic scale.

8.15

The absolute magnitude of Gliese 581 can be found from its apparent magnitude and parallax, using Eq. 8.8:

$$M_g = m_g + 5\log\pi_p'' + 5 \approx 9.0\,.$$

Since $M_\odot = 4.83$, the ratio of the intrinsic luminosities is:

$$\frac{L_g}{L_\odot} = 10^{-0.4(M_g - M_\odot)} \approx 1/48\,.$$

Since Gliese 581g has similar characteristics to the Earth, we require their surface temperature to be approximately the same. This amounts to equating the fluxes of their respective stars, as seen from the planets:

$$\frac{L_g}{4\pi a_g^2} = \frac{L_\odot}{4\pi a_\oplus^2} \Rightarrow a_{g,\text{au}} = \sqrt{\frac{L_g}{L_\odot}} \approx 0.144\,.$$

The period is found using Kepler's third law:

$$T_\text{years} = \sqrt{\frac{a_\text{au}^3}{M/M_\odot}} \approx 36\,\text{days}\,.$$

8.16

The distance of Altair to the Earth is $d = 1/\pi_p'' = 5.13\,\text{pc}$, i.e. $d_a/d_\odot \approx 10^6$. From Eq. 8.5:

$$\frac{F_a}{F_\odot} = 10^{-0.4(m_a - m_\odot)}\,.$$

The ratio of the fluxes can be found from the distances and intrinsic luminosities of the stars:

$$\frac{F_a}{F_\odot} = \frac{L_a}{L_\odot}\left(\frac{d_\odot}{d_a}\right)^2.$$

Substituting in the first equation and isolating $L_a/L_\odot$:

$$\frac{L_a}{L_\odot} = \left(\frac{d_a}{d_\odot}\right)^2 \cdot 10^{-0.4(m_a - m_\odot)}\,.$$

Because Altair is a main-sequence star with a mass in the range $1.4 - 2.1M_\odot$, looking at Tab. 7.1, its temperature its around 8000 K. According to Stefan-Boltzmann law:

$$L_a = 4\pi R_a^2 \sigma T_a^4\,,$$
$$L_\odot = 4\pi R_\odot^2 \sigma T_\odot^4\,.$$

Dividing the two equations, substituting for $L_a/L_\odot$ the expression previously found, and isolating $R_a/R_\odot$:

$$\frac{R_a}{R_\odot} = \frac{d_a}{d_\odot}\left(\frac{T_\odot}{T_a}\right)^2 \cdot 10^{-0.2(m_a - m_\odot)} \approx 1.642\,.$$

The ratio of the densities is:

$$\frac{\rho_a}{\rho_\odot} = \frac{M_a}{M_\odot}\left(\frac{R_\odot}{R_a}\right)^3 \approx 0.384\,.$$

From which it follows that:

$$\rho_a \approx 0.384 \cdot \frac{3}{4\pi}\frac{M_\odot}{R_\odot^3} \approx 437\,\mathrm{kg/m^3}\,.$$

8.17

The composite magnitude can be found using Eq. 8.9, with $n = 3$:

$$m_{\rm sys} = -2.5\log[10^{-0.4m_1} + 10^{-0.4m_2} + 10^{-0.4m_3}] = 3.11\,.$$

8.18

Let m_1 be the magnitude of the larger and colder star, and m_2 that of the smaller and hotter star. The composite magnitude is smallest when both stars are visible, while it is greatest when the hotter star is completely eclipsed. Hence, it immediately follows that $m_{\rm min} = m_1 = 4.85$, while m_2 is found using Eq. 8.9:

$$\begin{aligned} m_{\rm max} &= -2.5\log\left[10^{-0.4m_1} + 10^{-0.4m_2}\right] \\ 10^{-0.4m_2} &= 10^{-0.4m_{\rm max}} - 10^{-0.4m_1} \\ \Rightarrow m_2 &= -2.5\log\left[10^{-0.4m_{\rm max}} - 10^{-0.4m_1}\right] = 4.16\,. \end{aligned}$$

Therefore, the magnitudes of the two stars are $m_1 = 4.85$ and $m_2 = 4.16$. The ration of the maximum and minimum fluxes is:

$$\frac{F_{\rm max}}{F_{\rm max}} = 10^{-0.4(m_{\rm max} - m_{\rm min})} = 2.88\,.$$

8.19

From Eq. 8.9, we know that the composite magnitude is given by:

$$m_s = -2.5\log[10^{-0.4m_1} + 10^{-0.4m_2} + ... + 10^{-0.4m_n}]\,.$$

In this case $n = 24$, equal to the number of letters in the Greek alphabet. Furthermore, the magnitude of the n-th star is $m_n = 0.10 \cdot n$, hence the previous equation becomes:

$$\begin{aligned} m_s &= -2.5\log[10^{-0.4\cdot 0.10} + 10^{-0.4\cdot 0.20} + ... + 10^{-0.4\cdot 2.4}] \\ &= -2.5\log[(10^{-0.04}) + (10^{-0.04})^2 + ... + (10^{-0.04})^{24}] \\ &= -2.5\log\left[\frac{1 - 10^{-0.04\cdot 25}}{1 - 10^{-0.04}} - 1\right] \approx -2.41\,. \end{aligned}$$

where we have used the fact that:

$$1 + x + x^2 + ... + x^n = \frac{1 - x^{n+1}}{1 - x}\,.$$

8.20

The brightness of the image of a star is proportional to the number of photons captured, which in turn is proportional to the product of the flux of the star and the exposure time. Let F_1, F_2 be the fluxes of the two stars and t_1, t_2 the exposure times. To obtain the same luminosity, we need $F_1 t_1 = F_2 t_2$, i.e. $F_1/F_2 = t_2/t_1$. Using Eq. 8.5:

$$\frac{F_1}{F_2} = 10^{-0.4(m_1 - m_2)} .$$

Substituting $F_1/F_2 = t_2/t_1$:

$$t_2 = t_1 \cdot 10^{-0.4(m_1 - m_2)} = 1000\,\text{s} .$$

Indeed, since the ratio of the fluxes is 1/100, the exposure time for the second star must be 100 times longer, i.e. 1000 s.

8.21

Using Eq. 8.5, we obtain the ratio of the fluxes:

$$\frac{F_☾}{F_\odot} = 10^{-0.4(m_☾ - m_\odot)} .$$

From Eq. 8.12, we have:

$$\frac{F_☾}{F_\odot} = e^{-\tau} \Rightarrow \tau = -\ln \frac{F_☾}{F_\odot} .$$

Substituting the ratio of the fluxes found in the first equation, using the logarithm property $\ln x = \log x / \log e$, we find:

$$\tau = \frac{0.4}{\log e}(m_☾ - m_\odot) = 12.9 .$$

8.22

First, we compute the maximum distance at which a star, exploding as a supernova, would raise the temperature on Earth beyond $T = 333\,\text{K}$. Assuming the Earth to be a perfect black body, the power emitted in space is $P_{\text{out}} = 4\pi R_\oplus^2 T^4$, while the incident power is $P_{\text{in}} = F_{\text{tot}} \pi R_\oplus^2$. Equating P_{out} and P_{in}, we find that the fourth power of the temperature is directly proportional to the incident flux. Since the temperature on Earth was originally $T_0 = 14° = 287\,\text{K}$, it follows that:

$$\left(\frac{T}{T_0}\right)^4 = \frac{F + k}{k} = 1 + \frac{F}{k} = 1 + \frac{L/4\pi d^2}{L_\odot/4\pi d_\oplus^2} = 1 + \frac{L/L_\odot}{d_{\text{au}}^2} , \tag{14.5}$$

where d_{au} is the distance of the supernova, expressed in astronomical units, and $L/L_\odot$ its luminosity, expressed as a multiple of the Sun's luminosity. Isolating d_{au}:

$$d_{\text{max}} = \sqrt{\frac{L/L_\odot}{(T/T_0)^4 - 1}} \approx 3.5 \cdot 10^5\,\text{au} = 1.7\,\text{pc} . \tag{14.6}$$

On average, the number of stars at a distance $d < 1.7\,\text{pc}$ is:

$$N = \frac{4}{3}\pi d_{\text{max}}^3 \cdot n = 2.88 . \tag{14.7}$$

Since the Solar System has a lifetime of $t_\odot = 10^{10}$ years, and there is a supernova every $t = 30$ years among the $N_{\text{gal}} = 10^{11}$ stars that populate our galaxy, the probability that a supernova wipes out life on Earth during the lifetime of the Solar System is:

$$p = \frac{N}{N_{\text{gal}}} \frac{t_\odot}{t} = \frac{2.88}{10^{11}} \frac{10^{10}}{30} = \frac{2.88}{300} \approx 0.01 . \tag{14.8}$$

9. Cosmic Distance Ladder

9.1

From Eq. 9.3, we obtain the distance:

$$d = \frac{1}{\pi''_p} = 3.50 \text{ pc}.$$

We then substitute $F_{10d}/F_d = (d/10d)^2 = 1/100$ in Eq. 8.4, since the intrinsic luminosity is the same. The magnitude of the star at a distance 10 times greater is:

$$m_{10d} - m_d = -2.5 \log \frac{1}{100} \Rightarrow m_{10d} = m_d - 2.5 \cdot (-2) = 0.34 + 5 = 5.34.$$

Because $m_{10d} < 6$, the star is still visible.

9.2

Eq. 9.10 gives the ratio of the intrinsic luminosities of the two galaxies:

$$\frac{L_1}{L_2} = \left(\frac{v_{\text{rot},1}}{v_{\text{rot},2}}\right)^4.$$

Using Eq. 8.5, we obtain the ratio of their fluxes from the apparent magnitudes:

$$\frac{F_1}{F_2} = 10^{-0.4(m_1 - m_2)}.$$

But the flux depends on both the intrinsic luminosity and the distance:

$$\frac{F_1}{F_2} = \frac{L_1}{L_2}\left(\frac{d_2}{d_1}\right)^2.$$

Solving the last equation for d_2/d_1, substituting L_1/L_2 and F_1/F_2 given by the first two equations, we find:

$$\frac{d_2}{d_1} = \left(\frac{v_{\text{rot},2}}{v_{\text{rot},1}}\right)^2 \cdot 10^{-0.2(m_1 - m_2)} \approx 7.$$

9.3

Using Eq. 8.4:

$$\Delta M = (M + \Delta M) - M = -2.5 \log \frac{F_{M+\Delta M}}{F_M} = -2.5 \log \frac{d^2}{(d+\Delta d)^2} = 5 \log\left(1 + \frac{\Delta d}{d}\right).$$

Hence:

$$\frac{\Delta d}{d} = 10^{0.2 \cdot \Delta M} - 1 \approx 25.9\,\%.$$

9.4

Since spiral galaxies are approximately circular, but the image appears elliptic, it is clear that the galactic plane does not coincide with the observation line. Let i be the angle between the perpendicular to the plane of the galaxy and the observation line, and let r be the radius of the galaxy. The length of its apparent semi-major axis is $a = r$, while that of its semi-minor axis is $b = r \cos i$, hence $\cos i = b/a$. The rotational velocity of the galaxy is greater than the observed velocity by a factor of $\sin i$, therefore:

$$v_{\text{rot}} = v_{\text{oss}}/\sin i = v_{\text{oss}}/\sqrt{1 - (b/a)^2} = 366\,\text{km/s}.$$

Using the Tully-Fisher relation in the H band, we find:

$$M_{\rm H} = -9.5 \log v_{\rm rot} + 2.08 \approx -22.3 \,.$$

The distance of the galaxy is:

$$M_{\rm H} - m_h = -5 \log d + 5 \Rightarrow d = 10^{\frac{m_h - M_H}{5} + 1} \approx 10^9 \,\text{pc} = 1000\,\text{Mpc} \,.$$

Since z is small, we can use the non-relativistic formula:

$$zc = H_0\, d_{\rm Mpc} \Rightarrow H_0 = \frac{cz}{1000\,\text{Mpc}} = 60\,\text{km s}^{-1}\text{Mpc}^{-1} \,.$$

9.5

The distance of the galaxy can be obtained from Hubble's law:

$$d = \frac{zc}{H} = 440\,\text{Mpc} \,.$$

Since $z \ll 1$, we use Eq. 8.7, neglecting the decrease in brightness due to redshift:

$$M = m + 5 - 5 \log d \approx -20 \,.$$

We can now compare the luminosity of the galaxy to that of the Sun, using Eq. 8.5:

$$\frac{F}{F_\odot} = 10^{-0.4(M - M_\odot)} \approx 10^{10} \,.$$

The luminosity of this galaxy is 10^{10} times greater than the luminosity of the Sun. Since there is practically no gas and dust in elliptical galaxies, we can assume there is no absorption. Furthermore, assuming that the average mass of stars in the galaxy is equal to the mass of the Sun, and the average luminosity is the luminosity of the Sun, it follows that the mass of all stars in the galaxy is 10^{10} solar masses. Taking into account dark matter, the mass of a galaxy is around 5 times greater than the mass of visible matter (stars, in this case). Therefore, the final estimate of the total mass will be $5 \cdot 10^{10}$ solar masses.

9.6

From Hubble's law (Eq. 9.13), it follows that:

$$v = H_0 d \,,$$

where v is the velocity of an object and d its distance from Earth. Rearranging:

$$\frac{1}{H_0} = \frac{d}{v} = \frac{\text{distance}}{\text{velocity}} = \text{time} \,.$$

Hence, by calculating the reciprocal of Hubble's constant, we can estimate the age of the universe. Using $H_0 = 68\,\text{km}\,\text{s}^{-1}\,\text{Mpc}^{-1}$:

$$\begin{aligned}\tau \approx \frac{1}{H_0} &= \frac{1}{68\,\text{km}\,\text{s}^{-1}\,\text{Mpc}^{-1}} = \frac{1}{68/(206265 \cdot 149.6 \cdot 10^{12})\,\text{Mpc}\,\text{s}^{-1}\,\text{Mpc}^{-1}} \\ &= \frac{206265 \cdot 149.6 \cdot 10^{12}}{68}\,\text{s} = 4.54 \cdot 10^{17}\,\text{s} \\ &= \frac{4.54}{86400 \cdot 365.25} \cdot 10^{17}\,\text{years} = 14.4 \cdot 10^9\,\text{years} \,.\end{aligned}$$

Close enough to the true value of $\tau = 13.7 \cdot 10^9$ years. Since H_0 is known with a certain amount of error, i.e. $H_0 = (68 \pm 5.5)\,\text{km}\,\text{s}^{-1}\,\text{Mpc}^{-1}$, we should take this into account when estimating τ. The relative error on H_0, equal to $5.5/68 = 0.081$, is also the relative error on $1/H_0$, hence the absolute error on τ is $14.4 \cdot 10^9 \cdot 0.081 = 1.2 \cdot 10^9$ years. Therefore:

$$\tau = (14.4 \pm 1.2) \cdot 10^9\,\text{years} \,,$$

which contains the accepted value of $13.7 \cdot 10^9$ years.

9.7

At a distance r, the universe expands at a rate of $v(r) = Hr$. The mass of a spherical shell between r and $r + dr$ is $dm = \rho dV = 4\pi\rho r^2\, dr$. Hence, its kinetic energy is:

$$K = \frac{1}{2}\,dm\,v^2 = \frac{1}{2}(4\pi\rho\, r^2\, dr)(Hr)^2 = 2\pi H^2 \rho\, r^4\, dr\,.$$

The mass inside a sphere of radius r is $m = \rho V = 4/3\,\pi\rho\, r^3$, therefore the potential energy of this shell is:

$$U = -\frac{Gm\,dm}{r} = -\frac{G}{r}\left(\frac{4}{3}\pi\rho\, r^3\right)(4\pi\rho\, r^2\, dr) = -\frac{16}{3}G\pi^2\rho^2 r^4\, dr\,.$$

The total energy is then:

$$E = K + U = 2\pi\rho\left(H^2 - \frac{8}{3}G\pi\rho\right)r^4\, dr\,.$$

Setting $E = 0$, we find the corresponding density:

$$-\frac{8}{3}G\pi\rho_c + H^2 = 0 \Rightarrow \rho_c = \frac{3H^2}{8\pi G} \approx 10^{-26}\,\mathrm{kg/m^3}\,.$$

This is the so-called *critical density* of the universe, which can also be obtained from Einstein's theory of relativity. The critical density is the average density of matter required for the universe to just halt its expansion, but only after an infinite time. A universe with the critical density is said to be *flat*. If the density of matter in the universe is higher than the critical density (*closed universe*), self-gravity slows the expansion until it halts, and ultimately the universe re-collapses. Instead, if the density of matter is lower than the critical density (*open universe*), self-gravity is insufficient to stop the expansion, and the universe continues to expand forever (albeit at an ever decreasing rate). The critical density corresponds to approximately 10 hydrogen atoms per cubic metre. Its measurement is still a matter of debate, but experimental data suggests that our universe is flat.

9.8

The Lyman series results from the transition of an electron in an excited state ($n \geqslant 2$) to the ground state ($n = 1$). Therefore, the lowest wavelength is obtained from the most energetic transition, i.e. from $n = \infty$ to $n = 1$. This energy difference is, by definition, the ionization energy of hydrogen, hence:

$$\lambda_l = \frac{hc}{E_{\mathrm{ion}}} = 91.2\,\mathrm{nm}\,.$$

For the V band, $\lambda_{\max} = 600.5\,\mathrm{nm}$ and $\lambda_{\min} = 501.5\,\mathrm{nm}$. Using $z = \Delta\lambda/\lambda_l$, we find $z = 4.50$–5.58. Using the relativistic Doppler shift (Eq. 7.20), we have $v_{\mathrm{rel}} = (0.937$–$0.953)\,c$, i.e. $v_{\mathrm{rel}} = (281$–$286)\cdot 10^3$ km/s. Applying Hubble's law, the range of distances is 4130–4200Mpc.

9.9

The redshift parameter of the galaxy is $z = (\lambda_E - \lambda_\odot)/\lambda_\odot$. According to Wien's law, the wavelengths of maximum emission are related to the temperatures by $\lambda_E = k/T_E$ and $\lambda_\odot = k/T_\odot$, hence $\lambda_E/\lambda_\odot = T_\odot/T_E$. Therefore, the redshift parameter is:

$$z = \frac{\lambda_E - \lambda_\odot}{\lambda_\odot} = \frac{T_\odot}{T_E} - 1 \approx 0.18\,.$$

The galaxy recedes from us with a velocity of $v = c \cdot z = 5.4 \cdot 10^4$ km/s. By Hubble's law, its distance is $d_g = v/H_0 \approx 760\,\mathrm{Mpc}$. The photons emitted by the galaxy have energy $E = hc/\lambda_\odot$, while those arriving on Earth have energy $E = hc/\lambda_E$, hence the ratio of the emitted to the observed flux is $F_g/F_o = \lambda_E/\lambda_\odot = T_\odot/T_E$. Additionally, we know that the galaxy appears 1000 times fainter than ϵ Eridani, i.e. $F_o/F_E = 1/1000$. We then have:

$$\begin{aligned} \frac{L_g}{L_E} &= \frac{F_g}{F_E}\left(\frac{d_g}{d_E}\right)^2 = \frac{F_g}{F_o}\frac{F_o}{F_E}(d_g\,\pi''_{p,E})^2 \\ &= \frac{5780}{4900}\frac{1}{1000}(760\cdot 10^6\cdot 0.311)^2 \approx 6.6\cdot 10^{13}\,. \end{aligned}$$

The ratio between the luminosity of ϵ Eridani and that of the Sun can be obtained from their apparent magnitudes and parallaxes:

$$\begin{aligned} m_E - m_\odot &= -2.5\log\frac{F_E}{F_\odot} = -2.5\log\frac{L_E}{L_\odot}\left(\frac{d_\odot}{d_E}\right)^2 \\ \Rightarrow \frac{L_E}{L_\odot} &= \left(\frac{d_E}{d_\odot}\right)^2\cdot 10^{-0.4(m_E-m_\odot)} \approx 0.283\,. \end{aligned}$$

Because the galaxy mainly consists of Sun-like stars, the number of stars is equal to the ratio between the intrinsic luminosities of the galaxy and the Sun:

$$N = \frac{L_g}{L_\odot} = \frac{L_g}{L_E}\cdot\frac{L_E}{L_\odot} \approx (6.6\cdot 10^{13})\cdot 0.283 = 1.9\cdot 10^{13}\,.$$

9.10

The ratio of the fluxes during maximum and minimum brightness can be computed from the apparent magnitudes, using Eq. 8.5:

$$\frac{F_1}{F_2} = 10^{-0.4(m_1-m_2)}\,.$$

Since the star is at a constant distance from the Earth, the ratio of the fluxes is equal to the ratio of the intrinsic luminosities, given by Stefan-Boltzmann law:

$$\frac{F_1}{F_2} = \frac{L_1}{L_2} = \frac{4\pi R_1^2\sigma T_1^4}{4\pi R_1^2\sigma T_1^4} \Rightarrow \frac{F_1}{F_2} = \left(\frac{R_1}{R_2}\right)^2\left(\frac{T_1}{T_2}\right)^4.$$

According to Wien's law, $\lambda_1 = k/T_1$ and $\lambda_2 = k/T_2$, hence $T_1/T_2 = \lambda_2/\lambda_1$. Substituting T_1/T_2 and F_1/F_2 into the last equation:

$$\frac{R_1}{R_2} = \left(\frac{\lambda_1}{\lambda_2}\right)^2\cdot 10^{-0.2(m_1-m_2)} \approx 0.89\,.$$

which is the desired result.
After a period T, the surface of the star has covered a distance $R_2 - R_1$, at a velocity of v. Hence, $R_2 - R_1 = vT$, and substituting $R_1 = 0.89R_2$, we find:

$$\begin{aligned} R_2(1+0.89) = vT &\Rightarrow R_2 = 5.76\cdot 10^9\,\mathrm{m} \\ &\Rightarrow R_1 = 0.89R_2 = 5.12\cdot 10^9\,\mathrm{m}\,. \end{aligned}$$

Let $k = 1367\,\mathrm{W/m^2}$ be the solar constant and $m_\odot = -26.74$ the apparent magnitude of the Sun. Then:

$$\frac{F_2}{k} = 10^{-0.4(m_2-m_\odot)} \Rightarrow F_2 = 6.42\cdot 10^{-10}\,\mathrm{W/m^2}\,.$$

When the radius is maximum, the temperature of the star is:

$$T_2 = \frac{k}{\lambda_2} = \frac{0.002897}{649.1\cdot 10^{-9}}\,\mathrm{K} = 4463.1\,\mathrm{K}\,.$$

Hence, its intrinsic luminosity is:

$$L_2 = 4\pi R_2^2\sigma T_2^4 = 9.38\cdot 10^{27}\,\mathrm{W}\,.$$

While the flux is:

$$F_2 = \frac{L_2}{4\pi d^2} \Rightarrow d = R_2 T_2^2\sqrt{\frac{\sigma}{F}} = 34.94\,\mathrm{pc}\,.$$

10. Gravitation and Kepler's Laws

10.1

From Eq. 10.2, we can write:

$$g_\odot = \frac{GM_\odot}{R_\odot^2}, \qquad g_\oplus = \frac{GM_\oplus}{R_\oplus^2},$$

where $R_\odot$, $R_\oplus$ are the radii and $M_\odot, M_\oplus$ the masses of the Sun and Earth, respectively. Dividing the two above equations:

$$\frac{g_\odot}{g_\oplus} = \frac{M_\odot}{M_\oplus}\left(\frac{R_\oplus}{R_\odot}\right)^2 = 27.96\,.$$

10.2

Writing the mass as a function of the radius and density, i.e. $M = 4/3\,\pi r^3\rho$, from Eq. 10.2:

$$g = \frac{4\pi G}{3}\rho r\,.$$

Using the above equation to write g_{SW3} and $g_{♃}$, dividing one by the other:

$$\frac{g_{\mathrm{SW3}}}{g_{♃}} = \left(\frac{\rho_{\mathrm{SW3}}}{\rho_{♃}}\right)\left(\frac{r_{\mathrm{SW3}}}{r_{♃}}\right).$$

Substituting $\rho_{\mathrm{SW3}}/\rho_{♃} = 4$ and $r_{\mathrm{SW3}}/r_{♃} = 1/4$, we finally get:

$$\frac{g_{\mathrm{SW3}}}{g_{♃}} = 1\,.$$

Hence, the gravitational acceleration is the same as Jupiter's.

10.3

Since $\rho = 3m/(4\pi r^3)$, the ratio of the densities can be written as:

$$\frac{\rho_\oplus}{\rho_\odot} = \frac{M_\oplus}{M_\odot}\left(\frac{R_\odot}{R_\oplus}\right)^3.$$

We now need to express all the terms on the RHS as a function of the four known parameters $\theta_\odot, g, T_{\mathrm{sid}}, s$. Let us start with $R_\odot^3/M_\odot$. If $\theta_\odot$ is expressed in radians, then:

$$\theta_\odot \approx \frac{2R_\odot}{d_\oplus} \Rightarrow d_\oplus \approx \frac{2R_\odot}{\theta_\odot},$$

where $d_\oplus$ is the Earth-Sun distance. Substituting $d_\oplus$ in Kepler's third law:

$$T^2 = 4\pi^2\frac{8{R_\odot}^3}{GM_\odot\theta_\odot^3} \Rightarrow \frac{{R_\odot}^3}{M_\odot} = \frac{GT^2{\theta_\odot}^3}{32\pi^2}.$$

From the definition of s, it follows that:

$$360\cdot s = 2\pi R_\oplus \Rightarrow R_\oplus = \frac{180}{\pi}s\,.$$

Using 10.2 to express g:

$$g = \frac{GM_\oplus}{{R_\oplus}^2} \Rightarrow \frac{M_\oplus}{{R_\oplus}^2} = \frac{g}{G}.$$

Hence:

$$\frac{M_\oplus}{{R_\oplus}^3} = \frac{M_\oplus}{{R_\oplus}^2}\cdot\frac{1}{R_\oplus} = \frac{\pi\, g}{180\, s\, G}.$$

Substituting everything back into the first equation:

$$\frac{\rho_\oplus}{\rho_\odot} = \frac{GT^2{\theta_\odot}^3}{32\pi^2}\cdot\frac{\pi\, g}{180\, s\, G} = \frac{T^2{\theta_\odot}^3\, g}{5760\, s\,\pi} \approx 3.91\,.$$

10.4

The velocity in a circular orbit is given by Eq. 10.18:

$$v_c = \sqrt{\frac{GM}{r}} = \sqrt{\frac{GM}{R_\oplus + H}} = 9688.4\,\mathrm{m/s}\,.$$

The orbital period is given by Kepler's third law:

$$T = 2\pi\sqrt{\frac{(R_\oplus + H)^3}{GM}} = 5745.4\,\mathrm{s} = 1^h 35^m 45.4^s\,.$$

10.5

The orbital period of a geostationary satellite is equal to the period of rotation of the Earth about its axis ($T_{\mathrm{sid}} = 86164\,\mathrm{s}$), hence by Kepler's third law:

$$T_{\mathrm{sid}}^2 = \frac{4\pi^2}{GM}{a_{\mathrm{gs}}}^3 \Rightarrow a_{\mathrm{gs}} = \left(\frac{GM}{4\pi^2}T^2\right)^{1/3} \approx 42150\,\mathrm{km}\,.$$

Therefore, the distance from the Earth's surface is 35770 km. The velocity is:

$$v = \sqrt{\frac{GM}{a_{\mathrm{gs}}}} = \sqrt[3]{\frac{2\pi GM}{T_{\mathrm{sid}}}} \approx 3.07 \cdot 10^3\ \mathrm{m/s}$$

Geostationary satellites are often used for telecommunications, since an antenna pointed at a fixed point in the sky is sufficient to exchange data with a satellite. However, there is a delay of at least $35770/(3 \cdot 10^6) \approx 0.01$ seconds, so locally cables are used to transfer data. The geostationary orbit was first proposed by Herman Potočnik in 1928 and popularized by science fiction writer Arthur C. Clarke. In theory, to cover the entire equator, three satellites in coplanar geostationary orbits, distant 120° from each other, would be sufficient.

10.6

On Earth, the sky appears to move from east to west with the same angular velocity as the rotation of the Earth about its axis. Therefore, a satellite moves in the same direction as the apparent rotation of the sky if its orbital period is greater than the period of rotation of the Earth. For a geostationary satellite, the orbital period is exactly equal to the period of rotation of the Earth, and since the angular velocity decreases with the square root of the distance, we require the orbital radius to be greater than the radius of the geostationary orbit. Thus, the minimum orbital radius is $a = a_{\mathrm{gs}}$, where a_{gs} is the radius of the geostationary orbit (see previous exercise). We conclude that the minimum height from the ground is:

$$h_{\min} = a_{g,s} - R_\oplus \approx 35770\,\mathrm{km}\,.$$

10.7

Since a is already expressed in astronomical units, we convert T in years and apply Eq. 10.14:

$$T_{\mathrm{yr}}^2 = \frac{M}{M_\odot}a_{\mathrm{au}}^3 \Rightarrow M = M_\odot\frac{T_{\mathrm{yr}}^2}{a_{\mathrm{au}}^3} = 2.14 \cdot 10^{30}\,\mathrm{kg}\,.$$

10.8

The sum of the distances at perihelion and aphelion is equal to two times the semi-major axis:

$$a = \frac{d_p + d_a}{2} = 16\,\text{au}\,.$$

From Kepler's third law:

$$T_{\text{yr}}^2 = a_{\text{au}}^3 \Rightarrow T = \sqrt{2^{12}}\,\text{years} = 64\,\text{years}\,.$$

To calculate the area swept by the radius vector every year, we divide the area of the ellipse by the orbital period (measured in years), since, according to Kepler's second law, the area swept by the radius vector per unit time is constant. The area of the ellipse is $A = \pi\,a\,b$. To find b, we first calculate c:

$$c = \frac{d_a - d_p}{2} = 15.5\,\text{au}\,.$$

Which gives:

$$b = \sqrt{a^2 - c^2} = 3.97\,\text{au}\,.$$

Hence, the area swept by the radius vector per year is:

$$\frac{\text{d}A}{\text{d}t} = \frac{\pi\,a\,b}{T_{\text{yr}}} = 3.1\,\text{au}^2/\text{year} = 6.9 \cdot 10^{16}\,\text{km}^2/\text{year}\,.$$

10.9

The energy is given by Eq. 10.26:

$$E = -\frac{GmM}{2a}\,.$$

The Sun-Earth distances at aphelion and perihelion are $d_a = a(1+e)$ and $d_p = a(1-e)$, respectively. Denoting with v_a and v_p the velocities at aphelion and perihelion, applying conservation of energy:

$$\frac{1}{2}mv_a^2 - \frac{GmM}{a(1+e)} = -\frac{GmM}{2a} \Rightarrow v_a = \sqrt{\frac{GM}{a}}\sqrt{\frac{1-e}{1+e}}\,,$$

$$\frac{1}{2}mv_p^2 - \frac{GmM}{a(1-e)} = -\frac{GmM}{2a} \Rightarrow v_a = \sqrt{\frac{GM}{a}}\sqrt{\frac{1+e}{1-e}}\,.$$

To check the correctness of our result, let us verify that the angular momentum is indeed conserved. Using Eq. 10.12, we require that:

$$d_p \cdot v_p = d_a \cdot v_a\,.$$

Substituting v_a and v_p obtained previously, and dividing by $\sqrt{GM/a}$:

$$d_p\sqrt{\frac{1+e}{1-e}} = d_a\sqrt{\frac{1-e}{1+e}} \Rightarrow d_p(1+e) = d_a(1-e)\,.$$

But $d_p = a(1-e)$ and $d_a = a(1+e)$, hence the last equation is an identity.

10.10

The idea of this problem is to think of the trajectory of a body in free fall as the borderline case of an elliptical orbit that tends to a straight line (when the eccentricity goes to 1). Assuming that g is constant and using the equations of uniformly accelerated motion is incorrect because, at the distance of $r = 10^5$ km, $g/g_0 = 1/(1 + h/R_\oplus)^2 \approx 0.65$, so the gravitational acceleration varies considerably. At the time of observation, the asteroids are at rest, so they are located at the apogee of their orbits. At the moment of impact, neglecting Earth's radius, the asteroids will be in the focus of the orbit (which coincides with the perigee), therefore they will have travelled exactly half of the orbit. The major axis is equal to the initial distance from the centre of the Earth ($d + R_\oplus$). Hence, by Kepler's third law:

$$T_1 = \pi\sqrt{\frac{[(d_1 + R_\oplus)/2]^3}{GM}} = 1777047\,\mathrm{s}\,,$$

$$T_2 = \pi\sqrt{\frac{[(d_2 + R_\oplus)/2]^3}{GM}} = 1776772\,\mathrm{s}\,.$$

The time difference is $\Delta t = 264$ seconds. The distance along the Earth is then:

$$d = v \cdot \Delta t = \omega R_\oplus \Delta t = 123\,\mathrm{km}\,.$$

We have assumed that the radius of the Earth is small compared to the initial distance. Without this assumption, we need to solve the equations of motion numerically, and we would obtain $T_1 = 1776550\,\mathrm{s}$ and $T_2 = 1776290\,\mathrm{s}$, giving $\Delta t = 260$ seconds. So, even though the time intervals are both smaller by about 500 seconds, their difference does not change substantially. The approximate solution is therefore correct up to 2 km.

10.11

In vacuum, electromagnetic radiation travels at the speed of light c, hence the distance of perihelion and aphelion are, respectively:

$$d_p = (2 \cdot 60\,\mathrm{s}) \cdot c \approx 3.60 \cdot 10^{10}\,\mathrm{m} = 0.24\,\mathrm{au}\,,$$

$$d_a = (39 \cdot 60\,\mathrm{s}) \cdot c \approx 70.1 \cdot 10^{10}\,\mathrm{m} = 4.79\,\mathrm{au}\,.$$

The semi-major axis is then:

$$a_s = \frac{d_a + d_p}{2} = 2.52\,\mathrm{au}\,.$$

The focal distance is equal to the difference between the distances at aphelion and perihelion:

$$c_s = \frac{d_a - d_p}{2} = 2.28\,\mathrm{au}\,.$$

Hence, the eccentricity is:

$$e_s = \frac{c_s}{a_s} = 0.9\,.$$

The period of Seneca is given by Kepler's third law:

$$T_{s,\mathrm{yr}} = \sqrt{a_{s,\mathrm{au}}^3} = 4\,\mathrm{years}\,.$$

10.12

In a circular orbit, the centrifugal and gravitational forces are always equal, ensuring the satellite does not accelerate in the radial direction, so that its orbital radius stays constant. In an elliptical orbit, however, the two forces are different in general, and this causes the body to accelerate radially, moving towards or away from the centre. Let us calculate the radial acceleration of the body at perigee. The gravitational force is:

$$F_g = \frac{GmM}{r^2}\,,$$

where $r = d_p + R_\oplus$. The centrifugal force is:

$$F_c = m\frac{v_p^2}{r} = \left(1 + \frac{6}{1000}\right)^2 \frac{GmM}{r^2},$$

where we have used the fact that the velocity at perigee is 0.6 % greater than the velocity of a circular orbit at the same height, which is just $v = \sqrt{GM/r}$. Expanding the square and neglecting the factor $(6/1000)^2$, we obtain the total force in the radial direction:

$$F_{\text{tot}} = F_c - F_g \approx \frac{12}{1000}\frac{GmM}{r^2} \Rightarrow a_{\text{tot}} = \frac{12}{1000}\frac{GM}{(d_p + R_\oplus)^2} \approx 0.103\,\text{m/s}^2 .$$

The satellite undergoes approximately uniformly accelerated motion, starting with zero velocity at perigee, covering a distance of $s = d - d_p = 70\,\text{km}$ in the radial direction. The desired time is thus:

$$s = \frac{1}{2}a_{\text{tot}}t^2 \Rightarrow t = \sqrt{2s/g_{\text{tot}}} \approx 1200\,\text{s} = 20\,\text{minutes}.$$

10.13

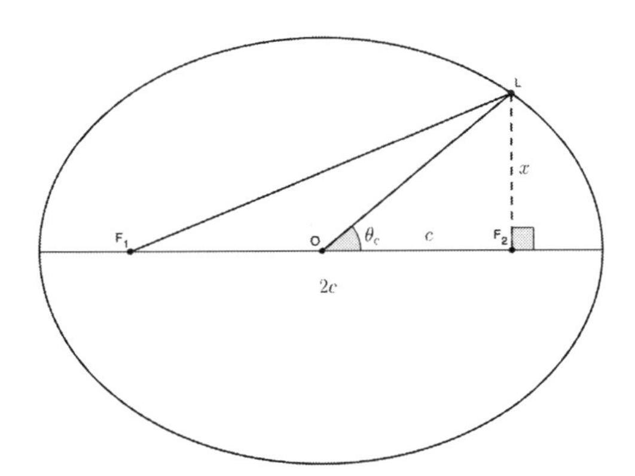

Figure 14.7: The length of the latus rectum can be derived by applying the Pythagorean theorem to the triangle F_1LF_2.

Solution 1. First, we find the length of the latus rectum. Looking at Fig. 14.7, let $\overline{LF_2} = x$. Since $\overline{F_1L} + \overline{F_2L} = 2a$, it follows that $\overline{F_1L} = 2a - \overline{F_2L}$. We also know that $\overline{F_1F_2} = 2c = 2ea$ and $\overline{OF_2} = c = ea$. Applying the Pythagorean theorem to the triangle F_1LF_2.

$$\begin{aligned}
\overline{LF_2}^2 + \overline{F_1F_2}^2 &= \overline{F_1L}^2 \\
x^2 + 4e^2a^2 &= (2a - x)^2 \\
4e^2a^2 &= 4a^2 - 4ax \\
\Rightarrow x &= a(1 - e^2).
\end{aligned}$$

The angle θ_c (in Fig. 14.7) is given by:

$$\tan\theta_c = \frac{\overline{LF_2}}{\overline{OF_2}} = \frac{x}{c} = \frac{1 - e^2}{e}.$$

Using Eq. A.7, the area of the sector of the ellipse between $-\theta_c$ and θ_c is:

$$A = 2A(\theta_c) = a\,b \arctan\left(\frac{a}{b}\tan\theta_c\right) = a\,b \arctan\left(\sqrt{1 - e^2}/e\right) = a\,b \arccos e,$$

where we have used the fact that $b = a\sqrt{1-e^2}$ and $\arctan(\sqrt{1-e^2}/e) = \arccos e$. The total area swept by the radius vector is then:

$$A_s = A - 2A_{\mathrm{OLF_2}} = a\,b\arccos e - a^2 e(1-e^2) = a\,b(\arccos e - e\sqrt{1-e^2})\,.$$

The total area of the ellipse is πab and the orbital time is T, thus Eq. 10.11 gives:

$$t = \frac{A_s}{A_{\mathrm{tot}}}T = \frac{T}{\pi}(\arccos e - e\sqrt{1-e^2})\,.$$

Solution 2. The length of the latus rectum can be found using Eq. 10.32, with $\theta = 90°$, which gives $r = a(1-e^2)$ (as before). Using Eq. 10.33, the eccentric anomaly for the latus rectum is:

$$a(1-e^2) = a(1 - e\cos E) \Rightarrow E = \arccos e\,.$$

The mean anomaly is then given by Eq. 10.34:

$$M = \arccos e - e\sin(\arccos e) = \arccos e - e\sqrt{1-e^2}\,.$$

where we have used $\sin(\arccos e) = \sqrt{1-e^2}$. Since M increases linearly with time and completes a full revolution after a period T, it follows that:

$$t = \frac{M}{\pi}T = \frac{T}{\pi}(\arccos e - e\sqrt{1-e^2})\,.$$

which is the same result obtained previously. Note that this method is generally faster.

10.14

Let a_f be the semi-major axis of Earth's orbit after the catastrophe, M the initial mass of the Sun and m the mass of the Earth. Earth's kinetic energy immediately before and after the catastrophe is the same, because there is no net force that changes its speed. Being proportional to the mass of the Sun, Earth's potential energy is instead halved after the catastrophe. Let us examine the two cases:

Case 1, Earth in aphelion As shown in Ex. 10.9, the velocity at aphelion is obtained by applying conservation of energy:

$${v_a}^2 = \frac{GM}{a}\frac{1-e}{1+e}\,.$$

The sum of the kinetic and potential energies is equal to the total energy of the elliptical orbit with semi-major axis a_f, and is given by Eq. 10.26:

$$\frac{1}{2}m{v_a}^2 - \frac{GM/2m}{a(1+e)} = -\frac{GM/2m}{2a_f}\,.$$

Substituting ${v_a}^2$, given by the first equation:

$$\begin{aligned}
-\frac{GM}{4a_f} &= \frac{GM}{2a}\frac{1-e}{1+e} - \frac{GM}{2a(1+e)}\\
\frac{1}{4a_f} &= \frac{1}{2a}\left(\frac{1}{1+e} - \frac{1-e}{1+e}\right)\\
\frac{1}{a_f} &= \frac{2}{a}\frac{e}{1+e}\\
\Rightarrow a_f &= a\frac{1+e}{2e}\,.
\end{aligned}$$

The period of the new orbit is given by Kepler's third law:

$$T_f{}^2 = \frac{4\pi^2 a_f{}^3}{GM/2}\,.$$

While the initial period, equal to one year, was given by:

$$T^2 = \frac{4\pi^2 a^3}{GM}\,.$$

Dividing the two equations:

$$T_{f,\mathrm{yr}}{}^2 = 2a_{f,\mathrm{au}}{}^3\,,$$

where the factor 2 appears because the mass of the Sun is halved after the catastrophe. Note that a_f can be easily expressed in astronomical units if we divide by a the equation for a_f. We then find:

$$T_f = 118.8\,\text{years}\,.$$

Case 2, Earth in perihelion The velocity of the Earth at perihelion is:

$$v_p{}^2 = \frac{GM}{a}\frac{1+e}{1-e}\,.$$

The energy of the Earth after the catastrophe is:

$$E = \frac{1}{2}mv_p{}^2 - \frac{GMm}{2a(1-e)} = \frac{GMm}{2a}\left(\frac{1+e}{1-e} - \frac{1}{1-e}\right).$$

From which we obtain:

$$E = \frac{GMm}{2a}\frac{e}{1-e}\,.$$

Since E is greater than zero, the Earth will follow an hyperbolic orbit, and its motion will not be periodic.

10.15

On average, a person can jump (lift their centre of mass) by 50 cm on the Earth, so their launch velocity v_l is:

$$mg\Delta h = \frac{1}{2}mv_l^2 \Rightarrow v_l = \sqrt{2gh}\,.$$

The escape velocity from the planet is given by Eq. 10.27:

$$v_{\mathrm{esc}} = \sqrt{\frac{2GM}{R}}\,.$$

Since the density of the planet is the same as the density of the Earth:

$$\frac{M}{M_\oplus} = \left(\frac{R}{R_\oplus}\right)^3.$$

Substituting M in v_{esc}:

$$v_{\mathrm{esc}} = \sqrt{\frac{2GM_\oplus}{R_\oplus{}^3}R^2} \Rightarrow v_{\mathrm{esc}} = \sqrt{\frac{2gR^2}{R_\oplus}}\,,$$

where we have used $g = GM_\oplus/R_\oplus^2$. Finally, setting $v_l = v_{\mathrm{esc}}$:

$$R = \sqrt{hR_\oplus} \approx 2\,\mathrm{km}\,.$$

We have retained only one significant digit, since v_l was an estimate.

10.16

The escape velocity is given by Eq. 10.27:

$$v_{\rm esc} = \sqrt{\frac{2GM}{R}}\,.$$

Assuming that, on average, the stars have the same mass as the Sun, $M = NM_\odot$. Substituting M in the previous equation and solving for N:

$$N = \frac{Rv_{\rm esc}^2}{2GM_\odot} \approx 8\cdot 10^4\,\text{stars}\,.$$

10.17

The Schwarzschild radius for a black hole of mass $M_\odot$ is given by Eq. 10.28:

$$R_s = \frac{2GM_\odot}{c^2} \approx 3\,\text{km}\,.$$

The density is:

$$\rho_m = \frac{M_\odot}{4/3\,\pi {R_s}^3} = \frac{3}{4\pi M_\odot^2}\left(\frac{c^2}{2G}\right)^3 = 1.8\cdot 10^{19}\,\text{kg/m}^3\,.$$

10.18

By symmetry, the gravitational field of a cylinder must point in the radial direction. If we consider a Gaussian surface with cylindrical shape whose axis coincides with that of the planet, the only contribution to the gravitational flux is given by the lateral surface, since the vector area of the base of the cylinder is perpendicular to the gravitational field. Therefore, the total flux of the gravitational field is:

$$\Phi = -2\,\pi\,l\,r\,g(r)\,,$$

where l is the height of the cylinder. From Eq. 10.4:

$$\Phi = -4\pi G M_{\rm int} = -4\pi^2\,G\,\rho\,R^2\,l\,.$$

Equating the two expressions for the flux:

$$g(r) = \frac{2\pi G\rho R^2}{r}\,.$$

In a circular orbit, the centripetal force is equal to the gravitational force:

$$m\frac{v_{c,1}^2}{r} = m\frac{2\pi G\rho R^2}{r} \Rightarrow v_{c,1} = \sqrt{2\pi\,G\,\rho\,R^2} = \sqrt{\frac{3}{2}\frac{GM_\oplus}{R_\oplus}} = 14.72\,\text{m/s}\,,$$

where we have substituted $\rho = M_\oplus/(\frac{4}{3}\pi R_\oplus^3)$ and $R = R_\oplus$, since the density and radius are the same as on Earth. The angular velocity of rotation is also the same as on Earth:

$$\frac{v_{c,1}}{h+R_\oplus} = \omega_e \Rightarrow h = \frac{v_{c,1}}{\omega_e} - R_\oplus = 195.5\cdot 10^3\,\text{km}\,.$$

The second cosmic velocity is infinite. Indeed, the gravitational field of the planet is constant, hence it is never possible to escape from it. As you travel farther and farther away from the planet, which is an infinitely long cylinder, imagine rescaling all distances. The infinite cylinder would still look identical, however your distance from it would be reduced.

10.19

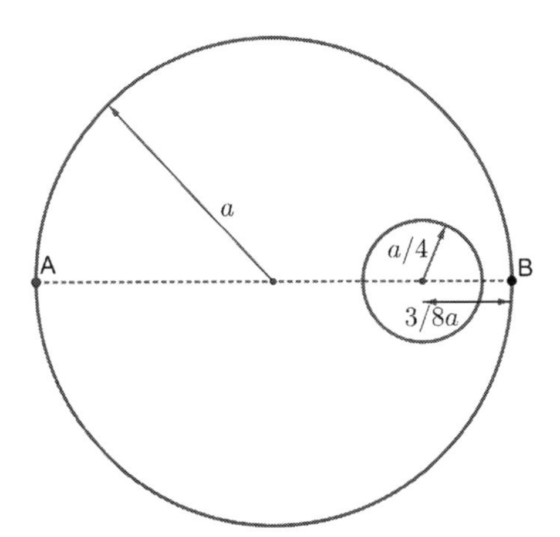

Figure 14.8: The total gravitational field is given by the sum of the gravitational fields of the large sphere and the small sphere, taken with negative mass.

Gravity satisfies the superposition principle, hence, in Fig. 14.8, the gravitational field at each point in space is the sum of the gravitational field produced by the sphere of radius a and by the sphere of radius $a/4$, taken with negative mass. In fact, in considering the gravitational field of the entire sphere a, we are overestimating the total gravitational field, since a portion of radius $a/4$ has been removed. Therefore, the total gravitational field can be calculated by subtracting the contribution of the cavity or, alternatively, by adding the contribution of the cavity, taken with negative mass. The calculation of the gravitational field is particularly simple on the diameter AB since the two contributions point in the same direction, and it is sufficient to subtract their modules. Let ρ be the density of the body and let M, m be the masses of the large and small spheres, respectively. Then:

$$M = \frac{4}{3}\pi\,\rho\,a^3\,, \qquad\qquad m = \frac{1}{48}\pi\,\rho\,a^3\,.$$

Summing the contributions, for $|x| \geqslant a$:

$$g(x) = \frac{GM}{x^2} - \frac{Gm}{(x - 5/8\,a)^2} = \frac{4\pi\rho G a^3}{3}\left[\frac{1}{x^2} - \frac{1}{(8x-5a)^2}\right].$$

Substituting $x_B = a$ and $x_A = -a$, we find:

$$\frac{g_B}{g_A} = \frac{1/9 - 1}{1/169 - 1} = \frac{169}{189}\,.$$

10.20

Since the projectile reaches high altitudes, the gravitational field is constant is not constant.

Part 1. To find the semi-major axis, we compute the total energy energy:

$$E = \frac{1}{2}m{v_0}^2 - \frac{GM_\oplus m}{R_\oplus} = \frac{GM_\oplus m}{2R_\oplus} - \frac{GM_\oplus m}{R_\oplus} = -\frac{GM_\oplus m}{2R_\oplus}\,.$$

Comparing with Eq. 10.26, we conclude that $a = R_\oplus$.

Part 2. We give two solutions: one uses conservation of energy and angular momentum, the other is based on purely geometrical arguments.

Solution 1. Initially, the angle between the instantaneous velocity and the radius vector is $\theta = 120°$. At the highest point of the orbit, the radius vector is perpendicular to the instantaneous velocity (if it weren't, the rocket would be ascending or descending, but then it wouldn't be located in the highest point of the orbit). Applying conservation of energy and angular momentum between the launch point and the point of maximum height:

$$-\frac{GM_\oplus m}{2R_\oplus} = -\frac{GM_\oplus m}{r_{\max}} + \frac{1}{2}mv^2 \Rightarrow v^2 = GM_\oplus\left(\frac{2}{r_{\max}} - \frac{1}{R_\oplus}\right),$$
$$m\,R_\oplus\,v_0\,\sin\theta = m\,r_{\max}\,.$$

Squaring both sides of the last equation and substituting v^2 found in the first equation:

$$R_\oplus{}^2\,\frac{GM_\oplus}{R_\oplus}\left(\frac{\sqrt{3}}{2}\right)^2 = r_{\max}{}^2\cdot GM_\oplus\left(\frac{2}{r_{\max}} - \frac{1}{R_\oplus}\right)$$
$$\Rightarrow \frac{3}{4}R_\oplus = 2r_{\max} - \frac{r_{\max}{}^2}{R_\oplus}\,,$$

which is a quadratic equation:

$$4r_{\max}{}^2 - 8r_{\max}R_\oplus + 3R_\oplus{}^2 \Rightarrow r_{\max} = \frac{4\pm\sqrt{16-12}}{4}R_\oplus\,,$$

with solutions $r_{\max,1} = R_\oplus/2$ (not acceptable since $r_{\max,1} < R_\oplus$) and $r_{\max} = 3R_\oplus/2$. The maximum distance from the surface of the Earth is therefore $h_{\max} = R_\oplus/2$.

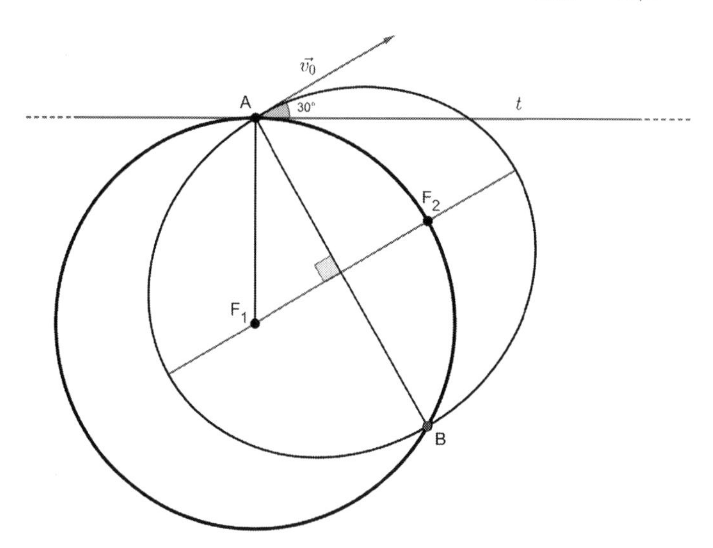

Figure 14.9: Projectile launched at 30° to the horizontal with the first cosmic velocity.

Solution 2. The sum of the distances of a point on the ellipse from the two foci is equal to the major axis. Considering the launch point A in Fig. 14.9, we have:

$$\overline{\mathrm{AF}_1} + \overline{\mathrm{AF}_2} = 2R_\oplus\,,$$

since $a = R_\oplus$. But $\overline{\mathrm{AF}_1} = R_\oplus$, hence $\overline{\mathrm{AF}_2} = \overline{\mathrm{AF}_1} = R_\oplus$. Being equidistant from the foci, A is on the minor axis of the ellipse. The orbit of the projectile is symmetric with respect to the launch and landing points, hence the major axis passes through F_1, F_2 and is perpendicular to AB, which is the minor axis. It follows that $\angle tAF_1 = 90°$, $\angle v_0At = 30°$ and $\angle v_0AB = 90°$ (since AB is the minor axis), hence:

$$\angle BAF_1 = \angle tAF_1 - \angle tAB = \angle tAF_1 - (\angle v_0AB - \angle v_0At) = 90° - (90° - 30°) = 30° .$$

At the same time, $\angle AF_1F_2 = 90° - \angle BAF_1 = 60°$, therefore, by symmetry, the triangle F_1AF_2 is equilateral. The semi-focal distance is then $c = \overline{F_1F_2}/2 = R_\oplus/2$. Thus, the maximum distance from the centre of the Earth is:

$$r_{\max} = 2a - c = 2R_\oplus - \frac{1}{2}R_\oplus = \frac{3}{2}R_\oplus ,$$

as we found earlier.

Part 3. The angle at the centre of the Earth is $\alpha = 2 \cdot \angle AF_1B = 120° = 2\pi/3$ radians. Hence, the range of the projectile is:

$$s = \frac{2\pi}{3}R_\oplus .$$

Part 4. By Kepler's second law, the flight time is given by the ratio of the area swept by the projectile to the total area of the elliptical orbit, multiplied by the time necessary to travel the full orbit. To calculate the total area we need the semi-minor axis:

$$b = \sqrt{a^2 - c^2} = R_\oplus\sqrt{1 - 1/4} = \frac{\sqrt{3}}{2}R_\oplus$$

$$\Rightarrow A_{\text{tot}} = \pi a\, b = \frac{\sqrt{3}\pi}{2}{R_\oplus}^2 .$$

The area swept by the radius vector during the flight is the sum of the area to the right of the minor axis AB, which is just half the are of the ellipse $A_1 = (\sqrt{3}\pi/4) \cdot {R_\oplus}^2$, and the area of triangle F_1AB:

$$A_2 = b \cdot c = \frac{\sqrt{3}}{2}R_\oplus \cdot \frac{1}{2}R_\oplus = \frac{\sqrt{3}}{4}{R_\oplus}^2 .$$

Hence, the area swept by the radius vector is:

$$A_s = A_1 + A_2 = \frac{\sqrt{3}}{4}(\pi + 1){R_\oplus}^2 .$$

The ratio of the areas is:

$$R = \frac{A_s}{A_{\text{tot}}} = \frac{\pi + 1}{2\pi} .$$

Therefore, the flight time is:

$$t = R \cdot T = (\pi + 1)\sqrt{\frac{{R_\oplus}^3}{GM_\oplus}} \approx 3335.7\,\text{s} = 55^m\, 36^s .$$

10.21

As in the previous problem, the energy of the probe is:

$$E = \frac{1}{2}m{v_0}^2 - \frac{GM_\oplus m}{R_\oplus} = \frac{GM_\oplus m}{2R_\oplus} - \frac{GM_\oplus m}{R_\oplus} = -\frac{GM_\oplus m}{2R_\oplus} ,$$

from which it follows that the semi-major axis is $a = R_\oplus$. Looking at Fig. 14.10 (overleaf), it is clear that F_1PF_2E is a square, and PE must be the minor axis of the ellipse. The solution

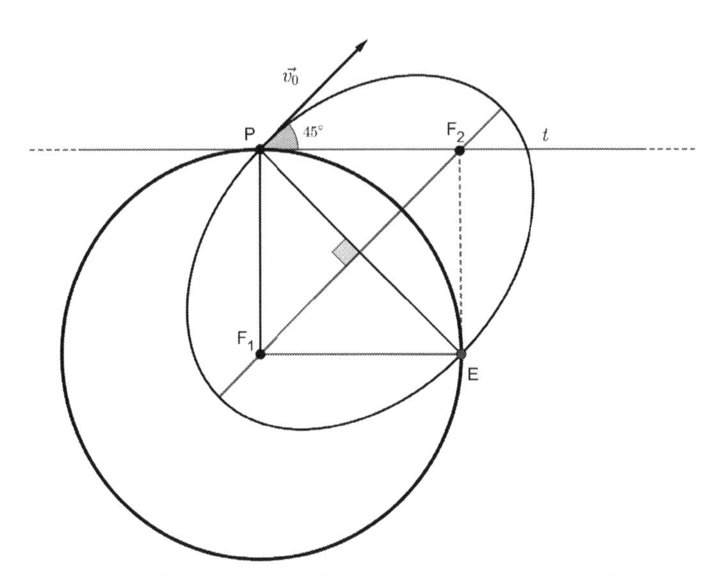

Figure 14.10: Probe launched at 45° to the horizontal from the north pole, landing at the equator.

is similar to the previous problem, only with $\theta_0 = 45°$. Applying conservation of energy and angular momentum between the launching point and the point of maximum height:

$$\frac{1}{2}m{v_0}^2 - \frac{GmM}{R_\oplus} = \frac{1}{2}mv^2 - \frac{GmM}{r_{\max}},$$

$$\frac{1}{\sqrt{2}}R_\oplus v_0 = r_{\max}v\,.$$

Squaring the second equation and substituting v^2 obtained from the first equation:

$$-\frac{1}{2R_\oplus} = \frac{R_\oplus}{4r^2_{\max}} - \frac{1}{r_{\max}},$$

where we used ${v_0}^2 = GM/R_\oplus$. It follows that:

$$\frac{1}{2R_\oplus}{r_{\max}}^2 - r_{\max} + \frac{1}{4}R_\oplus = 0 \Rightarrow r_{\max} = R_\oplus\left(1 \pm \frac{1}{\sqrt{2}}\right).$$

The only acceptable solution is for $r_{\max} \geqslant R_\oplus$, hence:

$$r_{\max} = R_\oplus(1 + 1/\sqrt{2})\,.$$

The maximum height with respect to the surface of the Earth is $h_{\max} = R_\oplus/\sqrt{2}$.
To find the time of flight, we compute the total area swept by the radius vector, which is just the sum of two contributions:

$$A_s = A_{\mathrm{F_1PE}} + A_{\text{half ellipse}} = \frac{1}{2}\left({R_\oplus}^2 + \frac{\pi}{\sqrt{2}}{R_\oplus}^2\right) = \frac{R^2_\oplus}{2}\left(1 + \frac{\pi}{\sqrt{2}}\right).$$

From the second and third Kepler's laws, we conclude that:

$$T^2 = \frac{(1+\frac{\pi}{\sqrt{2}})R^2_\oplus/2}{\frac{\pi}{\sqrt{2}}{R_\oplus}^2}\cdot\frac{4\pi^2{R_\oplus}^3}{GM_\oplus}$$

$$\Rightarrow T = \sqrt{2\pi(\pi+\sqrt{2})\frac{{R_\oplus}^3}{GM_\oplus}} \approx 4320s = 1^h12^m\,.$$

10.22

In the case of minimum speed, the asteroid initially orbits the Sun at the same distance and in the same direction as the Earth. We can therefore assume that the initial relative velocity between the asteroid and the Earth is zero. When the asteroid is close enough to the Earth (at a distance smaller than the Hill sphere, about 1.5 million kilometres), it stars accelerating towards the Earth because of their mutual gravitational attraction.
Since the radius of the Hill sphere is very large, and since the asteroid has a small but non-zero velocity when it enters the Hill sphere, we can very well assume that the asteroid starts at rest, from an infinite distance, and accelerates towards Earth. Then, the impact velocity is exactly equal to the escape velocity from Earth:

$$\frac{1}{2}mv^2 - \frac{GmM_\oplus}{R_\oplus} = 0 \Rightarrow v = \sqrt{\frac{2GM_\oplus}{R_\oplus}} = 11.2\,\mathrm{km/s}\,.$$

If the asteroid hits the Earth at the equator, in the same direction as the rotation of the Earth, the impact velocity will be approximately 0.5 km/s smaller. Hence, the minimum velocity of an asteroid that impacts the Earth is $v_{\min} \approx 10.7$ km/s.
In the case of the maximum velocity, the orbit of the asteroid is a parabola. In fact, since the asteroid belongs to the Solar System, it cannot follow an hyperbolic orbit, therefore its maximum energy is $E = 0$. The direction of the asteroid's orbit must be opposite to the direction of motion of the Earth around the Sun and the vertex of the parabolic orbit must coincide with the collision point. When the asteroid is closest to the Sun, its velocity is:

$$\frac{1}{2}mv_1^2 - \frac{GmM_\odot}{d_\oplus} = 0 \Rightarrow v_1 = \sqrt{\frac{2GM_\odot}{d_\oplus}}\,.$$

Since the orbital velocity of the Earth is $v_\oplus = \sqrt{GM_\odot/d_\oplus}$, at this moment, the relative velocity between the asteroid and the Earth is:

$$v_r = \sqrt{\frac{2GM_\odot}{d_\oplus}} + \sqrt{\frac{GM_\odot}{d_\oplus}} = (1+\sqrt{2})\sqrt{\frac{GM_\odot}{d_\oplus}}\,.$$

As before, we can assume that the asteroids starts at infinity. This time, however, the initial velocity is v_r. Applying conservation of energy:

$$\frac{1}{2}mv^2 - \frac{GmM_\oplus}{R_\oplus} = \frac{1}{2}mv_r^2 \Rightarrow v = \sqrt{(3+2\sqrt{2})\frac{GM_\odot}{d_\oplus} + 2\frac{GM_\oplus}{R_\oplus}} = 72.8\,\mathrm{km/s}\,.$$

If the asteroid hits the Earth at the equator, in the opposite direction as the rotation of the Earth, the impact velocity is 0.5 km/s greater, hence $v_{\max} \approx 73.3$ km/s. Taking into account the eccentricity of Earth's orbit, we obtain an even greater velocity of impact if the Earth is at perihelion. In this case, using $d_\oplus = a(1-e)$ in the equation for v_r, we find $v_r = 72.7$ km/s. Adding in quadrature the contribution of the Earth's gravitational acceleration, we then obtain $v_{\max} = \sqrt{72.7^2 + 11.2^2} + 0.5 \approx 74$ km/s.

10.23

Solution 1. The geometric solution makes use of the fact that, according to Archimedes' theorem, the area of a parabolic segment is 4/3 of the area of the inscribed triangle. Hence, in Fig. 14.11 (overleaf), the area of the parabolic segment, on the left of AB, is 4/3 of the area of triangle AOB. Let p be the distance of the comet at perihelion. Since the orbit is parabolic, $E = 0$, and applying conservation of energy we find the velocity at perihelion $v_p = \sqrt{2GM/p}$. Therefore, the angular momentum of the comet is $L = mpv_p = m\sqrt{2GMp}$.

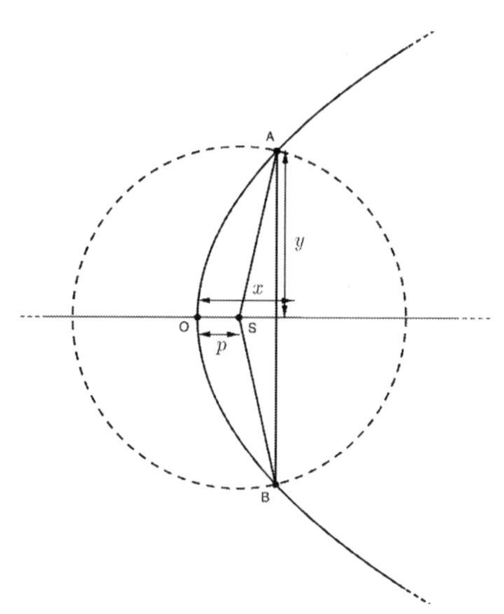

Figure 14.11: The dashed line represents the orbit of the Earth; the heavy solid line, the parabolic orbit of the comet, with closest approach at a distance p from the Sun (S).

From Kepler's second law, the constant rate at which the radius vector sweeps out area is:

$$\frac{\mathrm{d}A}{\mathrm{d}t} = \frac{L}{2m} = \sqrt{\frac{GMp}{2}}\,.$$

Let the x-axis be the axis of symmetry of the parabolic orbit, and let the origin coincide with its vertex (O). The equations of the orbits of the comet and Earth are, respectively:

$$y^2 = 4px\,, \qquad (x-p)^2 + y^2 = 1\,.$$

Substituting $y^2 = 4px$ in the equation of the Earth's orbit:

$$4px + (x-p)^2 = 1 \Rightarrow x^2 + 2px + p^2 - 1 = 0\,,$$

which is a quadratic equation, with the only acceptable ($|x| < 1$) solution $x = 1 - p$. Substituting back into the equation of the parabola, we obtain $y = 2\sqrt{p(1-p)}$. Hence, the area of the inscribed triangle is $A_t = yx = 2\sqrt{p}(1-p)^{3/2}$, and that of the parabolic segment $A_p = 8/3\sqrt{p}(1-p)^{3/2}$. The area swept by the radius vector is then:

$$\begin{aligned} A_s = A_p - A_{\mathrm{ASB}} &= \frac{8}{3}\sqrt{p}\,(1-p)^{3/2} - 2\sqrt{p(1-p)}\,(1-2p) \\ &= \frac{2}{3}\sqrt{p(1-p)}\,[4(1-p) - 3(1-2p)] = \frac{2}{3}\sqrt{p(1-p)}\,(1+2p)\,. \end{aligned}$$

The time taken to cover this area is:

$$\begin{aligned} T_{\mathrm{comet}} &= \frac{A_s}{dA/dt} = \frac{2/3\sqrt{p(1-p)}\,(1+2p)}{\sqrt{GMp/2}} \\ &= \frac{2}{3}\sqrt{\frac{1}{GM}}\sqrt{2(1-p)}\,(1+2p) = \frac{1}{3\pi}\sqrt{2(1-p)}\,(1+2p)\,\text{years}\,, \end{aligned}$$

since $\sqrt{1/GM} = T_{\mathrm{yr}}/2\pi$, where T_{yr} is the duration of one year (remember we set $d_\oplus = 1$). To obtain the maximum time, we require $dT/dp = 0$:

$$2\sqrt{1-p} - \frac{1}{2}\frac{1+2p}{\sqrt{1-p}} = 0 \Rightarrow 4(1-p) = 1 + 2p \Rightarrow p = 1/2\,.$$

Hence:

$$T_{\text{comet, yr}} = \frac{2}{3\pi}\,\text{years} \approx 77\,\text{days}\,.$$

Solution 2. Writing $\dot{r} = dr/dt = v_r$ in Eq. 10.25, we have:

$$E = \frac{1}{2}m\dot{r}^2 + \frac{L^2}{2mr^2} - \frac{GmM}{r}\,.$$

Since the orbit is parabolic, the energy is $E = 0$. The angular momentum can be found by considering the moment when the comet is in perihelion, at a distance p from the Sun. At this point, $\dot{r} = 0$, hence the above equation becomes:

$$0 = \frac{L^2}{2mp^2} - \frac{GmM}{p} \Rightarrow L = m\sqrt{2GMp}\,.$$

Isolating $\dot{r}$ from the first equation, we find:

$$\dot{r} = \frac{\mathrm{d}r}{\mathrm{d}t} = \sqrt{\frac{2}{m}\left(\frac{GmM}{r} - \frac{L^2}{2mr^2}\right)}\,.$$

The time the comet spends inside Earth's orbit is obtained by separating variables and integrating both sides:

$$\int_p^{d_\oplus} \frac{dr}{\sqrt{\frac{2}{m}\left(\frac{GmM}{r} - \frac{L^2}{2mr^2}\right)}} = \int_0^{T/2} dt \Rightarrow T = 2\int_p^{d_\oplus} \frac{dr}{\sqrt{\frac{2}{m}\left(\frac{GmM}{r} - \frac{L^2}{2mr^2}\right)}}\,,$$

where $d_\oplus$ is the radius of Earth's orbit. But $L^2 = 2GMm^2p$, hence:

$$T = 2\int_p^{d_\oplus} \frac{dr}{\sqrt{\frac{2}{m}\left(\frac{GmM}{r} - \frac{GmMp}{r^2}\right)}} = \sqrt{\frac{2}{GM}}\int_p^{d_\oplus} \frac{r}{\sqrt{r-p}}\,dr\,.$$

The above expression can be easily integrated by parts, yielding:

$$\int_p^{d_\oplus} \frac{r}{\sqrt{r-p}}\,dr = \frac{2}{3}d_\oplus^{3/2}\left(1 + \frac{2p}{d_\oplus}\right)\sqrt{1 - \frac{p}{d_\oplus}}\,.$$

The time, as a function of p, is therefore:

$$T = \frac{2\sqrt{2}}{3}\sqrt{\frac{d_\oplus^3}{GM}}\left(1 + \frac{2p}{d_\oplus}\right)\sqrt{1 - \frac{p}{d_\oplus}}\,.$$

Since $T_{\text{yr}} = 2\pi\sqrt{d_\oplus^3/GM}$, we recover the previous solution.

10.24

Suppose we want to produce the maximum gravitational field at the origin of our coordinate system, along the x-axis. An infinitesimal element of the planet, of volume dV and position vector r, which makes and angle ψ with the x-axis, produces a gravitational field of:

$$dg = G\frac{\rho\,dV}{r^2}\,,$$

where ρ is the density of the planet. The component of the gravitational field along the x-axis is $dg_x = dg\cos\psi$, hence the specific contribution of this element is:

$$\frac{\mathrm{d}g_x}{\mathrm{d}V} = G\rho\frac{\cos\psi}{r^2}\,.$$

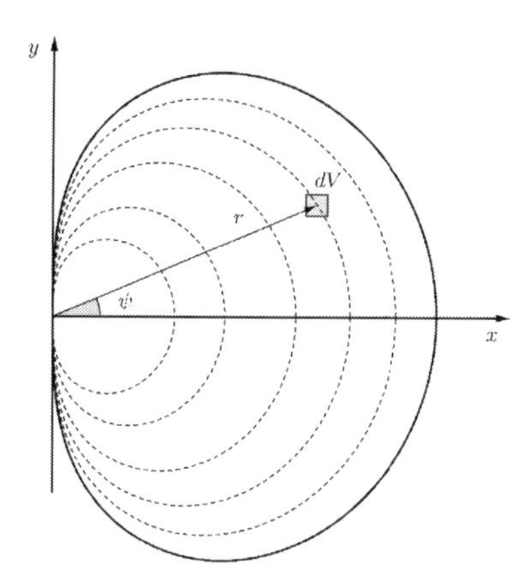

Figure 14.12: The dashed line represents the "level surfaces"; the heavy solid line, the shape of the planet.

The contribution is the same for all parts of the planet for which the fraction $\cos\psi/r^2$ has the same value. Such "level surfaces" (Fig. 14.12) can be described in a polar coordinate system (with the x-axis as the polar axis) by the equation:

$$r(\psi) = a\sqrt{\cos\psi}\,.$$

The larger the value of a, the farther (on average) the surface is from the centre and the smaller its specific contribution. Imagine that the shape of a planet is described by the above equation. Changing its surface in any way must involve moving some element from the interior to the exterior of the planet, which would inevitably lower its contrition to the gravitational field. This proves that $r(\psi) = a\sqrt{\cos\psi}$ is indeed the shape giving the greatest gravitational field at the origin. The value of a can be related to the volume V of the planet. Transforming the polar coordinates (r, ψ) into Cartesian coordinates, we have:

$$x = r\cos\psi = a(\cos\psi)^{3/2}\,, \qquad y = r\sin\psi = a(\cos\psi)^{1/2}\sin\psi\,.$$

In terms of these, the volume of the planet (as a solid of revolution) is:

$$V = \int_0^a \pi y(x)^2\,dx = -\frac{3\pi}{2}a^3\int_{\pi/2}^0 (\cos\psi)^{3/2}\sin^3\psi\,d\psi = \frac{4\pi}{15}a^3\,.$$

Expressing a in terms of V, the equation describing the shape of the planet becomes:

$$r(\psi) = \sqrt[3]{\frac{15V}{4\pi}}\sqrt{\cos\psi}\,.$$

The total gravitational field at the origin is obtained by integrating the contribution of all the "level shells", with a ranging from 0 to $\sqrt[3]{15V/(4\pi)}$. These shells have an infinitesimal volume of $dV = d(4\pi/15a^3) = 4/5\pi a^2\,da$, therefore:

$$g_x = \int dg_x = \int \frac{G\rho}{a^2}\,dV = \int_0^{\sqrt[3]{\frac{15V}{4\pi}}} \frac{G\rho}{a^2}\frac{4}{5}\pi a^2\,da = \frac{4}{5}\pi G\rho\sqrt[3]{\frac{15V}{4\pi}} = \frac{4}{5}\sqrt[3]{\frac{15}{4}}\left(\frac{\pi}{V}\right)^{2/3} GM\,.$$

On the other hand, the gravitational field produced by a sphere is $g = GM/R^2$, where $R = \sqrt[3]{3V/(4\pi)}$. Therefore, the ratio of the maximum gravitational field produced by a planet and by a spherical planet with the same mass and density is:

$$\frac{g_{\text{max}}}{g_{\text{sph}}} = \frac{4}{5}\sqrt[3]{\frac{15}{4}}\left(\frac{\pi}{V}\right)^{2/3}\frac{GM}{GM\left(4\pi/3V\right)^{2/3}} = \frac{4}{5}\sqrt[3]{\frac{135}{64}} \approx 1.026\,.$$

The maximum gravitational field is only 2.6% greater than that of a simple sphere.

10.25

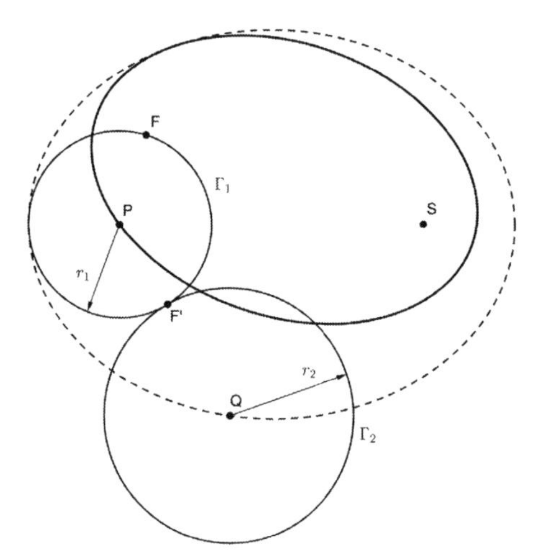

Figure 14.13: The dashed line represents the envelope; the heavy solid line, the orbit of a piece of debris.

Since the fragments fly off with the same velocities and, at the moment of impact, they are all at the same distance $\overline{\text{SP}}$ from the Sun, it follows that they have the same energies, hence the major axes of their orbits have a common value of $2a$. In Fig. 14.13, the heavy solid line represents the possible orbit for a piece of debris; it passes through P and has the Sun (S) as one of its foci. The situation has cylindrical symmetry about the line SP, therefore it is sufficient to work in an arbitrary plane that contains it. Since, for any point on an ellipse, the sum of the distances from the two foci is equal to the length of the major axis, it follows that the distance of the second focus (F), as measured from P, is also fixed: it lays on a circle (Γ_1) with centre in P and radius $r_1 = 2a - r$.
Consider now an arbitrary point Q on the plane. If an elliptical trajectory passes through Q, its focus must be on a circle (Γ_2), centred in Q, with radius $r_2 = 2a - \overline{\text{SQ}}$. Therefore, the second focus (F) must belong to both Γ_1 and Γ_2. Three cases are now possible:

1. if circle Γ_2 intersects circle Γ_1, then point Q is on two different "eligible" ellipses, and the two intersection points give the second foci of the two corresponding trajectories.
2. if circle Γ_1 and Γ_2 have no common point, then none of the possible orbits of the debris pieces can pass through Q; hence, Q lies outside the required envelope.
3. if circles Γ_1 and Γ_2 are tangent (at point F′ in Fig. 14.13), only a single elliptical trajectory passes through Q; this means that Q lies on the required envelope.

In the third case, it follows that:

$$\overline{\text{PQ}} = r_1 + r_2 = (2a - \overline{\text{SP}}) + (2a - \overline{\text{SQ}}), \tag{14.9}$$

that is:

$$\overline{\text{PQ}} + \overline{\text{SQ}} = 4a - \overline{\text{SP}} = \text{const.} \tag{14.10}$$

Therefore, the points of the envelope themselves lie on an ellipse (the dashed line in Fig. 14.13), with foci at P and S. Because of the cylindrical symmetry, the three-dimensional envelope of the orbits of the pieces of debris is a spheroid (ellipsoid of revolution), with major axis of length $4a - \overline{\text{SP}}$, and foci at P and S.

11. Motion of the Planets

11.1

For a superior planet we use Eq. 11.3:

$$T_{\rm syn} = \frac{T_{\rm yr}T_p}{T_p - T_{\rm yr}} = 398.85\,\text{days}\,.$$

Dividing numerator and denominator by $T_{\rm yr}$, we find:

$$\frac{T_{p,\rm yr}}{T_{p,\rm yr} - 1} = \frac{398.85}{325.25}\,.$$

Hence:

$$T_{p,\rm yr} = \frac{398.85/325.25}{398.85/325.25 - 1} = 11.87\,\text{years}\,,$$

which is Jupiter's orbital period

11.2

We don't know whether this is a superior or inferior planet, hence we consider two cases. For an inferior planet, we use Eq. 11.2:

$$\frac{T_{\rm yr}T_p}{T_{\rm yr} - T_p} = 584 \Rightarrow T_{p,\rm yr} = \frac{584/365.25}{1 + 584/365.25} = 0.615\,\text{years}\,.$$

For a superior planet, we use Eq. 11.3:

$$\frac{T_{\rm yr}T_p}{T_p - T_{\rm yr}} = 584 \Rightarrow T_{p,\rm yr} = \frac{584/365.25}{584/365.25 - 1} = 2.67\,\text{years}\,.$$

Since there is no planet with an orbital period of 2.67 years, while the orbital period of Venus is exactly 0.615 years, we conclude that the planet under consideration is Venus.

11.3

The orbital periods can be found using Kepler's third law:

$$T_i = 2\pi\sqrt{\frac{(R_\oplus + h_i)^3}{GM_\oplus}} = 5304\,\text{s}\,, \qquad T_s = 2\pi\sqrt{\frac{(R_\oplus + h_s)^3}{GM_\oplus}} = 5548\,\text{s}\,,$$

where i and s are the inferior and superior satellites, respectively. Using Eq. 11.2:

$$T_{\rm syn} = \frac{T_iT_s}{T_s - T_i} = 120600\,\text{s} = 1^d 9^h 30^m\,.$$

11.4

The maximum elongation of an inferior planet is always smaller than 90°. Since the elongation is greater than 90° in this case, it must be a superior planet.

11.5

In the evening, the Sun sets in the west. Planets that rise in the east have an angular distance from the Sun (i.e. elongation) around 180°. Since the elongation of inferior planets is always less than 90°, it must be a superior planet.

11.6

For an inferior planet, we use Eq. 11.2, setting $T_{\rm syn} = T_p$:

$$T_p = \frac{T_p T_{\rm yr}}{T_{\rm yr} - T_p} \Rightarrow T_{\rm yr} - T_p = T_{\rm yr} \Rightarrow T_p = 0\,,$$

which is clearly impossible. For a superior planet, Eq. 11.3 gives:

$$T_p = \frac{T_p T_{\rm yr}}{T_p - T_{\rm yr}} \Rightarrow T_p - T_{\rm yr} = T_{\rm yr} \Rightarrow T_p = 2T_{\rm yr}\,.$$

Hence, it must be a superior planet with an orbital period of two years.

11.7

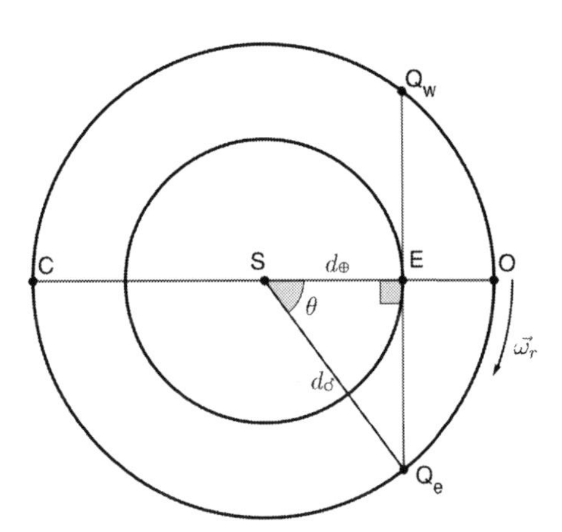

Figure 14.14: Apparent motion of Mars.

The synodic period of Mars can be calculated using Eq. 11.3:

$$T_{\rm syn,yr} = \frac{T_{♂,\rm yr}}{T_{♂,\rm yr} - 1} = 2.135\,\text{years}\,.$$

The angle between eastern quadrature and opposition is (see Fig. 14.14):

$$\theta = \arccos\left(d_\oplus/d_♂\right) = \arccos\left(1/d_{♂,\rm au}\right). \tag{14.11}$$

Let $\omega_r = 2\pi/T_{\rm syn}$ be the angular velocity of Mars. Then, the time taken to arrive in eastern quadrature starting from opposition is:

$$\Delta t = \frac{\theta}{\omega_r} = \frac{\arccos\left(1/d_{♂,\rm au}\right)}{2\pi} T_{\rm syn} \approx 0.29\,\text{years} = 105^d 22^h 8^m\,.$$

11.8

The greatest and the smallest maximum elongations of Mercury are observed when the planet is in aphelion and perihelion, respectively:

$$\frac{d_p}{d_\oplus} = \sin\left(\epsilon_{\max,2}\right), \qquad \frac{d_a}{d_\oplus} = \sin\left(\epsilon_{\max,1}\right).$$

From Eq. 10.7, we have:

$$e = \frac{d_a - d_p}{d_a + d_p} = \frac{\sin\left(\epsilon_{\max,1}\right) - \sin\left(\epsilon_{\max,2}\right)}{\sin\left(\epsilon_{\max,1}\right) + \sin\left(\epsilon_{\max,2}\right)} = 0.206\,,$$

where we neglected the eccentricity of Earth's orbit, since it is 10 times smaller than Mercury's.

11.9

Using Kepler's third law (Eq. 10.14), we estimate the semi-major axis of the asteroid's orbit:

$$a_{\mathrm{au}} = T_{\mathrm{yr}}^{2/3} = 15.21^{2/3} \approx 2.5\,\mathrm{au}\,.$$

From Eq. 8.5, we obtain the ratio of the maximum and minimum fluxes perceived on Earth:

$$\frac{F_{\max}}{F_{\min}} = 10^{-0.4\Delta m} = 10\,.$$

The flux depends not only on the distance of the asteroid from the Sun, but also on the distance between the observer and the asteroid. Let e be the eccentricity of the asteroid's orbit, R its radius and A its albedo. Then, the fluxes at perihelion and aphelion, for an observer on Earth, are:

$$F_p = \frac{L_\odot}{4\pi a^2(1-e)^2}\frac{\pi R^2 A}{2\pi[a(1-e)-d_\oplus]^2}\,,$$
$$F_a = \frac{L_\odot}{4\pi a^2(1+e)^2}\frac{\pi R^2 A}{2\pi[a(1+e)-d_\oplus]^2}\,,$$

where we have assumed that the asteroid only reflects light on the hemisphere in the direction of the incoming rays. A perfect sphere would reflect light isotropically, so in this case 2π should be substituted with 4π. In reality, this coefficient depends on the surface of the asteroid, and could be as small as π. We have also assumed that the orbit of the Earth is a circle. This is a good assumption, since we expect the eccentricity of the orbit of the asteroid to be much greater (as can be verified at the end). Expressing the distances in astronomical units, taking the ratio of the above equations:

$$\frac{F_{\max}}{F_{\min}} = \frac{F_p}{F_a} = \left(\frac{1+e}{1-e}\frac{a_{\mathrm{au}}(1+e)-1}{a_{\mathrm{au}}(1-e)-1}\right)^2 = 10$$
$$\Rightarrow \frac{1+e}{1-e}\frac{a_{\mathrm{au}}(1+e)-1}{a_{\mathrm{au}}(1-e)-1} = \sqrt{10}\,.$$

Rearranging, we obtain a quadratic equation for the eccentricity:

$$e^2 a(1-\sqrt{10}) + e(2a-1)(1+\sqrt{10}) + (a-1)(1-\sqrt{10}) = 0\,,$$

which leads to $e \approx 0.21$ (an order of magnitude larger than the eccentricity of Earth's orbit)

11.10

Because Mars is visible on the meridian when the Sun is setting, it is in eastern quadrature and it rises at midday. Looking at Fig. 14.14, initially the angle formed by Mars (M) with the SO direction (as seen from S) is:

$$\angle\mathrm{OSM} = -\arccos\left(1/d_{♂,\mathrm{au}}\right) = -\arccos\left(1/1.524\right) = -49^\circ\,.$$

Since the sidereal period of Mars is $T_♂ = 1.881$ years, after a year it will have covered an angle of $360^\circ/1.881 = 191.4^\circ$ to the west, as seen from the Sun. Hence, $\angle\mathrm{OSM} = 191.4^\circ - 49^\circ = 142.4^\circ$. To find the angle formed with the CE direction, as seen from E, we can either apply the sine and cosine rules to triangle SEM, or note that:

$$\tan\left(\angle\mathrm{CEM}\right) = \frac{d_{♂,\mathrm{au}}\sin\left(\angle\mathrm{CSM}\right)}{d_{♂,\mathrm{au}}\cos\left(\angle\mathrm{CSM}\right)+1}$$
$$= \frac{d_{♂,\mathrm{au}}\sin\left(180^\circ-\angle\mathrm{OSM}\right)}{d_{♂,\mathrm{au}}\cos\left(180^\circ-\angle\mathrm{OSM}\right)+1}\,.$$

Hence:

$$\angle\text{CEM} = \arctan \frac{\sin(180^\circ - 142.4^\circ)}{\cos(180^\circ - 142.4^\circ) + 1} = 22.8^\circ\,.$$

At the autumnal equinox, the Sun rises exactly at 6 am; therefore Mars, being further west by 22.8°, will rise $22.8^\circ/360^\circ \cdot 24^h \approx 1.5^h$ earlier, i.e. at approximately 4 : 30 am, local time.

12. Orbital Manoeuvres

12.1

Let us compute the desired velocity increments using Eqs. 12.2 and 12.3. In this case, $R_2/R_1 = 5.2$ and $v_0 = \sqrt{GM_\odot/d_\oplus} = 29.787$ km/s, therefore:

$$\Delta v_1 = v_P - v_0 = v_0\left[\sqrt{2\left(1 - \frac{1}{1 + R_2/R_1}\right)} - 1\right] = 8.792\,\text{km/s}\,,$$

$$\Delta v_2 = v_0\sqrt{\frac{R_1}{R_2}}\left[1 - \sqrt{\frac{2}{1 + R_2/R_1}}\right] = 5.643\ \text{km/s}\,.$$

The transfer time can be obtained from Eq. 12.4:

$$T = \sqrt{\frac{\pi^2}{8GM_\odot}(R_1 + R_2)^3} = 2.73\,\text{years}\,.$$

12.2

Summing Eqs. 12.2 and 12.3, setting $\alpha = R_2/R_1$:

$$v_0\left[\frac{1}{\alpha} - 1 + \frac{\sqrt{2}}{\sqrt{1+\alpha}}\left(-\frac{1}{\sqrt{\alpha}} + \sqrt{\alpha}\right)\right] = \Delta v_{\text{tot}}\,,$$

where $v_0 = \sqrt{GM/R_1}$ is the velocity of the initial circular orbit. Solving for the semi-major axis in Eq. 12.4:

$$R_1(1+\alpha) = \sqrt[3]{\frac{8GMT^2}{\pi^2}}\,.$$

Substituting R_1, thus obtained, in the first equation and simplifying:

$$\left[\sqrt{1+\alpha}\left(\frac{1}{\sqrt{\alpha}} - 1\right) + \sqrt{2\alpha} - \sqrt{\frac{2}{\alpha}}\right] = \Delta v_{\text{tot}}\left(\frac{2\sqrt{2}T}{GM\pi}\right)^{1/3}.$$

Inserting the numerical values on the RHS:

$$\sqrt{1+\alpha}\left(\frac{1}{\sqrt{\alpha}} - 1\right) + \sqrt{2\alpha} - \sqrt{\frac{2}{\alpha}} \approx 0.5\,.$$

It can be verified that $\alpha \approx 2$. Hence:

$$R_1 = \frac{1}{1+\alpha}\sqrt[3]{\frac{8GMT^2}{\pi^2}} \approx 0.01\,\text{au}\,,$$

from which $R_2 = \alpha R_1 \approx 0.02\,\text{au}$.

12.3

It is better to fire the engine at take-off. Indeed, looking at the result derived in Pr. 12.4, it is best to maximize the rate of fuel consumption.

12.4

Since the rope is in equilibrium, the centrifugal force F_c due to the rotation of the Earth is equal to the gravitational force F_g. Let λ be the linear density of the rope. Let us consider an infinitesimal element of rope, of length dr and mass $dm = \lambda\, dr$, at a distance r from the centre of the Earth. The gravitational force acting on this element is:

$$dF_g(r) = g(r) \cdot dm = \frac{GM_\oplus \lambda}{r^2}\, dr\,.$$

The gravitational force on the rope is the sum (integral) of the contributions from all infinitesimal elements:

$$F_g = \int_{R_\oplus}^{h+R_\oplus} \frac{GM_\oplus \lambda dr}{r^2} \Rightarrow F_g = GM_\oplus \lambda \Big(\frac{1}{R_\oplus} - \frac{1}{R_\oplus + h} \Big)\,,$$

where h is the total length of the rope. The centrifugal force on an infinitesimal element is:

$$dF_c(r) = \omega^2 r\, dm = \lambda \omega^2 r\, dr\,,$$

where ω is the angular velocity of rotation of the Earth. The total centrifugal force is then:

$$F_c = \int_{R_\oplus}^{R_\oplus + h} \lambda \omega^2 r\, dr \Rightarrow F_c = \frac{1}{2}\lambda\omega^2 \Big(2R_\oplus h + h^2\Big)\,.$$

Setting $F_c = F_g$, we obtain:

$$\Big(\frac{h}{R_\oplus}\Big)^2 + 3\frac{h}{R_\oplus} + 2 - \frac{2GM_\oplus}{\omega^2 R_\oplus^3} = 0\,.$$

Denoting with $g = GM_\oplus / R_\oplus^2$ the gravitational acceleration on the surface of the Earth and solving the quadratic equation in $h/R_\oplus$, we find:

$$\frac{h}{R_\oplus} = \frac{1}{2}\Big[-1 + \sqrt{\frac{2g}{\omega^2 R_\oplus} - 1}\Big] \approx 22.5\,.$$

The rope must be approximately 22.5 times longer than the Earth's radius, which corresponds to a length of 143000 km. The tension is maximum at the point where the gravitational and centrifugal forces acting on an infinitesimal element are the same. Equating dF_g and dF_c, it is clear this happens at the distance of the geostationary orbit $r_{\rm gs} \approx 42000$ km. The maximum tension is then $\sigma_{\rm max}/\rho \approx 4.8 \cdot 10^7$ N m/kg. This is much greater than the maximum sustainable tension of any construction material. For steel, the breaking point is about $2.6 \cdot 10^5$ N m/kg, for carbon it is $1.7 \cdot 10^6$ N m/kg. One possible way to overcome this would be using a rope with a variable cross-section.

12.5

The energetically optimal trajectory is an ellipse, with its perihelion and aphelion tangent to the orbit of the Earth and the planet. Since the probe returned on Earth at the same initial position, the entire flight must have taken an integer number of years. Of course, this is impossible in the case of an inner planet, since it would take less than a year. Therefore, we only need to consider the case of an outer planet. The time of flight is twice the value given in Eq. 12.4:

$$T^2 = \frac{\pi^2}{2GM_\odot}(d_\oplus + d_p)^3\,. \tag{14.12}$$

The length of a year can be written as $T_{\rm yr} = 2\pi\sqrt{d_\oplus^3/GM_\odot}$. Since $T = n\,T_{\rm yr}$, were n is an integer, it follows that:

$$n = \frac{1}{2\sqrt{2}}(1 + \frac{d_p}{d_\oplus})^{3/2}\,. \tag{14.13}$$

Writing $d_{p,\mathrm{au}} = d_p/d_\oplus$, isolating $d_{p,\mathrm{au}}$:

$$d_{p,\mathrm{au}} = 2n^{2/3} - 1\,. \qquad (14.14)$$

Trying every integer, we see that $n = 12$ gives $d_p = 9.48\,\mathrm{au}$, which is exactly the distance of Saturn from the Sun.

12.6

Writing the density as mass divided by volume:

$$\frac{M_p}{4/3\pi R_p^3} = \left(2320 + 3180\frac{R_p}{R_\oplus}\right)\mathrm{kg/m^3} \Rightarrow \frac{M_p}{R_p} = 9718R_p^2 + 13320\frac{R_p^3}{R_\oplus}\,.$$

From Eq. 10.27, the escape velocity from the planet is:

$$v_{\mathrm{esc}} = \sqrt{\frac{2GM_p}{R_p}} = \sqrt{2G\left(9718R_p^2 + 13320\frac{R_p^3}{R_\oplus}\right)}\,. \qquad (14.15)$$

Assuming that friction with the atmosphere is negligible and that the rocket expels all its fuel instantaneously (hence ignoring the effect of gravity, Pr. 12.4), we have:

$$v_{\mathrm{esc}} = \Delta v = v_e \ln\frac{m_i}{m_f} = 4460\cdot\ln\left(\frac{1}{1-0.96}\right) = 14356\,\mathrm{m/s}\,. \qquad (14.16)$$

Substituting for v_{esc}:

$$\begin{aligned}
\sqrt{2G\left(9718R_p^2 + 13320\frac{R_p^3}{R_\oplus}\right)} &= 14356 \\
9718\left(\frac{R_p}{R_\oplus}\right)^2 + 13320\left(\frac{R_p}{R_\oplus}\right)^3 &= \frac{14356^2}{2GR_\oplus^2} \\
9718\left(\frac{R_p}{R_\oplus}\right)^2 + 13320\left(\frac{R_p}{R_\oplus}\right)^3 &= 37979\,,
\end{aligned}$$

which can be solved numerically, giving $R_p = 1.21R_\oplus$. If the expulsion of fuel is not instantaneous, we need to use Eq. 15.18 (see Pr. 12.4). Assuming the burn lasts for 10 minutes, the decrease in Δv is approximately $0.25\,\mathrm{km/s}$. This leads to $R_p = 1.19R_\oplus$. The atmosphere may reduce the Δv by another $2\,\mathrm{km/s}$, giving $R_p = 1.06R_\oplus$.

13. Binary Stars

13.1

Using Eq. A.38, we compute the following Taylor expansions:

$$\begin{aligned}
\cos(\delta + \Delta\delta) &= \cos\delta - \sin\delta\Delta\delta - \cos\delta\,\frac{\Delta\delta^2}{2} + \ldots\,, \\
\sin(\delta + \Delta\delta) &= \sin\delta + \cos\delta\Delta\delta - \sin\delta\,\frac{\Delta\delta^2}{2} + \ldots\,, \\
\cos\Delta\alpha &= 1 - \frac{\Delta\alpha^2}{2} + \ldots\,.
\end{aligned}$$

In Eq. 13.1, the term in square brackets can then be written as:

$$\begin{aligned}
&\cos\delta\left(\cos\delta - \sin\delta\Delta\delta - \cos\delta\frac{\Delta\delta^2}{2}\right)\left(1 - \frac{\Delta\alpha^2}{2}\right) + \sin\delta\left(\sin\delta + \cos\delta\Delta\delta - \sin\delta\frac{\Delta\delta^2}{2}\right) \\
&= 1 - \frac{\Delta\delta^2}{2} - \frac{\Delta\alpha^2}{2}\left(\cos^2\delta - \sin\delta\cos\delta\Delta\delta - \cos^2\delta\frac{\Delta\delta^2}{2}\right).
\end{aligned}$$

If $\delta \approx 90°$, the term in round brackets is an infinitesimal of second order, hence the term containing $\Delta\alpha^2$ can be neglected, since it is an infinitesimal of order higher than $\Delta\delta^2$. Otherwise, the terms $\sin\delta\cos\delta\Delta\delta$ and $-\cos^2\delta\,\Delta\delta^2/2$ inside the round brackets can be neglect, since much smaller than $\cos^2\delta$. Hence, in general:

$$d^2 = d_1^2 + d_2^2 - 2d_1d_2\left[1 - \frac{\Delta\delta^2}{2} - \frac{\Delta\alpha^2}{2}\cos^2\delta\right],$$

proving Eq. 13.2.

13.2

Using Eq. 9.3, we find $d_m = 1/\pi''_{p,m} = 25.40\,\text{pc}$ and $d_a = 1/\pi''_{p,a} = 25.06\,\text{pc}$. The differences in their right ascensions and declinations are $\Delta\alpha = 78^s = 0.00567\,\text{rad}$ and $\Delta\delta = 226'' = 0.0011\,\text{rad}$, respectively. Using Eq. 13.1:

$$d = \sqrt{d_m^2 + d_a^2 - 2d_md_a\left[1 - \frac{\Delta\delta^2}{2} - \frac{\Delta\alpha^2}{2}\cos^2\delta_m\right]} \approx 0.351\,\text{pc} \approx 1.14\,\text{ly}$$

Hence, Mizar and Alcor are close enough to form a binary system, the first ever discovered. Modern telescopes have since found that Mizar is itself a pair of binaries (Mizar Aa, Mizar Ab, Mizar Ba, Mizar Bb), revealing what was once thought of as a single star to be four stars orbiting each other. Alcor has been sometimes considered a fifth member of the system, orbiting far away from the Mizar quadruplet. Recently, astronomers have discovered that Alcor is also a binary, gravitationally bound to the Mizar system. The whole group is therefore a sextuplet.

13.3

Using Eqs. 13.11 and 13.12, we know that, when the larger star covers the smaller entirely:

$$\frac{L_{\min,1}}{L_0} = \frac{{R_1}^2{T_1}^4}{{R_1}^2{T_1}^4 + {R_2}^2{T_2}^4}.$$

Instead, when the smaller star partially eclipses the larger star:

$$\frac{L_{\min,2}}{L_0} = \frac{({R_1}^2 - {R_2}^2){T_1}^4 + {R_2}^2{T_2}^4}{{R_1}^2{T_1}^4 + {R_2}^2{T_2}^4}.$$

Substituting $R_2 = R_1/4$ and $T_2 = T_1/3$, we find:

$$\frac{L_{\min,1}}{L_0} = \frac{1296}{1297},$$
$$\frac{L_{\min,2}}{L_0} = \frac{1216}{1297}.$$

Therefore, the difference in magnitudes is:

$$\Delta m = -2.5\log\frac{L_{\min,1}}{L_{\min,2}} = -2.5\log\frac{1296}{1216} \approx -0.069\,.$$

13.4

The *transit method* consists in simultaneously monitoring the brightness of thousands of stars, with the aim of identifying small periodic variations, which could indicate the presence of orbiting bodies, known as *extrasolar planets*.
Let $R_s \approx R_\odot = 6.96\cdot10^5\,\text{km}$, $R_r = 0.2R_\odot \approx 1.4\cdot10^5\,\text{km}$, $r_g = R_{♃} = 7.0\cdot10^4\,\text{km}$ and $r_t = R_\oplus = 6400\,\text{km}$ be the radii of a Sun-like star, of a red dwarf, of a gas giant and of an

Earth-like planet, respectively. During a transit, the brightness of the system decreases by a factor equal to the ratio between the cross-sectional areas of the orbiting planet and the host star. In the case of a Sun-like star:

$$f_{s,g} = \left(\frac{r_g}{R_s}\right)^2 \approx 0.01 \Rightarrow 1\,\%\,,$$

$$f_{s,t} = \left(\frac{r_t}{R_s}\right)^2 \approx 0.0000845 \Rightarrow 0.0084\,\%\,.$$

While, for a red dwarf:

$$f_{r,g} = \left(\frac{r_g}{R_n}\right)^2 \approx 0.25 \Rightarrow 25\,\%\,,$$

$$f_{r,t} = \left(\frac{r_t}{R_n}\right)^2 \approx 0.00209 \Rightarrow 0.209\,\%\,.$$

The decrease in brightness due to the transit of a gas giant is about 100 times (or 5 magnitudes) greater than that due to an Earth-like planet. Given the limited sensitivity of the tools available (about 0.1 %), it is clear that gas giants are much more likely to be discovered than Earth-like planets, which instead can only be found when orbiting a red dwarf. By measuring the maximum variation in the brightness of the star, and estimating its radius from the spectral class, it is possible to obtain the radius of the planet. Furthermore, using the *radial velocity method*, it is possible to measure the mass of the planet, thus obtaining an estimate of its density. The period of variation of the brightness of the star gives the time of revolution, from which it is then possible to derive the semi-major axis of the orbit. However, it is very rare for the orbital plane of a system to coincide with our line of sight. In fact, if we consider an Earth-like planet orbiting around a Sun-like star, the angle of inclination can be at most $\theta_{\text{rad}} < R_s/d_\oplus \approx 0.00465\,\text{rad}$ for a transit to be observable. The probability of this happening is $2\pi \cdot (2\theta_{\text{rad}})/4\pi = \theta_{\text{rad}}$, that is, only 1 in 200.

13.5

Denote with $\alpha = R_1/R_2$ the ratio of the radii of the larger to the smaller star and with $\beta = T_1/T_2$ the ratio of their temperatures. Using Eqs. 13.11 and 13.12, in conjunction with Eq. 8.5, we find Δ_1 and Δ_2:

$$\Delta_1 = -2.5 \log \frac{L_{\text{min, 1}}}{L_{\text{min, 2}}} = -2.5 \log \frac{\alpha^2\beta^4}{(\alpha^2 - 1)\beta^4 + 1}\,,$$

$$\Delta_2 = -2.5 \log \frac{L_{\text{min, 1}}}{L_0} = -2.5 \log \frac{\alpha^2\beta^4}{\alpha^2\beta^4 + 1}\,.$$

Raising both equations to exponents of 10:

$$\frac{\alpha^2\beta^4}{(\alpha^2 - 1)\beta^4 + 1} = 10^{-0.4\Delta_1}\,,$$

$$\frac{\alpha^2\beta^4}{\alpha^2\beta^4 + 1} = 10^{-0.4\Delta_2}\,.$$

Solve for $\alpha^2\beta^4$ in the second equation:

$$\alpha^2\beta^4 = \frac{10^{-0.4\Delta_2}}{1 - 10^{-0.4\Delta_2}}\,.$$

Solving for β^4 in the first equation, writing it as a function of $\alpha^2\beta^4$:

$$\frac{\alpha^2\beta^4}{\alpha^2\beta^4 + 1 - \beta^4} = 10^{-0.4\Delta_1}$$

$$\Rightarrow \beta^4 = 1 - \alpha^2\beta^4(10^{0.4\Delta_1} - 1)\,.$$

Hence, inserting the expression for $\alpha^2\beta^4$ obtained earlier:

$$\beta^4 = 1 - \frac{10^{-0.4\Delta_2}}{1 - 10^{-0.4\Delta_2}}(10^{0.4\Delta_1} - 1) \Rightarrow \beta \approx 0.533\,.$$

Substituting the numerical value of β in the equation for $\alpha^2\beta^4$, we find α:

$$\alpha^2 = \frac{10^{-0.4\Delta_2}}{1 - 10^{-0.4(\Delta_2 - \Delta_1)}} \Rightarrow \alpha \approx 1.289\,.$$

If we had not know that $R_1 > R_2$, $T_1 < T_2$ we would have had to examine two cases, according to which minimum was the deepest.

13.6

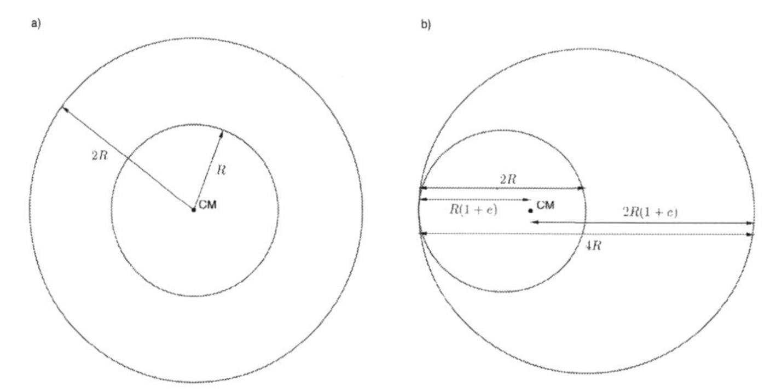

Figure 14.15: a) Initial shape of the orbits, b) smallest eccentricity for which they intersect.

The bodies must be diametrically opposite at all times, relative to their centre of mass, and their distances must obeys the relation $m\,d_m = 2m\,d_{2m}$, i.e. $d_m = 2d_{2m}$. Therefore, their orbits must have coincident apse lines and equal eccentricities. Furthermore, when one body is at minimum distance from the centre of mass, the other must also be at minimum distance, hence their orbits are flipped relative to the minor axis. From Fig. 14.15 b), we see that the orbits intersect if:

$$R(1+e) + 2R(1+e) = 4R \Rightarrow e = \frac{1}{3}$$

13.7

Let's calculate the velocity of the Sun with respect to the centre of mass of the system, using Eq. 13.6 and taking $i = 90°$, since the orbit of the planet is on our line of sight.

$$v_1 = m_2\left[\frac{2\pi G}{(m_1 + m_2)^2 T}\right]^{1/3}.$$

The frequency shift is given by the Doppler formula:

$$\Delta f = \frac{v_1}{c} f = \frac{v_1}{\lambda} = \frac{m_2}{\lambda}\left[\frac{2\pi G}{(m_1 + m_2)^2 T}\right]^{1/3}.$$

In the case of the Sun-Jupiter system, we take $m_1 = M_\odot$ and $m_2 = M_{♃}$. Since $M_{♃} \ll M_\odot$, we can neglect the mass of Jupiter at denominator, obtaining:

$$\Delta f = \frac{M_{♃}}{\lambda}\left[\frac{2\pi G}{M_\odot{}^2 T}\right]^{1/3}.$$

If we observe with a telescope in the visible region of the electromagnetic spectrum, $\lambda \approx 550\,\text{nm}$. The orbital period of Jupiter can be measured from the periodicity with which f varies. Using $T = 11.86\,\text{years} = 3.74 \cdot 10^8\,\text{s}$:

$$\Delta f \approx 2.49 \cdot 10^7\,\text{Hz}\,,$$

which corresponds to a wavelength shift of:

$$\Delta\lambda = \frac{\lambda^2}{c}\Delta f = 2.51 \cdot 10^{-5}\,\text{nm}\,.$$

Hence, the resolution needed is $\lambda/\Delta\lambda \approx 2 \cdot 10^7$.

13.8

Optical binaries Regardless of the orientation of the system, at some point in time the segment connecting the stars will be perpendicular to the line of sight. Since its length is 1 au, and the limiting angle is $0.1''$, such a segment is visible from a distance of 10 pc. At this distance, the apparent magnitude of each star is equal to its absolute magnitude, which is equal to that of the Sun ($M_\odot = 4.83$), and is more than enough for observations with an astrograph. As a result, it will be possible to resolve all binary systems inside a sphere of radius 10 pc. The number of such systems is:

$$N_\text{o} = n\left(\frac{4}{3}\pi {d_\text{o}}^2\right) \approx 4\,\text{stars}\,.$$

Eclipsing binaries Eclipsing systems can be registered with a photometer. Its accuracy is 0.001 magnitudes, so we can assume that only a small partial eclipse is sufficient for registration, i.e. the apparent disks of the stars need only touch each other slightly. In this case, the angle between the line connecting the centres of the stars and the direction to the observer is:

$$\gamma = \frac{2R}{d_\oplus} = 0.0094\,\text{rad}\,.$$

Such eclipses can be observed if the perpendicular to the plane of rotation of the stars lays on a thin ring with angular width 2γ, centred on the plane perpendicular to the direction of the observer. This ring covers an area of $2\pi(2\gamma)$, whilst the area of the whole sphere is 4π, hence the probability that the stars eclipse each other is:

$$p_\text{e} = \frac{4\pi\gamma}{4\pi} = \gamma\,.$$

In addition, the binary system must be observable with a photometer. Since it consists of two Sun-like stars, each with absolute magnitude $M = 4.83$, the total magnitude of the system is:

$$M = -2.5\log 2 \cdot 10^{-0.4M} = M - 2.5\log 2 = 4.83 - 0.75 \approx 4\,.$$

The distance at which the system will have an apparent magnitude of 15 is:

$$M - m = -5\log d_\text{e} + 5 \Rightarrow d_\text{e} = 10^{-0.2(M-m)+1} \approx 1600\,\text{pc}\,,$$

from which we obtain:

$$N_\text{e} = n\gamma\left(\frac{4}{3}\pi {d_\text{e}}^3\right) = 1.6 \cdot 10^5\,\text{stars}\,.$$

Spectroscopic binaries To determine the number of spectroscopic binaries, we find the velocities of the stars about the centre of rotation. From Kepler's third law (Eq. 10.20), the orbital period is:

$$T = 2\pi\sqrt{\frac{d_\oplus^3}{2GM}}\,.$$

Since each star is at a distance $d_\oplus/2$ from the centre of rotation, their linear velocities are:

$$v = \frac{2\pi d_\oplus/2}{T} = \frac{d_\oplus}{2}\sqrt{\frac{2GM}{d_\oplus^3}} = \sqrt{\frac{GM}{2d_\oplus}}\,.$$

The stars travel in opposite direction, hence their maximum relative velocity is twice as high. If the perpendicular to the plane of rotation of the binary system is inclined by an angle i to the direction of the observer, the observed relative speed is:

$$v_{\mathrm{rel}} = 2v\sin i\,.$$

The Doppler shift implies $\Delta\lambda/\lambda = v_{\mathrm{rel}}/c$, hence the resolution limit requires the inclination to be greater than:

$$10^{-5} = \frac{2\sin i}{c}\sqrt{\frac{GM}{2d_\oplus}} \Rightarrow i = \arcsin\left(\frac{c}{2}\sqrt{\frac{2d_\oplus}{GM}}\right)\cdot 10^{-5} = 0.071\,\mathrm{rad} \approx 4^\circ\,.$$

As a result, the area suitable for observation covers almost the whole sphere, with the exclusion of a patch of area $2\pi(2i)$. Hence, the spectroscopic binary can be observed with a probability of:

$$p_{\mathrm{s}} = \frac{4\pi - 4\pi i}{4\pi} = 1 - i = 0.93\,.$$

The magnitude of the system must be smaller than 12, hence the limiting distance is:

$$d_{\mathrm{s}} = 10^{-0.2(M-m)+1} \approx 400\,\mathrm{pc}\,.$$

The number of such systems is therefore:

$$N_{\mathrm{s}} = np_{\mathrm{s}}\left(\frac{4}{3}\pi {d_{\mathrm{s}}}^3\right) = 2.5\cdot 10^5\,\mathrm{stars}\,.$$

This gives an idea of how few binary systems are optical binaries. Instead, the number of eclipsing and spectroscopic binaries are on the same order of magnitude, and are approximately 10^5 more frequent than optical binaries. If the stars eclipse each other, the photometer can be used in conjunction with a spectrometer to better determine the characteristics of the binary.

15

Problem Solutions

1. Celestial Coordinate Systems

1.1 The seasons

The temperature of a location on the surface of the Earth is proportional to the flux incident on it. The flux depends on both the distance to the Sun and on the inclination of the solar rays. Let us examine which of the two effects gives rise to the greatest variation in temperature. Let a be the semi-major axis of the Earth's orbit and e its eccentricity. The maximum and minimum Sun-Earth distances are:

$$d_{\max} = (1+e)a\,, \qquad d_{\min} = (1-e)a\,.$$

Since the flux is proportional to the inverse of the distance squared, the ratio of the fluxes at aphelion and perihelion is:

$$\frac{F_{\mathrm{a}}}{F_{\mathrm{p}}} = \left(\frac{1-e}{1+e}\right)^2 \approx 0.94\,.$$

Let us now examine how the flux varies with the altitude of the Sun. If the Sun's rays form an angle h relative to the ground, those passing through a section of area A (perpendicular to the direction of the rays) are distributed on Earth on a surface of area $A/\sin h$. The angle h is just the altitude of the Sun. The declinations of the Sun during the summer and winter solstices are ϵ and $-\epsilon$, respectively, where $\epsilon = 23\,^\circ 27'$ is the obliquity of the ecliptic. For an observer in the northern hemisphere, the corresponding altitudes of the Sun are:

$$h_{\mathrm{s}} = 90\,^\circ - \phi + \epsilon\,, \qquad h_{\mathrm{w}} = 90\,^\circ - \phi - \epsilon\,.$$

The flux is inversely proportional to the area on which the Sun's rays are distributed:

$$\frac{F_{\mathrm{s}}}{F_{\mathrm{w}}} = \frac{\sin\left(90\,^\circ - \phi + \epsilon\right)}{\sin\left(90\,^\circ - \phi - \epsilon\right)}\,.$$

For intermediate latitudes ($\phi \approx 45\,^\circ$), we find:

$$\frac{F_{\mathrm{s}}}{F_{\mathrm{w}}} = 2.5\,.$$

As the Sun-Earth distance varies, the flux arriving on Earth changes at most by 6 %, while, as the declination of the Sun changes, the flux varies by approximately 250 %, at intermediate latitudes. Of course, the change in temperature is mitigated by the atmosphere. The Earth is close to aphelion during the summer solstice and close to perihelion during the winter solstice. Hence, the winter in the northern hemisphere is milder than the winter in the southern hemisphere, while the summer is slightly warmer.

1.2 Rotation of the Earth

Buildings rotate with the same angular velocity ω as the Earth. While the speed of the Earth's surface is $\omega R_{\oplus} \cos\phi$, the speed of an object launched from the top of a building with height h is $\omega(R_{\oplus} + h)\cos\phi$, where ϕ is the latitude of the place. The relative velocity is therefore $\omega h \cos\phi$ and remains constant during the fall, if we neglect the curvature of the Earth and friction with the air. Since the vertical motion is uniformly accelerated, $h = gt^2/2$, hence the object hits the ground after a time $t = \sqrt{2h/g}$. The horizontal distance travelled is therefore $d = \omega h\sqrt{2h/g}\cos\phi$. A body dropped from the Leaning Tower of Pisa ($h = 57\,\mathrm{m}$, $\phi = 44°$) travels a horizontal distance of $d \approx 1\,\mathrm{cm}$.
The above calculation is not entirely correct, as it assumes that the horizontal speed of the body is constant. In reality, since the object moves tangentially and the Earth is curved, the gravitational force gradually moves away from the initial vertical, acquiring a component in the tangential direction and decelerating the body accordingly. This correction adds a factor of 2/3 in front of the expression previously obtained. We can show this by either considering an inertial or an accelerating frame.
Inertial frame If the ball has moved a distance x to the right, then the westward component of gravity is approximately $g(x/R_{\oplus})$. To the leading order, the distance travelled horizontally is $x = R_{\oplus}\omega t\cos\phi$, so the sideways acceleration is $a = -\omega g t\cos\phi$. Integrating the last expression, using $(R_{\oplus} + h)\omega\cos\phi$ as the initial speed, we obtain the horizontal speed:

$$v(t) = (R + h)\omega\cos\phi - \frac{1}{2}\omega g t^2 \cos\phi$$

Integrating again, we find the horizontal distance as a function of time:

$$x(t) = (R + h)\omega t\cos\phi - \frac{1}{6}\omega g t^3 \cos\phi$$

Subtracting off the distance travelled by the base of the tower:

$$\Delta x = h\omega t\cos\phi - \frac{1}{6}\omega g t^3 \cos\phi$$

For small deflections, we can substitute $t = \sqrt{2h/g}$ in the above equation:

$$\Delta x = h\omega\sqrt{\frac{2h}{g}}\cos\phi - \frac{1}{6}\omega g\frac{2h}{g}\sqrt{\frac{2h}{g}}\cos\phi \Rightarrow \Delta x = \frac{2}{3}h\omega\sqrt{\frac{2h}{g}}\cos\phi$$

Accelerating frame In general, the Coriolis force is given by $F_c = -2m\vec{\omega} \times \vec{v}$. In this case, the angle between $\vec{\omega}$ and $\vec{v}$ is $\pi/2 - \phi$, hence the Coriolis force is directed eastward with magnitude $2m\omega(gt)\cos\phi$, where gt is the vertical speed at time t. Note that the ball is deflected eastward, with acceleration $2\omega g t$, independent of which hemisphere it is in. Integrating the last expression, we obtain the eastward speed (which is initially zero) as a function of time $v(t) = \omega g t^2$. Integrating again, we find a total displacement of $x = \omega g t^3/3$. Substituting $t = \sqrt{2h/g}$, we recover the previous result.

1.3 Foucault's Pendulum

If the pendulum is located at the north pole, the solution is simple: for an inertial observer hovering above the north pole, the pendulum's plane of oscillation appears stationary (because of the conservation of angular momentum), while the Earth below it rotates anticlockwise with a period of T_{sid}. Hence, for an observer at the poles, the plane of the pendulum rotates clockwise with period $T = T_{\mathrm{sid}}$. On the other hand, for an observer at latitude ϕ, the rotation of the Earth can be decomposed into a component pointing along the vertical line, and another pointing along the horizon (Fig. 15.1). The component along the vertical line

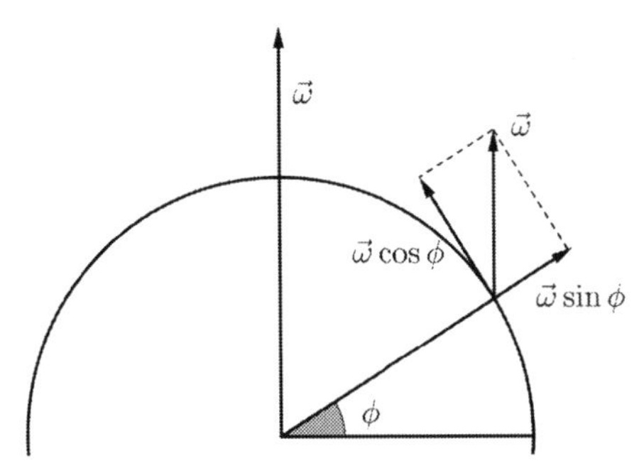

Figure 15.1: Components of the angular velocity of Earth's rotation for an observer at latitude ϕ.

is $\omega_F = \omega \sin\phi$. We are only interested in this component, since the component along the horizon just modifies the effective gravitational acceleration experienced by the pendulum. Hence, since $\omega = 2\pi/T_{\text{sid}}$, we finally get $T = T_{\text{sid}}/\sin\phi$. For an observer on the equator, the pendulum's plane of oscillation appears stationary.

1.4 Dip of the horizon with refraction

The effective curvature of the Earth is simply diminished by the curvature of the ray: the Earth's curvature is $1/R$, the ray's curvature is k/R, and the difference of the curvatures is $R_{\text{eff}} = (1-k)/R$. The value of k mostly depends on the temperature gradient. It turns out that in normal conditions, k ranges between 1/6 to 1/7 at sea level, and is smaller on sunny days or at higher elevations. On the other hand, when there is a strong temperature inversion, k can reach, or exceed, unity. The case $k = 1$ corresponds to horizontal rays that orbit the Earth indefinitely. Values larger than 1 correspond to ducting conditions; if the observer is inside the duct, a pseudo-horizon appears above the astronomical one, a remarkable phenomenon that really is observed.

2. Transformation of Coordinates

2.1 Upper and Lower Culmination

Consider a location at latitude ϕ and a star of declination δ.
For an inertial observer, the Earth's axis is fixed (because of the conservation of angular momentum), and the direction of the star is also fixed, since the celestial equatorial coordinates are independent of time. Looking at Fig. 15.2 (overleaf), we see that the altitudes of the star at lower and upper culmination are:

$$h_{\min} = \delta - (90^\circ - \phi) = \delta + \phi - 90^\circ\,,$$
$$h_{\max} = \delta + (90^\circ - \phi) = 90^\circ - \phi + \delta\,.$$

where, by convention, $h_{\max}$ is measured from the south and $h_{\min}$ from the north.
Now consider the reference system of an observer on Earth. Since the direction of the Earth's axis remains constant, the polar star appears fixed in the sky, while all other stars circle around it. A star with declination δ describes a circle with radius equal to the angle between the polar star and itself, i.e. $90^\circ - \delta$. The polar star forms an angle ϕ with the northern horizon. Looking at Fig. 15.3 (overleaf), the altitudes of lower and upper culmination are:

$$h_{\min} = \phi - (90^\circ - \delta) = \delta + \phi - 90^\circ\,,$$
$$h_{\max} = 180^\circ - \phi - (90^\circ - \delta) = 90^\circ - \phi + \delta\,.$$

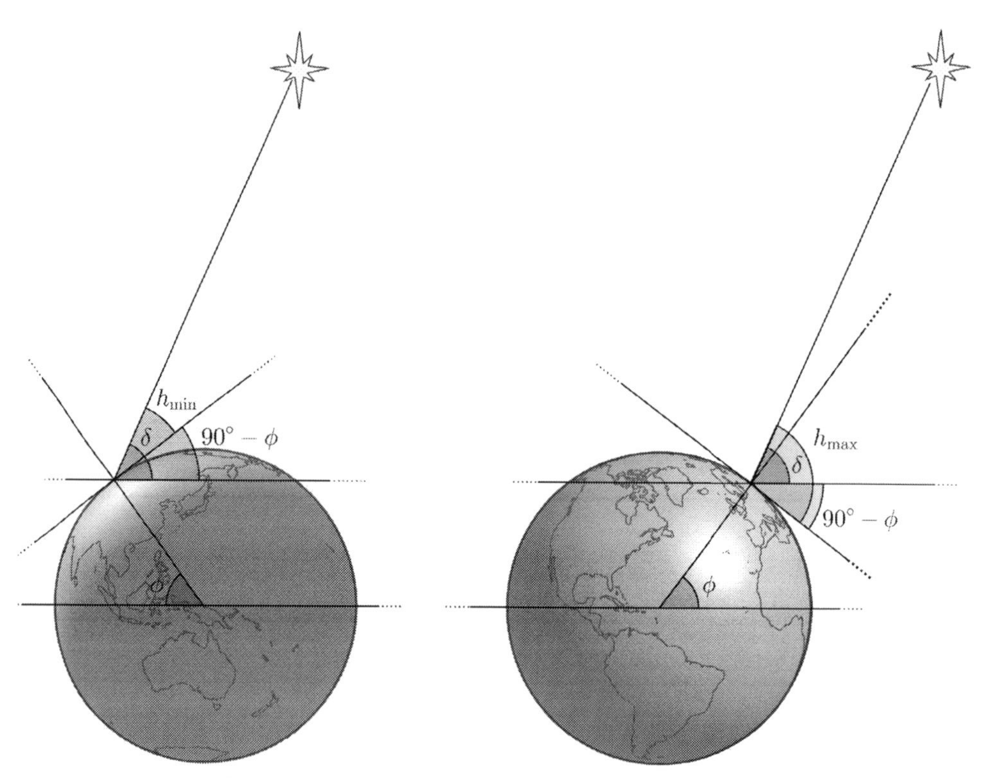

Figure 15.2: Inertial reference system of the Earth.

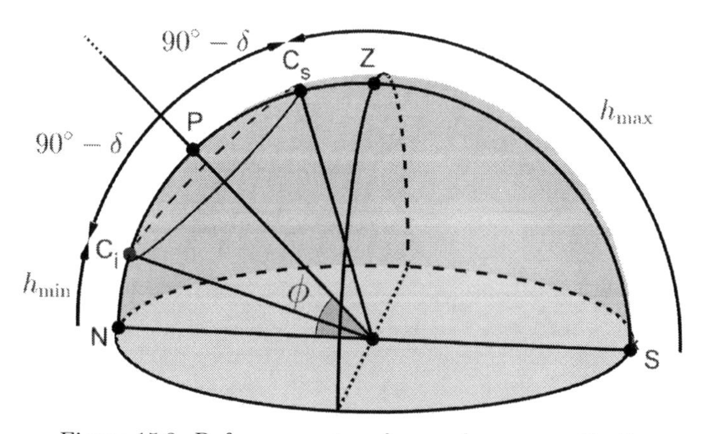

Figure 15.3: Reference system for an observer on Earth.

3. Perturbation of Coordinates

3.1 Precession of the Equinoxes

Let us first calculate the contribution of the Sun. The origin of our coordinate system is the centre of the Earth. Let $\vec{R}$ be the position vector of the Sun, and $\vec{r}$ the position vector of an infinitesimal element of the Earth, with mass dm. The gravitational force between the Sun and the element is:

$$d\vec{F} = \frac{GM}{|\vec{R}-\vec{r}|^3}(\vec{R}-\vec{r})\,dm\,.$$

The torque about the centre of the Earth is then:

$$d\vec{\tau} = \vec{r}\times d\vec{F} = \frac{GM}{|\vec{R}-\vec{r}|^3}(\vec{R}\times\vec{r})\,dm \approx \frac{GM}{R^3}\left(1+3\frac{\vec{R}\cdot\vec{r}}{R^2}\right)(\vec{R}\times\vec{r})\,dm\,,$$

where we have used $\vec{r}\times\vec{r} = 0$ and $(1+x)^n \approx 1+nx$ for $x \ll 1$, with $x = |\vec{r}|/|\vec{R}|$ and $n = 3$. The total torque is obtained by integrating the last equation over the whole ellipsoid. Separating the vector relation into components, we first compute τ_x:

$$\tau_x = \frac{GM}{R^3}\int\left(1+3\frac{r_xR_x+r_yR_y+r_zR_z}{R^2}\right)(r_yR_z - r_zR_y)\,dm\,.$$

When solving the integral, we see that all factors of the form $\int r_x\,dm$ and $\int r_xr_y\,dm$ are equal to zero. Indeed, by symmetry, there are as many elements with $r_x \geqslant 0$ as there are with $r_x \leqslant 0$. Hence, summing (integrating) over the whole ellipsoid, every element simplifies with its symmetric about the yz plane, giving a net zero contribution. The only non-zero factors are those of the form $\int r_x^2\,dm$:

$$\tau_x = \frac{3GM}{R^5}R_yR_z\left[\int r_y^2dm - \int r_z^2dm\right].$$

The integrals $\int r_y^2dm, \int r_z^2dm$ are the moments of inertia about the principal axes y and z, hence:

$$\tau_x = \frac{3GM}{R^5}R_yR_z[I_{yy}-I_{zz}]\,.$$

We then obtain τ_y and τ_z by cyclically permuting x, y, z:

$$\tau_y = \frac{3GM}{R^5}R_zR_x[I_{zz}-I_{xx}]\,, \qquad \tau_z = \frac{3GM}{R^5}R_xR_y[I_{xx}-I_{yy}]\,.$$

It is possible to write the position of the Sun $\vec{R}$ as a function of its ecliptic longitude. Looking at Fig. 15.4 (overleaf), we see that the components of $\vec{R}$ are:

$$\begin{cases} R_x = R\sin\lambda\cos\epsilon \\ R_y = R\cos\lambda \\ R_z = R\sin\lambda\sin\epsilon\,. \end{cases}$$

The torque in the y direction is then:

$$\tau_y = \frac{3GM}{R^3}\sin^2\lambda\sin\epsilon\cos\epsilon[I_{zz}-I_{xx}]\,.$$

Denoting with $\langle\tau_y\rangle$ the time average of τ_y and with $\langle\sin^2\lambda\rangle$ the time average of the square of the sine of the ecliptic longitude, we have:

$$\langle\tau_y\rangle = \frac{3GM}{R^3}\sin\epsilon\cos\epsilon[I_{zz}-I_{xx}]\langle\sin^2\lambda\rangle \Rightarrow \langle\tau_y\rangle = \frac{3GM}{2R^3}\sin\epsilon\cos\epsilon[I_{zz}-I_{xx}]\,,$$

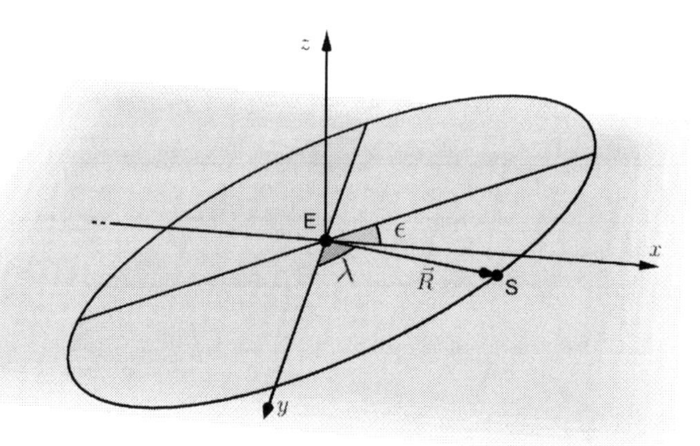

Figure 15.4: The position of the Sun $\vec{R}$ can be written as a function of its ecliptic longitude λ, which can be assumed to increase uniformly throughout a year.

where we have assumed that λ increases linearly with time, i.e. that the Earth's orbit can be approximated as circular, so that:

$$\langle \sin^2 \lambda \rangle = \frac{1}{T} \int_0^T \sin^2 \left(\frac{2\pi}{T} t \right) dt = \frac{1}{2} \,.$$

During a year, the torque is maximum at the solstices and minimum at the equinoxes. From the equation for $\langle \tau_y \rangle$, we see that the average torque throughout a year is equal to the average of the torque at a solstice and at an equinox. We can also show that τ_x, τ_z are proportional to $\sin \lambda \cos \lambda$, but since $\langle \sin \lambda \cos \lambda \rangle = 0$, we have $\langle \tau_x \rangle = \langle \tau_z \rangle = 0$. From the formula for $\langle \tau_y \rangle$, we see that the torque is proportional to M/R^3. Therefore, the ratio of the contribution of the Moon and the contribution of the Sun is:

$$\frac{M_{☾}}{R_{☾}^3} \cdot \frac{R_{\odot}^3}{M_{\odot}} = \frac{M_{☾}}{M_{\odot}} \left(\frac{R_{\odot}}{R_{☾}} \right)^3 \approx 2.17 \,.$$

In fact, this number is slightly different, since the orbit of the Moon is inclined by $i = 5°9' \approx 0.09$ radians with respect to the ecliptic. Therefore, the position of the Moon ranges from $\epsilon + i$ to $\epsilon - i$, and it can be shown that this leads to a correction factor of $1 - 0.09^2 \approx 0.992$. Hence, the combined effect of the Sun and the Moon is equal to 3.15 the effect of the Sun alone. Setting $I_{zz} = I_p$ and $I_{xx} = I_e$, the total torque acting on the Earth is:

$$\tau_{\text{tot}} = 3.15 \cdot \frac{3GM_{\odot}}{2R_{\odot}^3} \sin \epsilon \cos \epsilon \, [I_p - I_e] \,.$$

As a result of this torque, the angular momentum vector of the Earth describes a double cone, in which only the component of the angular momentum perpendicular to the ecliptic changes with time. The value of this component is $L_{\perp} = I_p \, \omega \sin \epsilon$, hence we have:

$$\frac{\mathrm{d}\vec{L}}{\mathrm{d}t} = \vec{\Omega}_{\text{pr}} \times \vec{L} \Rightarrow \langle \tau_{\text{tot}} \rangle = \frac{2\pi}{T_{\text{pr}}} L_{\perp} \,.$$

Isolating T_{pr}:

$$T_{\text{pr}} = \frac{1}{3.15} \cdot \frac{4\pi\omega R_{\odot}^3}{GM_{\odot}^3 \cos \epsilon} \frac{I_p}{I_p - I_e} \approx 25300 \text{ years} \,.$$

This is the precession period of the Earth's axis, with an error of about 2%.
The value of $I_p/(I_p - I_e)$ can be estimated from the eccentricity of the Earth. It is well known

that the moments of inertia of an ellipsoid of revolution along the polar and equatorial axes are:

$$I_p = \frac{2}{5}Ma^2 , \qquad I_e = \frac{1}{5}M(a^2 + b^2) .$$

Hence:

$$\frac{I_p - I_e}{I_e} = \frac{a^2 - b^2}{2a^2} .$$

Since the equatorial and polar radii are $a = R_\oplus = 6378\,\text{km}$ and $b = R_p = 6357\,\text{km}$, respectively; we conclude that $(I_p - I_e)/I_e \approx 0.00335 \approx 1/300$.

4. Observation and Instruments

4.1 Combination of thin lenses

The image formed by the first lens is located at the same position as if the second lens were not present. It serves as the object for the second lens, whose image is the final image of the system. Let p be the distance of the object from the first lens and let f_1, f_2 be the focal lengths of the first and second lenses. Then, according to Eq. 4.6:

$$\frac{1}{p} + \frac{1}{q_1} = \frac{1}{f_1} ,$$

where q_1 is the distance of the image from the first lens. Treating the image formed by the first lens as the object for the second lens, we see that the object distance for the second lens must be $p_2 = -q_1$ (the lenses are in contact and assumed to be infinitesimally thin). Let q be the distance of the final image from the second lens. We then have:

$$-\frac{1}{q_1} + \frac{1}{q} = \frac{1}{f_2} .$$

Summing the two equations, q_1 simplifies, and we are left with:

$$\frac{1}{p} + \frac{1}{q} = \frac{1}{f_1} + \frac{1}{f_2} = \frac{1}{f_{\text{eff}}} \Rightarrow f_{\text{eff}} = \frac{f_1 f_2}{f_1 + f_2} .$$

4.2 Lens maker's equation

Assuming all angles are small in Fig. 15.5 (overleaf), we see that α, β, γ are given by:

$$\alpha = \frac{d}{p}, \quad \beta = \frac{d}{R_1}, \quad \gamma = \frac{d}{q_1} . \tag{15.1}$$

The angles of incidence and refraction are $\theta_i = \alpha + \beta$ and $\theta_r = \beta - \gamma$; hence, by Snell's law:

$$\begin{aligned} n_1\theta_1 &= n_2\theta_2 \\ n_1(\alpha + \beta) &= n_2(\beta - \gamma) \\ n_1 d\Big(\frac{1}{p} + \frac{1}{R_1}\Big) &= n_2 d\Big(\frac{1}{R_1} - \frac{1}{q_1}\Big) . \end{aligned}$$

Therefore, we find:

$$\frac{n_1}{p} + \frac{n_2}{q_1} = \frac{n_2 - n_1}{R_1} .$$

Now, suppose the back of the refractive material has a radius of curvature R_2, and that the overall lens is infinitesimally thin. As in Pr. 4.1, the image formed by R_1 acts as the object for R_2, so $p_2 = -q_1$. Let q be the distance of the final image from R_2, then:

$$-\frac{n_2}{q_1} + \frac{n_1}{q} = \frac{n_1 - n_2}{R_2} .$$

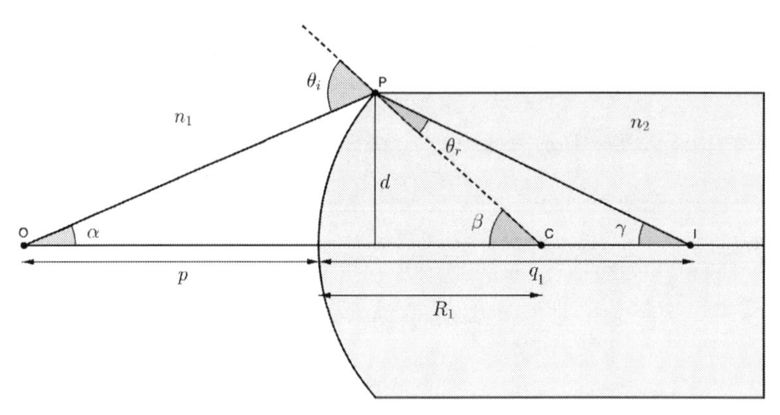

Figure 15.5: Refraction of light rays originating from a point O at a distance p from a spherical surface with radius R_1. The image forms at a distance q_1.

Summing the two above equations, we obtain:

$$\frac{n_1}{p} + \frac{n_1}{q} = (n_2 - n_1)\left(\frac{1}{R_2} - \frac{1}{R_1}\right).$$

Setting $n_1 = 1$ (i.e. the index of refraction of air) and $n_2 = n$, we recover Eq. 4.5:

$$\frac{1}{p} + \frac{1}{q} = (n-1)\left(\frac{1}{R_2} - \frac{1}{R_1}\right).$$

4.3 Airy Disk

The source (S) produces spherical waves with a strength a_S, and is at a distance s from the aperture element at (x, y). The wave arriving at the element is therefore:

$$\psi_1(r) = \frac{a_s e^{iks}}{s}.$$

The aperture, denoted by Σ, can change the amplitude or phase of the incident radiation, and its transmission properties can be described by the aperture function, $h(x, y)$. Usually $h = 0$ or 1 for obstructing or open areas respectively. By *Huygens' principle*, the aperture element can now be considered to act as a source of secondary, spherical, wavelets with a strength and phase given by:

$$a_{\rm ap} \propto \psi_1 h(x, y) dx\, dy.$$

Next, we need to calculate the amplitude of the secondary wavelet reaching the observation point (P), which is a distance r away from the aperture element. This amplitude is given by:

$$d\psi_p \propto a_{\rm ap} \frac{e^{ikr}}{r}.$$

In reality, Huygens' principle is not exact, since the simple idea of spherical wavelets would lead to a backward-propagating wave-front as well as a forward-propagating one, which is not observed experimentally. To fix Huygens' principle, we should multiply the previous equation by the *obliquity factor* $K(\theta)$, which describes the fall-off in intensity of the wavelets with angle θ away from the forward direction. However, assuming angles are small (*Fraunhofer regime*), $K(\theta) \approx 1$. To calculate the total amplitude at point P, we finally sum over all aperture elements:

$$\psi_p \propto \int\int_\Sigma h(x, y) \frac{a_s e^{iks}}{s} \frac{e^{ikr}}{r}\, dx\, dy.$$

Consider the diffraction pattern in a plane at a distance L from the aperture (Fig. 4.9). We denote the coordinates of a point P in this plane by (x_0, y_0). We will assume that the source S is a large distance behind the aperture (and centred on the aperture) so that $s \to \infty$ and the aperture is illuminated with a plane wave at normal incidence. Using the coordinate system shown in Fig. 4.9, we find the distance r from the aperture element at (x, y) to point P:

$$\begin{aligned} r^2 &= L^2 + (x - x_0)^2 + (y - y_0)^2 \\ &= L^2 + x_0^2 + y_0^2 - 2(xx_0 + yy_0) + x^2 + y^2 \\ &= R^2\left(1 - 2\frac{xx_0 + yy_0}{R^2} + \frac{x^2 + y^2}{R^2}\right), \end{aligned}$$

where $R^2 = L^2 + x_0^2 + y_0^2$. Using the binomial expansion, we obtain:

$$r \approx R - \frac{xx_0 + yy_0}{R},$$

where we have assumed that the last term can be neglected, i.e. that:

$$k\frac{x^2 + y^2}{2R} \ll \pi,$$

The wave amplitude at P is then given by the integral:

$$\psi_p \propto \int\int_\Sigma h(x, y) \exp\left\{-ik\frac{xx_0 + yy_0}{R}\right\} dx\, dy .$$

Changing to a polar coordinate system, we substitute $dx\, dy$ with $\rho\, d\rho\, d\phi$ and x, y with $\rho \cos\phi$ and $\rho \sin\phi$, respectively:

$$\psi_p \propto \int_0^{D/2}\int_0^{2\pi} \rho \exp\left\{-ik\rho\frac{x_0 \cos\phi + y_0 \sin\phi}{R}\right\} d\phi\, d\rho .$$

Since we expect the diffraction patter to be circular symmetric, it suffices to study the intensity along a given direction, say x. Then, setting $y_0 = 0$ in the previous equation, and substituting $x_0 = R\sin\xi$, we have:

$$\begin{aligned} \psi_p &\propto \int_0^{D/2}\int_0^{2\pi} \rho e^{-ik\rho \sin\xi \cos\phi} d\phi\, d\rho \\ &= 2\pi \int_0^{D/2} \rho J_0(k\rho \sin\xi)\, d\rho \\ &= \frac{2\pi}{k^2 \sin^2\xi} \int_0^{kD/2 \sin\xi} k\rho J_0(k\rho\sin\xi)\, d(k\rho\sin\xi) \\ &= \frac{2\pi}{k^2 \sin^2\xi} \frac{Dk\sin\xi}{2} J_1\left(\frac{kD\sin\xi}{2}\right). \end{aligned}$$

Which can be written as:

$$\psi_p \propto \frac{J_1(\pi D \sin\xi/\lambda)}{\pi D \sin\xi/\lambda},$$

where we have substituted $k = 2\pi/\lambda$. We know that the first root of $J_1(x)$ is at $x = 3.8317$. Hence, the first minimum in the intensity is at:

$$\frac{\pi D \sin\xi}{\lambda} = 3.8317 \Rightarrow \xi = \frac{3.8317}{\pi}\frac{\lambda}{D}$$

$$\xi \approx 1.2197\frac{\lambda}{D}.$$

As we wanted to show. Since ψ_p must be circular symmetric, its general form is:

$$\psi_p \propto \frac{J_1(\pi D\sqrt{\sin^2\xi + \sin^2\theta}/\lambda)}{\pi D\sqrt{\sin^2\xi + \sin^2\theta}/\lambda}.$$

To calculate the diffraction pattern of an elliptical aperture, let us define x', y', such that $x' = \rho\cos\phi$, $y' = (b/a)\,\rho\sin\phi$ and $x'^2 + y'^2 = a^2/\rho^2$ is a circle of radius a/ρ. The amplitude at point P$=(x_0, y_0)$ on the diffraction screen is:

$$\begin{aligned}
\psi_p(x_0, y_0) &\propto \int\int_\Sigma h(x,y)\exp\Big\{-ik\frac{xx_0 + yy_0}{R}\Big\}dx\,dy \\
&= \int_0^a\int_0^{2\pi}\exp\left\{-ik\rho\frac{x_0\cos\phi + \frac{b}{a}y_0\sin\phi}{R}\right\}\frac{b}{a}\rho\,d\phi\,d\rho \\
&\propto \int_0^a\int_0^{2\pi}\rho\exp\left\{-ik\rho\frac{x_0\cos\phi + \frac{b}{a}y_0\sin\phi}{R}\right\}d\phi\,d\rho
\end{aligned}$$

We can then write:

$$\begin{aligned}
\psi_p\Big(x_0, \frac{a}{b}y_0\Big) &\propto \int_0^a\int_0^{2\pi}\rho\exp\Big\{-ik\rho\frac{x_0\cos\phi + y_0\sin\phi}{R}\Big\}d\phi\,d\rho \\
&= \frac{J_1(\pi a\sqrt{x_0^2 + y_0^2}/(\lambda R))}{\pi a\sqrt{x_0^2 + y_0^2}/(\lambda R)},
\end{aligned}$$

which follows from the circular symmetric solution found earlier. Therefore, substituting $(a/b)\,y_0 \to y_0$:

$$\begin{aligned}
\psi_p(x_0, y_0) &\propto \frac{J_1(\pi a\sqrt{x_0^2 + (b/a)^2y_0^2}/(\lambda R))}{\pi a\sqrt{x_0^2 + (b/a)^2y_0^2}/(\lambda R)} \\
&= \frac{J_1(\pi\sqrt{a^2x_0^2 + b^2y_0^2}/(\lambda R))}{\pi a\sqrt{a^2x_0^2 + b^2y_0^2}/(\lambda R)}.
\end{aligned}$$

Hence, we obtain:

$$\psi_p(x_0, y_0) \propto \frac{J_1(\pi\sqrt{a^2\sin^2\xi + b^2\sin^2\theta}/\lambda)}{\pi a\sqrt{a^2\sin^2\xi + b^2\sin^2\theta}/\lambda}.$$

5. Time System

5.1 Effect of ecliptic obliquity on the equation of time

The ecliptic longitude of the Sun is $\beta_\odot = 0°$. Substituting $\beta_\odot$ in the system of equations 2.12, we have:

$$\begin{cases}
\sin\alpha\cos\delta = \cos\epsilon\sin\lambda \\
\cos\alpha\cos\delta = \cos\lambda \\
\sin\delta = \sin\epsilon\sin\lambda\,.
\end{cases}$$

Dividing the first equation by the second, discarding the first:

$$\begin{cases}
\tan\alpha = \cos\epsilon\tan\lambda \\
\cos\alpha\cos\delta = \cos\lambda \\
\sin\delta = \sin\epsilon\sin\lambda\,.
\end{cases}$$

Differentiating the first equation with respect to time, assuming ϵ is constant:

$$\frac{1}{\cos^2\alpha}\frac{d\alpha}{dt} = \frac{\cos\epsilon}{\cos^2\lambda}\frac{d\lambda}{dt}\,.$$

Substituting $\cos\alpha = \cos\lambda/\cos\delta$, obtained from the second equation:

$$\frac{d\alpha}{dt} = \frac{\cos\epsilon}{\cos^2\delta}\frac{d\lambda}{dt}\,,$$

which is what we wanted to show. This formula shows that the angular velocity of the Sun varies throughout the year because:

- $d\lambda/dt$ is not constant, but is greatest at perihelion and smallest at aphelion;
- $\cos\epsilon/\cos^2\delta$ varies from a minimum of $\cos\epsilon$ when $\delta = 0°$, to a maximum of $1/\cos\epsilon$ when $\delta = \epsilon$.

5.2 Equation of time

Both the energy and the angular momentum are conserved:

$$E = -\frac{GmM}{r} + \frac{1}{2}mv^2\,, \qquad L = m\,r\,v_t = m\,r\,\omega^2\,, \tag{15.2}$$

where v_t is the tangential velocity and $\omega = d\theta/dt$ is the angular velocity. The total velocity v is obtained by solving the equation for the energy:

$$v = \sqrt{\frac{2}{m}\left(E + \frac{GmM}{r}\right)}\,.$$

In an elliptical orbit, the energy is:

$$E = -\frac{GmM}{2a}\,.$$

The angular momentum can be calculated by considering the instant when the body is in perihelion:

$$\begin{aligned}
L = mr_pv_p &= mr_p\sqrt{\frac{2}{m}\left(E + \frac{GmM}{r_p}\right)}\\
&= ma(1-e)\sqrt{\frac{2}{m}\left(-\frac{GmM}{2a} + \frac{GmM}{a(1-e)}\right)}\\
&= m(1-e)\sqrt{GMa}\sqrt{-1 + \frac{2}{1-e}}\\
&= m\sqrt{1-e^2}\sqrt{GMa}\,.
\end{aligned}$$

Thus, the angular velocity is:

$$\begin{aligned}
\frac{d\theta}{dt} = \frac{L}{mr^2} &= m\sqrt{1-e^2}\sqrt{GMa}\,\frac{(1+e\cos\theta)^2}{a^2(1-e^2)^2}\\
&= \sqrt{\frac{GM}{a^3}}\frac{(1+e\cos\theta)^2}{(1-e^2)^{3/2}}\,,
\end{aligned}$$

as we wanted to show. The angular velocity of the Sun on the ecliptic is thus:

$$\frac{d\lambda}{dt} = \sqrt{\frac{GM}{a^3}}\frac{(1+e\cos\theta)^2}{(1-e^2)^{3/2}}\,.$$

From Pr. 5.1, the projection of this motion on the celestial equator is:

$$\frac{d\alpha}{dt} = \frac{\cos\epsilon}{\cos^2\delta}\frac{d\lambda}{dt} \,.$$

The declination of the Sun can be obtained from the ecliptic longitude, using Eq. 2.12c:

$$\sin\delta = \sin\epsilon\sin\lambda \,.$$

Hence, we have:

$$\frac{d\alpha}{dt} = \frac{\cos\epsilon}{1-\sin^2\epsilon\sin^2\lambda}\sqrt{\frac{GM}{a^3}\frac{\left[1+e\cos\left(\lambda-\gamma\right)\right]^2}{(1-e^2)^{3/2}}} \,,$$

where we have substituted $\theta = \lambda - \gamma$. Since $\theta = 0°$ corresponds to the direction of perihelion and $\lambda = 0°$ to the direction of the vernal point, γ is the angular distance between the perihelion and the vernal point. The Earth transits through perihelion on the 5th of January, while the spring equinox is on the 21st of March, i.e. 76 days later. Hence $\gamma = (76/365.25) \cdot 360° \approx 75°$. Denoting with $T_s = 2\pi(d\alpha/dt)^{-1}$, the length of the true solar day is:

$$\frac{1}{T_{\rm sol}} = \frac{1}{T_{\rm sid}} - \frac{1}{T_s} \Rightarrow T_{\rm sol} = \frac{T_{\rm sid}T_s}{T_s - T_{\rm sid}} \,.$$

The difference between the length of a given day and the mean solar day is therefore $T_{\rm sol} - 86400\,\rm s$. For a given day, the equation of time is the sum of all these differences from the 1st of January up to that day. We can then draw the analemma by calculating the altitude and the azimuth of the true Sun at each day of the year, using the system of Eqs. 2.4.
In the following plots, each point represents the position of the Sun seen at a given day. The equation of time and the analemma for an observer at latitude $\phi = 45°$, at time $12:00$, is shown in Figs. 15.6 a) and b), respectively. In Figs. 15.7 a), b) and c) we see the shape of the analemma for an observer at latitude $\phi = 45°$, at observation times $10:00$, $12:00$ and $14:00$, respectively. In Figs. 15.8 a), b) and c) we compare the shape of the analemma, at time $10:00$, for observers at latitudes 90°, 45° and 0°, respectively. Finally, in Figs. 15.9 a), b) and c), we see the equation of time if the eccentricity of Earth's orbit were 10 times smaller, the current value, and 10 times larger, respectively. You can find my code for this exercise on the web page of the book; have fun experimenting with different cases!

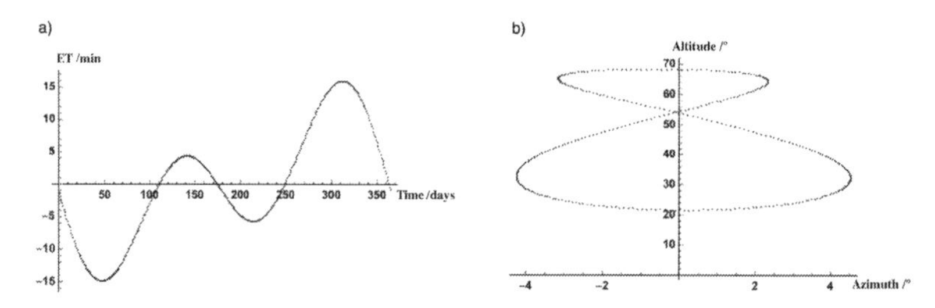

Figure 15.6: a) Equation of time, b) analemma at latitude $\phi = 45°$ and time 12:00.

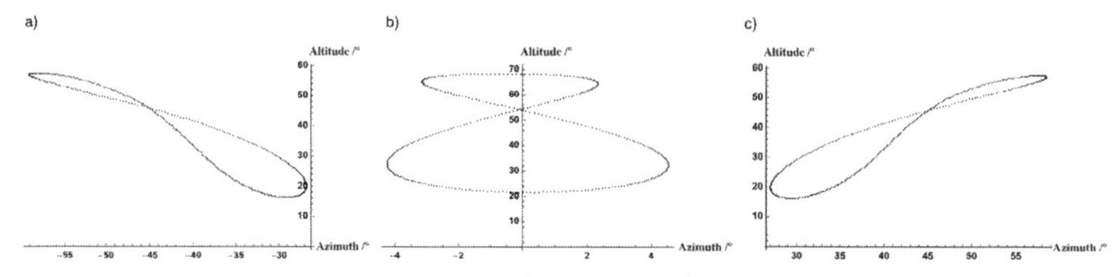

Figure 15.7: Analemma at latitude 45° and times a) 10:00 , b) 12:00 and c) 14:00.

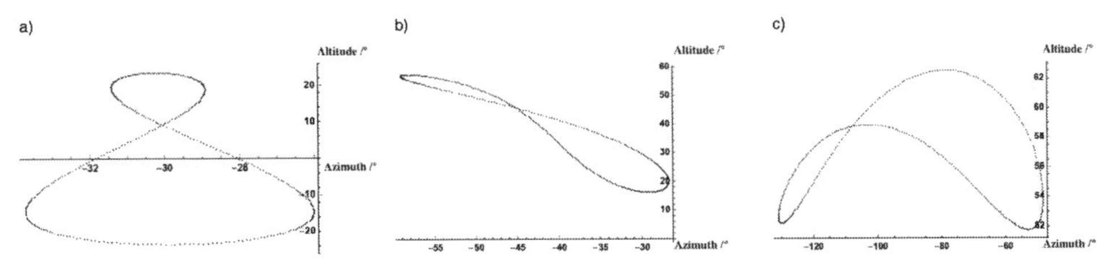

Figure 15.8: Analemma at time 10:00, for observers at latitudes a) 90°, b) 45° and c) 0°.

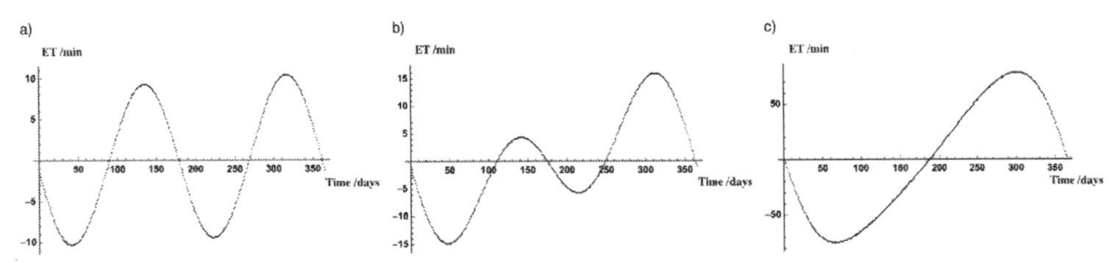

Figure 15.9: The equation of time if the eccentricity were a) 10 times smaller, b) the current value and c) 10 times larger.

6. The Moon

6.1 Meton cycle

Lunar cycles depend on the relative position between the Sun and the Moon. The position of the Sun relative to the Earth varies with a periodicity of one tropical year ($t_{\text{tr}} = 365.24^d$). The position of the Moon is the same, relative to the Earth, every synodic period ($t_{\text{syn}} = 29.53^d$). The length of a Meton cycle is therefore the smallest interval of time that contains an integer number of tropical years and synodic months. For n, m integers, we must have:

$$\frac{x}{29.53} = n\,, \qquad \frac{x}{365.24} = m\,.$$

Dividing the first equation by the second:

$$\frac{n}{m} = 12.36844\,.$$

We try several values of m and check whether n is approximatively integer. Note that varying m is easier than varying n, since m is smaller by one order of magnitude, and we therefore need to try fewer cases. By inspection, we find $m = 19$, which gives $n = 235.0003 \approx 235$. Hence, the Meton cycle is equal to 19 tropical years, or 235 synodic months.

6.2 Anomalistic month

Let $\omega_a = 2\pi/T_a$ and $\omega_{\text{sid}} = 2\pi/T_{\text{sid}}$ be the angular velocities of rotation of the apse line and of the Moon, respectively, where $T_a = 8.8504\,\text{years} = 3232.665\,\text{days}$ and $T_{\text{sid}} = 27.322\,\text{days}$. Since both the apse line and the Moon rotate in the anti-clockwise direction, the relative angular velocity is $\omega_r = \omega_{\text{sid}} - \omega_a$. Then:

$$\frac{1}{T_{\text{an}}} = \frac{1}{T_{\text{sid}}} - \frac{1}{T_a} \Rightarrow T_{\text{an}} = \frac{T_a T_{\text{sid}}}{T_a - T_{\text{sid}}} \approx 27.554\,\text{days}\,.$$

The anomalistic month is 5^h35^m longer than the sidereal month.

6.3 Measuring the distance of the Moon

As already found in Ex. 6.5, the distances between an observer on the Earth and the Moon at zenith or on the horizon are, respectively:

$$d_z = d_☾ - R_\oplus\,, \qquad d_h = \sqrt{d_☾^2 - R_\oplus^2}\,.$$

The angular diameter of the Moon in these two cases is:

$$\tan\alpha_z = \frac{D_☾}{d_z} \Rightarrow \alpha_{z,\mathrm{rad}} \approx \frac{D_☾}{d_☾ - R_\oplus}\,,$$

$$\tan\alpha_h = \frac{D_☾}{d_h} \Rightarrow \alpha_{h,\mathrm{rad}} \approx \frac{D_☾}{\sqrt{d_☾^2 - R_\oplus^2}}\,.$$

Dividing the two equations:

$$\left(\frac{\alpha_{z,\mathrm{rad}}}{\alpha_{h,\mathrm{rad}}}\right)^2 \approx \frac{(d_☾ - R_\oplus)^2}{d_☾^2 - R_\oplus^2}\,.$$

Denoting with $\alpha = \alpha_{z,\mathrm{rad}}/\alpha_{h,\mathrm{rad}} = 237/233$ and solving the second order equation for $d_☾$, we finally obtain:

$$d_☾ = R_\oplus \frac{1+\sqrt{1-(1-\alpha^4)}}{\alpha^2 - 1} \approx 375000\,\mathrm{km}\,,$$

which is 9000 km smaller than the true value.

6.4 Distance of the Moon, another method

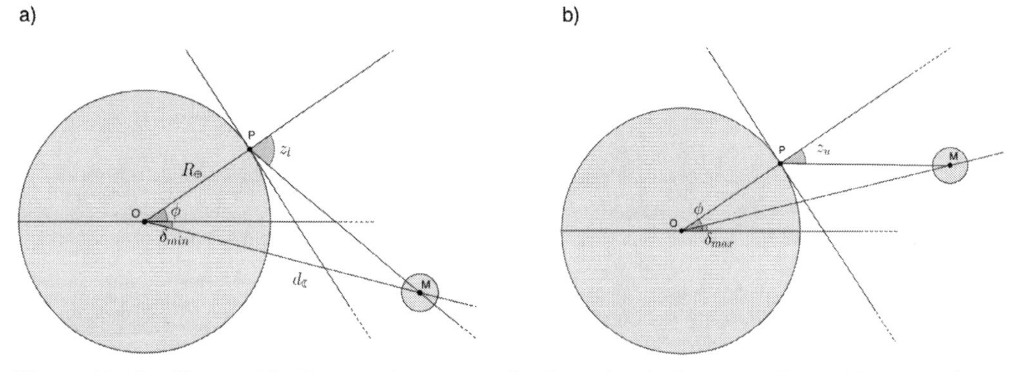

Figure 15.10: The zenith distance is measured when the declination of the Moon is a) minimum and b) maximum, respectively, from a place of latitude ϕ.

The apparent position of the Moon changes because of parallax. In Fig. 15.10 b), consider triangle OPM. We see that $\angle\mathrm{POM} = \phi - \delta_{\max}$ and $\angle\mathrm{OPM} = 180° - z_u$, so it follows that $\angle\mathrm{PMO} = 180° - (\phi - \delta_{\max} + 180° - z_u) = z_u - \phi + \delta_{\max}$. Using the law of sines (Eq. A.13):

$$\frac{\overline{\mathrm{OP}}}{\sin(\angle\mathrm{PMO})} = \frac{\overline{\mathrm{OM}}}{\sin(\angle\mathrm{OPM})}\,.$$

Substituting $\overline{\mathrm{OP}} = R_\oplus$ and $\overline{\mathrm{OM}} = d_☾$, using the reduction formulae (Appendix A.3):

$$\frac{d_☾}{R_\oplus} = \frac{\sin z_u}{\sin(z_u - \phi + \delta_{\max})}\,.$$

Similarly, looking at Fig. 15.10 a), we obtain a second equation for $d_{☾}/R_{\oplus}$. It is sufficient to swap $\delta_{\max}$ with $-|\delta_{\min}|$ and z_u with z_l in the previous formula:

$$\frac{d_{☾}}{R_{\oplus}} = \frac{\sin z_l}{\sin\left(z_l - \phi - |\delta_{\min}|\right)}\,.$$

Since the orbit of the Moon is symmetric with respect to the celestial equator, $\delta_{\max} = |\delta_{\min}| = \delta$. The problem then reduces to solving the system of equations:

$$\begin{cases} \dfrac{d_{☾}}{R_{\oplus}} = \dfrac{\sin z_u}{\sin\left(z_u - \phi + \delta\right)} \\ \dfrac{d_{☾}}{R_{\oplus}} = \dfrac{\sin z_l}{\sin\left(z_l - \phi - \delta\right)} \end{cases}.$$

Dividing the two equations, we find:

$$\frac{\sin z_u}{\sin z_l} = \frac{\sin\left(z_u - \phi + \delta\right)}{\sin\left(z_l - \phi - \delta\right)}\,.$$

Since δ is the only unknown, the above equation can be solved numerically, giving $\delta \approx 28.6°$. Finally, substituting the value of δ in one of the above equations, we find $d_{☾}/R_{\oplus} \approx 60.1$. Knowing that the obliquity of the ecliptic is $\epsilon = 23°27'$, we then obtain the angle between the plane of the lunar orbit and the ecliptic: $i = \delta - \epsilon = 5.15 = 5°9'$.

6.5 Lunar influence

Let $x_{☾}$ and $x_{\oplus}$ be the respective distances of the Moon and the Earth from the centre of mass of the Earth-Moon system. Then:

$$M_{\oplus}x_{\oplus} = M_{☾}\,x_{☾} \;\Rightarrow\; x_{☾} = \frac{M_{\oplus}}{M_{☾}}x_{\oplus}\,.$$

The Earth-Moon distance is $d_{☾} = x_{\oplus} + x_{☾} = 3.844\cdot 10^5$ km. Hence, substituting $x_{☾}$ in the equation for $d_{☾}$, we find:

$$d_{☾} = \left(1 + \frac{M_{\oplus}}{M_{☾}}\right)x_{\oplus} \;\Rightarrow\; x_{\oplus} = \frac{d_{☾}}{1 + M_{\oplus}/M_{☾}} \approx 4670\,\mathrm{km}\,.$$

Suppose there is a first quarter Moon during the vernal equinox. Then, the centre of mass of the Earth-Moon system starts at a distance $x_{\oplus}$ behind the Earth. Since there are approximately 12.5 synodic months in a year, at the end of the year there will be a last quarter Moon. At that moment, the centre of mass will be at a distance $x_{\oplus}$ in front of the Earth. Therefore, the centre of mass will have travelled an additional distance of $2x_{\oplus}$, compared to the average. Since the orbital velocity of the centre of mass is constant, the time taken to cover this additional distance is:

$$\Delta t = \frac{2x_{\oplus}}{v_{\oplus}} = 2x_{\oplus}\sqrt{\frac{d_{\oplus}}{GM_{\odot}}} \approx 310\,\mathrm{s} = 5^m 10^s\,.$$

The next year, there is a last quarter Moon during the vernal equinox, hence the centre of mass of the system starts at a distance $x_{\oplus}$ in front of the Earth. At the end of the year there will be a first quarter Moon, hence the centre of mass will now be at a distance $x_{\oplus}$ behind the Earth. Therefore, the centre of mass has travelled a distance smaller by $2x_{\oplus}$, compared to the average, and the duration of this year will be Δt shorter. Since the year before was Δt longer than the average, the difference in length of the two years is $2\Delta t \approx 10^m 20^s$.

6.6 Longitudinal libration

Let us first find an estimate for the amplitude of longitudinal libration. The velocity of the Moon at aphelion is (see Ex. 10.9):

$$v = \sqrt{\frac{1+e}{1-e}}\sqrt{\frac{GM_\oplus}{a_☾}}\,.$$

Therefore, the angular velocity at aphelion is:

$$\omega_{\text{rev}} = \frac{v}{(1-e)a_☾} = \sqrt{\frac{1+e}{(1-e)^3}}\sqrt{\frac{GM_\oplus}{a_☾^3}}\,.$$

Since the Moon rotates with angular velocity $\omega_{\text{rot}}^2 = GM_\oplus/a_☾^3$ about its axis (equal, on average, to that of revolution), after a time t we will observe an additional angle of the Moon:

$$2A = (\omega_{\text{rev}} - \omega_{\text{rot}})t = \sqrt{GM_\oplus/a_☾^3}\left(\sqrt{\frac{1+e}{(1-e)^3}} - 1\right)t\,.$$

Using the approximation for small e (since $e = 0.0549 \ll 1$), we find:

$$2A \approx \frac{2\pi}{T_{\text{sid}}}\Big[(1+e/2)(1+3e/2) - 1\Big]t = \frac{4\pi}{T_{\text{sid}}}et\,.$$

Since this is just an approximation, we can loosely chose t. To get rid of some numerical factors, let us take $t = T_{\text{sid}}/\pi$, in which case $A \approx 2e = 6^\circ$, approximately. This should be considered as an order-of-magnitude estimation.
Another solution, which does not rely on a guess for t, starts with Eq. 5.9:

$$\frac{\mathrm{d}\theta}{\mathrm{d}t} = \sqrt{\frac{GM_\oplus}{a_☾^3}}\frac{(1+e\cos\theta)^2}{(1-e^2)^{3/2}}\,.$$

The relative velocity of the lunar surface as seen from Earth is:

$$\left(\frac{\mathrm{d}\theta}{\mathrm{d}t}\right)_{\text{rel}} \approx \frac{\mathrm{d}\theta}{\mathrm{d}t} - \omega_{\text{rot}} = \sqrt{\frac{GM_\oplus}{a_☾^3}}\left[\frac{(1+e\cos\theta)^2}{(1-e^2)^{3/2}} - 1\right].$$

The relative velocity is zero for:

$$\frac{(1+e\cos\theta_0)^2}{(1-e^2)^{3/2}} - 1 = 0 \Rightarrow \cos\theta_0 = \frac{(1+e)^{3/4} - 1}{e}\,.$$

Using the approximation for small e, we get $\cos\theta_0 \approx 3/4\,e$. Since $\omega_r = dA/dt = (dA/d\theta)(d\theta/dt)$:

$$\int_0^A dA = \int_0^{\theta_{\max}}\left[1 - \frac{(1-e^2)^{3/2}}{(1+e\cos\theta)^2}\right]d\theta\,.$$

For small e, the integral becomes:

$$A = 2e\int_0^{\cos^{-1}\theta_0}\cos\theta\,d\theta = 2e\sqrt{1-\left(\frac{3}{4}e^2\right)} \approx 2e\,,$$

to first order. Hence, we recover the previous result.

7. Electromagnetic radiation

7.1 Radiation emitted by accelerated charge

Let C be the units of charge, T of time, L of length and M of mass. Writing down the physical dimensions of the quantities under consideration:

$$[\epsilon_0] = \frac{C^2T^2}{L^3M}\,, \qquad [c] = \frac{L}{T}\,, \qquad [q] = C\,, \qquad [a] = \frac{L}{T^2}\,.$$

Our aim is to combine these quantities to get another, whose physical dimension is power:

$$[P] = \frac{ML^2}{T^3}\,.$$

Let α, β, γ and δ be the exponents with which ϵ_0, c, q and a appear in the final formula for P. The problem reduces to solving a system of four equations:

$$\begin{cases} L: -3\alpha + \beta + \delta = 2 & (15.3a) \\ T: 2\alpha - \beta - 2\delta = -3 & (15.3b) \\ M: -\alpha = 1 & (15.3c) \\ C: \gamma + 2\alpha = 0\,. & (15.3d) \end{cases}$$

For instance, to get the first equation, we note that the length L appears with exponent -3 in ϵ_0 (so we add -3α); with exponent 1 in c (so we add β), and finally with exponent 1 in a (so we add δ). The length appears in the expression for P with exponent 2, hence the sum of the previous terms must equal 2, and we recover Eq. 15.3a.
From Eq. 15.3c, we immediately obtain $\alpha = -1$ and, substituting the value of α in Eq. 15.3d, we find $\gamma = 2$. Substituting the value of α and γ in Eqs. 15.3a and 15.3b, we have:

$$\begin{cases} \beta + \delta = -1 & (15.4a) \\ -\beta - 2\delta = -1\,. & (15.4b) \end{cases}$$

Summing Eqs. 15.4a and 15.4b, we find $-\delta = -2$, i.e. $\delta = 2$. Substituting the value of δ in any of the two above equations, we find $\beta = -3$. We know that the numerical constant is $1/6\pi$, hence we can write:

$$P = \frac{1}{6\pi}\frac{q^2a^2}{\epsilon c^3}$$

7.2 Maxwell distribution

According to Boltzmann's law, the velocity distribution $g(v_x)$ is proportional to:

$$g(v_x) \propto e^{-\frac{mv_x^2}{2k_BT}} \Rightarrow g(x) = k\,e^{-\frac{mv_x^2}{2k_BT}}\,,$$

where k is a constant. The probability that a particle has a velocity anywhere between $-\infty$ and ∞ is 1. Requiring $g(v_x)$ to satisfy this condition, for a certain k, is called *normalizing* the function. We change variable to $u = \sqrt{m/(2k_BT)}v_x$, hence $dv_x = du\sqrt{2k_BT/m}$. Integrating $g(v_x)$, we find:

$$k\sqrt{\frac{2k_BT}{m}}\int_{-\infty}^{\infty} e^{-u^2}du = k\sqrt{\frac{2\pi k_BT}{m}}\,,$$

where we have used the Gaussian integral $\int_{-\infty}^{\infty} e^{-u^2}du = \sqrt{\pi}$. The above integral must be equal to 1, therefore $k = \sqrt{m/(2\pi k_BT)}$. The probability function $g(v_x)$, and similarly $g(v_y), g(v_z)$, is given by:

$$g(v_x) = \sqrt{\frac{m}{2\pi k_BT}}e^{-\frac{mv_x^2}{2k_BT}}\,.$$

Let $n(v_x, v_y, v_z)$ be the velocity distribution function, such that $n(v)dv_x\, dv_y\, dv_z$ is equal to the probability that a particle's velocity is in the range (v_x, v_y, v_z) to $(v_x + dv_x, v_y + dv_y, v_z + dv_z)$. Then:

$$n(v) = g(v_x)\, g(v_y)\, g(v_z) = \left(\frac{m}{2\pi k_B T}\right)^{3/2} e^{-\frac{m(v_x^2+v_y^2+v_z^2)}{2k_B T}} ,$$

where $v_x^2 + v_y^2 + v_z^2 = v^2$. In a coordinate system where v_x, v_y and v_z are the three axes, the number of particles with velocities in the range v to $v + dv$ is given by the product of $n(v)$ with the volume of the spherical shell with radius v and thickness dv, i.e. $4\pi v^2\, dv$:

$$f(v)\, dv = 4\pi v^2 n(v)\, dv = 4\pi\left(\frac{m}{2\pi k_B T}\right)^{3/2} v^2 e^{-\frac{mv^2}{2k_B T}}\, dv ,$$

which is Maxwell distribution law.

7.3 Natural width of emission lines

When an atom passes from the excited state to the ground state, it emits a photon with energy $E = hf = h/(2\pi) \cdot (2\pi/T) = \hbar\omega$, where ω is the angular frequency. If the time between two successive emissions is Δt, the atom undergoes $1/\Delta t$ transitions per second, hence the power emitted is $E/\Delta t = \hbar\omega/\Delta t$. As shown in Pr. 7.1, the power emitted by an accelerated electron is on the order of $P = e^2 a^2/\epsilon_0 c^3$. The acceleration of an electron is on the order of $\omega^2 r$, where r is the radius of the hydrogen atom ($r = 50 \cdot 10^{-12}$ m). Equating P and $E/\Delta t$, we then obtain:

$$\Delta t = \frac{\hbar c}{e^2}\left(\frac{c}{\omega r}\right)^2 \frac{1}{\omega} .$$

In the case of visible radiation $c/\omega \approx 10^{-7}$ m, hence $\Delta t \approx 4 \cdot 10^8 \omega^{-1}$. Using $\Delta E = \hbar\Delta\omega$, and substituting ΔE in Heisenberg's uncertainty principle, we find:

$$\frac{\Delta E}{E} = \frac{\Delta\omega}{\omega} \approx 2 \cdot 10^{-9} ,$$

hence $\Delta E = \hbar\Delta\omega \approx 5 \cdot 10^{-28}$ J. Note that radio waves have a natural width which is approximately 10^{10} times smaller than for visible light, while the time between two successive emissions is longer by 10^{15}. The natural width of the emission lines increases with increasing pressure. Indeed, the higher the pressure, the shorter the average time between two consecutive collisions. Collisions interrupt the emission process, shortening its characteristic time, and increasing, according to Heisenberg's uncertainty principle, the uncertainty in E. Thanks to the pressure dependence of the natural width of the spectral lines, we are able to estimate the surface gravity of stars, by simply observing their spectra. This is the basis of the Yerkes classification system.

7.4 Doppler Broadening

As shown in Sec. 7.6, in the non-relativistic limit, the wavelength emitted by an atom with velocity v_x is $\lambda = \lambda_0(1 - v_x/c)$, where λ_0 is the wavelength emitted at rest. Solving for v_x:

$$v_x = c\left(1 - \frac{\lambda}{\lambda_0}\right) .$$

According to Boltzmann law, the number of atoms emitting photons between λ and $\lambda + d\lambda$ is proportional to $\exp\left\{-\frac{E}{k_B T}\right\}$, where $E = \frac{1}{2}mv_x^2 = \frac{1}{2}mc^2(\lambda_0 - \lambda)^2/\lambda_0^2$. Hence, the intensity is:

$$I(\lambda) \propto \exp\left\{-\frac{mc^2(\lambda_0 - \lambda)^2}{2\lambda_0^2 k_B T}\right\} .$$

The wavelength at which $I(\lambda)$ is half the maximum value satisfies:

$$\frac{I(\lambda_0 + \Delta\lambda/2)}{I(\lambda_0)} = \frac{1}{2}.$$

Substituting for $I(\lambda)$, previously found, we obtain the desired result.
This formula is valid when the gas can be treated as ideal, therefore the temperature and pressure should not be too high or too low. Furthermore, if the pressure is too high, the natural width of the emission lines would overwhelm the Doppler broadening. Other phenomena that can be observed on emission lines are the *Stark effect*, which consists in the separation of spectral lines caused by an electric field due to the matter in the state of plasma, or the *Zeeman effect*, equivalent to the Stark effect for the magnetic field.

7.5 Wien and Stefan-Boltzmann laws

To compute the distribution of P_λ, we note that the energy contained in the interval $[\lambda, \lambda+d\lambda]$ is the same as the energy is the interval $[f, f+df]$, where $f = c/\lambda$. Hence, $P_f\, df = P_\lambda\, d\lambda$, or:

$$P_\lambda = P_f \frac{\mathrm{d}f}{\mathrm{d}\lambda}.$$

But $df/d\lambda = -c/\lambda^2$, and substituting $f = c/\lambda$ in P_f, we obtain:

$$P_\lambda = \frac{2\pi hc^2}{\lambda^5} \frac{1}{e^{\frac{hc}{k_B T\lambda}} - 1}.$$

The maximum of P_λ can be found by setting $dP_\lambda/d\lambda = 0$. Since the numerator is constant, it is simpler to set the derivative of the denominator to zero. Denoting with $x = \lambda T = cT/f$ and $k = hc/k_B = 0.0143878\,\text{mm K}$, we have:

$$5\lambda^4(e^{\frac{hc}{k_B x}} - 1) - \lambda^4 \frac{hc}{k_B x} e^{\frac{hc}{k_B x}} = 0$$
$$\Rightarrow e^{k/x}(5 - k/x) = 5.$$

Solving this equation numerically, we find $x = 0.201405\,k$, hence $\lambda T = 2.89777\,\text{mm K}$, which is Wien's law. The maximum of P_f can be found setting $dP_f/f = 0$. As before, using $x = \lambda T = cT/f$ and $k = hc/k_B = 0.0143878\,\text{mm K}$:

$$\frac{f^2}{(e^{\frac{k}{x}} - 1)^2}\left[3\left(e^{\frac{k}{x}} - 1\right) - \frac{k}{x}e^{\frac{k}{x}}\right] = 0.$$

The term inside the brackets must be equal to zero. Thus, we need to numerically solve:

$$\left(3 - \frac{k}{x}\right)e^{\frac{k}{x}} = 3,$$

which gives $x = 0.354429\,k$ as the only real solution. Hence, the maximum of P_f is at $\lambda T = 5.09945\,\text{mm K}$. Integrating the power over all frequencies:

$$P = \int_0^\infty P_f\, df = \frac{2\pi h}{c^2} \int_0^\infty \frac{f^3}{e^{\frac{hf}{k_B T}} - 1} df.$$

Using the substitution $u = h/(k_B T)f$, with $df = k_B T/h\, du$:

$$P = \frac{2\pi h}{c^2}\left(\frac{k_B T}{h}\right)^4 \int_0^\infty \frac{u^3}{e^u - 1} du.$$

The last integral is just $\pi^4/15$:

$$P = \frac{2\pi^5 k_B^4}{15c^2h^3}T^4 .$$

This equation can be written in the form:

$$P = \sigma T^4 , \qquad \text{where} \qquad \sigma = \frac{2\pi^5 k_B^4}{15c^2h^3} .$$

7.6 Ultraviolet catastrophe

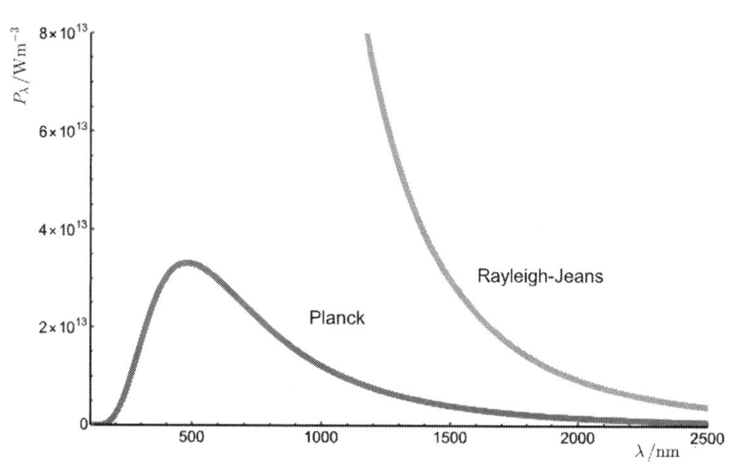

Figure 15.11: Rayleigh-Jeans and Planck laws for a body at the temperature of the Sun (T = 5778 K). In the ultraviolet part of the electromagnetic spectrum, Planck's function decreases after having reached a maximum, while Rayleigh-Jeans increases to infinity.

In the previous problem, we showed that:

$$P_\lambda = \frac{2\pi hc^2}{\lambda^5}\frac{1}{e^{\frac{hc}{k_BT\lambda}} - 1} .$$

For large λ, $hc/(k_BT\lambda)$ tends to zero, hence $e^{\frac{hc}{k_BT\lambda}}$ tends to $1 + hc/(k_BT\lambda)$. Therefore:

$$P_\lambda = \frac{2\pi hc^2}{\lambda^5}\frac{k_BT\lambda}{hc} = \frac{2\pi ck_BT}{\lambda^4} . \tag{15.5}$$

However, for $\lambda \to 0$, $P \to \infty$. Of course, a body cannot emit an infinite amount of energy. It is for this reason that classical physics, when applied to the study of the black body, leads to the so-called "ultraviolet catastrophe", being this the part of the electromagnetic spectrum where the Rayleigh-Jeans law significantly deviates from the true distribution.

7.7 Why do stars emit energy?

Backing our arguments with data directly measurable on Earth, let us show that the first two models cannot be correct, while the third one is reasonable. Let k = 1367 W/m^2 be the solar constant and $L_\odot = 4\pi d_\oplus^2 \cdot k \approx 3.8 \cdot 10^{26}$ W the power emitted by the Sun.

Hypothesis 1 An asteroid arriving from infinity has an initial energy of zero. Applying conservation of energy, we find the kinetic energy of the asteroid when it hits the surface of the Sun:

$$K + U = 0 \Rightarrow K = \frac{GmM_\odot}{R_\odot} .$$

The power produced in this way is:

$$P = \frac{\Delta K}{\Delta t} = \frac{GM_\odot}{R_\odot}\frac{\Delta m}{\Delta t}.$$

Since $P = L_\odot$, the mass of the Sun increases at the rate of:

$$\frac{\Delta m}{\Delta t} = L_\odot \frac{R_\odot}{GM_\odot}.$$

Hence, every year, the mass of the Sun would increase by:

$$\Delta m_{\rm yr} = 365 \cdot 86400 \cdot L_\odot \frac{R_\odot}{GM_\odot} \approx 6.3 \cdot 10^{22}\,\text{kg}.$$

The duration of a year and the semi-major axis of the Earth depend on the mass of the Sun. Let da, dM and dT be the variations in the semi-major axis, the mass of the Sun and the length of a year, after a time interval dt. At $t = 0$, the semi-major axis of the Earth is a, the mass of the Sun is M and the length of a year is T. Therefore, at $t = 0$, the kinetic energy of the Earth is:

$$K = \frac{GmM}{2a}.$$

When $t = dt$, the mass of the Sun increases by dM, thus the potential energy of the Earth decreases, but the kinetic energy remains the same (similarly to Ex. 10.14). Therefore, the total energy is now:

$$E = -\frac{Gm(M+dM)}{a} + \frac{GmM}{2a} = \frac{GmM}{2a}\left[1 - 2\left(1 + \frac{dM}{M}\right)\right] \approx -\frac{GmM}{2a(1-2\,dM/M)},$$

where we have used the approximation $(1+x) \approx 1/(1-x)$ for $x = dM/M \ll 1$. The total energy can be written as $E = -GmM/2(a+da)$, where $a + da$ is the new semi-major axis. Comparing this expression for E with the above equation:

$$a + da = a\left(1 - 2\frac{dM}{M}\right) \Rightarrow \frac{da}{a} = -2\frac{dM}{M}.$$

According to Kepler's third law, the new orbital period $(T + dT)$ is:

$$\begin{aligned}
(T+dT)^2 &= \frac{4\pi^2}{G(M+dM)}(a+da)^3 \\
T^2 + 2TdT &\approx \frac{4\pi^2}{GM}a^3\left(1 - \frac{dM}{M}\right)\left(1 + 3\frac{da}{a}\right) \\
T^2 + 2TdT &\approx T^2\left(1 - \frac{dM}{M}\right)\left(1 - 6\frac{dM}{M}\right) \\
2T\,dT &\approx -7T^2\frac{dM}{M} \\
\frac{dT}{T} &\approx -\frac{7}{2}\frac{dM}{M}.
\end{aligned}$$

We can then find by how much the period changes in a year:

$$\Delta T_{\rm year} = -\frac{7T}{2M}\Delta m_{\rm year} \approx -1.4\,\text{s}.$$

Hence, if the Sun got its energy solely from the impact of asteroids, the orbital period of the Earth would decrease by around 1.4 s every year. Integrating the above equation:

$$\int_{T_0}^{T(t)} \frac{dT}{T} = -\frac{7}{2}\int_{M_0}^{M(t)} \frac{dM}{M}$$
$$\ln\frac{T(t)}{T_0} = -\frac{7}{2}\ln\frac{M(t)}{M_0}$$
$$T(t) = T_0\Big(\frac{M(t)}{M_0}\Big)^{-7/2}$$
$$T(t) \approx T_0\Big(1+\frac{\Delta m_{\text{yr}}}{M_0}t_{\text{yr}}\Big)^{-7/2}.$$

After just 6.9 million years, the duration of a year would halve. Of course, this is not observed.

Hypothesis 2 In Pr. 10.4 we obtained the potential energy of a star:

$$U = -\frac{3GM^2}{5R}.$$

According to the contraction hypothesis, when R decreases, the potential energy decreases, thus the star must radiate the energy lost into space. In fact, according to the Virial theorem (Sec. 10.6), the magnitude of the kinetic energy is always equal to half the magnitude of the potential energy. Hence, while the star contracts, only half of the potential energy is emitted as radiation, while the rest goes to increasing the thermal velocity of atoms inside the star. The total energy that the star can emit according to this process is:

$$E_{\text{tot}} = \frac{3GM^2}{10R} \approx 1.14\cdot 10^{41}\,\text{J}.$$

Assuming the solar luminosity stays roughly constant, the lifetime of the Sun would be:

$$t_{\text{life}} = \frac{E_{\text{tot}}}{L_\odot} \approx 9.5\cdot 10^6\,\text{years}.$$

If this hypothesis were true, the Sun would only be able to radiate energy for a little more than 9 million years. However, we have ample evidence that shows the Solar System has existed for more than 5 billion years, hence this hypothesis must also be wrong. At the time when stars were believed to emit energy by this process, one of the objections to Darwin's theory of evolution was that it could only take place over a couple billion years, whereas the Solar System was thought to be much younger than that.
It is interesting to note that, if this hypothesis were true, the mean temperature of the Sun would increase every year by:

$$\frac{3}{2}nR\frac{\Delta T}{\Delta t} = L_\odot$$
$$\frac{3}{2}\frac{M_\odot}{m_p}k_B\frac{\Delta T}{\Delta t} = L_\odot$$
$$\Rightarrow \frac{\Delta T}{\Delta t} = \frac{2L_\odot m_p}{3M_\odot k_B} \approx 0.5\,\text{K/year},$$

where m_p is the mass of a proton (we assumed that the Sun is entirely made up of hydrogen and neglected the mass of an electron). Again, this variation is not measured.

Hypothesis 3 According to the process of nuclear fission, 2 protons and 2 neutrons are needed to form a helium atom. However, the mass of reagents and products is not the same, but it is smaller by 0.72 %, compared to the mass of a helium atom. This mass difference is directly converted into energy, according to the equation $E = mc^2$. Assuming that around 10 % of the total mass of the Sun can be converted into helium through this reaction, the total energy the Sun can emit is:

$$E_{\text{tot}} = M_{\odot} \cdot 0.1 \cdot 0.0072 \cdot c^2 = 1.3 \cdot 10^{44}\ \text{J}\,.$$

Hence, its lifetime is:

$$t_{\text{life}} = \frac{E_{\text{tot}}}{L_{\odot}} \approx 10 \cdot 10^9 \text{ years}\,.$$

Which sounds plausible.

7.8 Inside a star

Hydrostatic equilibrium In Sec. 10.2, we find that the gravitational acceleration at a distance r from the centre of a star is:

$$g(r) = \frac{Gm(r)}{r^2}\,.$$

Consider a cylindrical element, at a distance r from the centre of the star, with base area dA, thickness dr and mass $dm = \rho\, dA\, dr$. The gravitational force acting on it is:

$$F_g = g(r)\, dm = \frac{Gm(r)}{r^2} \rho\, dA\, dr\,.$$

The force due to the radiation pressure is always directed radially and is opposite to the force of gravity. Let dP be the pressure difference between the top and bottom of the cylinder. The cylinder experiences an outwards force of:

$$F_p = dP\, dA\,.$$

Equating the two forces:

$$dP\, dA = \frac{Gm(r)}{r^2} \rho\, dA\, dr \Rightarrow \frac{\mathrm{d}P}{\mathrm{d}r} = \frac{Gm(r)\rho}{r^2}\,.$$

Mass distribution Consider a spherical shell of radius r and thickness dr, with mass $dm = (4\pi r^2\, dr)\rho$. Then:

$$\frac{\mathrm{d}m}{\mathrm{d}r} = 4\pi r^2 \rho\,.$$

Energy production The contribution to the total luminosity from an infinitesimal mass element dm is $dL_r = \epsilon\, dm$. The mass between r and $r + dr$ is $dm = 4\pi\rho r^2\, dr$, hence:

$$\frac{\mathrm{d}L_r}{\mathrm{d}r} = 4\pi r^2 \rho\epsilon\,.$$

Temperature Consider a cylindrical element between r and $r + dr$, with base area dA. The radiation pressure is directly proportional to the incident flux:

$$dP_r = (k\rho\, dr)\frac{F_r}{c} = \frac{L_r}{4\pi r^2 c} k\rho\, dr\,,$$

where L_r is the luminosity of the star at a distance r from the centre, which can be calculated using the result from the previous part. At the same time, the pressure in a photon gas is $P = aT^4/3$. If dT is the temperature difference between the bottom and the top of the cylinder, then the pressure difference is:

$$dP_r = -d\left(\frac{1}{3}aT^4\right) = -\frac{4}{3}aT^3\, dT\,.$$

Equating the two expressions:

$$\frac{L_r}{4\pi r^2 c} k\rho\, dr = -\frac{4}{3}aT^3 dT \Rightarrow \frac{\mathrm{d}T}{\mathrm{d}r} = -\frac{3}{4ac}\frac{k\rho}{T^3}\frac{L_r}{4\pi r^2}$$

8. Flux and magnitude

8.1 Olbert paradox

Let $\overline{L}$ be the average luminosity of a star and n the average number density of stars in the universe. The number of stars between r and $r + dr$ is:

$$dN_r = n \cdot 4\pi r^2 \, dr \, .$$

Hence, the total luminosity produced by stars in this spherical shell is:

$$dL_r = \overline{L} \cdot dN_r = 4\pi n \overline{L} r^2 \, dr \, .$$

Therefore, the flux arriving on Earth is:

$$dF_r = \frac{dL_r}{4\pi r^2} = 4\pi n \overline{L} \, dr \, .$$

As we see, the flux does not depend on the distance of the spherical shell from the Earth. Thus, the fluxes arriving on Earth from spherical shells of radii R and r (with $r \neq R$) are the same. Since there are an infinite number of these shells, the flux arriving on Earth should be infinite, therefore the night sky should be as bright as daylight! This reasoning is also valid if we assume that stars are grouped into galaxies, since we only used the average luminosity and density of stars (which is certainly non-zero). Olbert believed that the answer to this paradox is that space is opaque. At that time, thermodynamics had yet to be fully developed, but today we know that, if this hypothesis were true, the interstellar medium would absorb part of the light emitted by stars, re-emitting it in the form of black body radiation: the energy arriving on Earth would be the same as if space were completely transparent. In reality, the speed of light is finite and, because light from distant regions of the universe has not yet had time to travel to the Earth, the observable universe is finite. At the same time, the universe expands according to Hubble's law, so radiation from distant stars is fainter, due to the Doppler effect.

8.2 Earth's temperature

Assuming the Sun to be a perfect black body, the flux arriving on Earth is given by the solar constant:

$$k = \frac{L_\odot}{4\pi {d_\oplus}^2} = \frac{\sigma {R_\odot}^2 {T_\odot}^4}{{d_\oplus}^2} \, .$$

To obtain the power incident on the Earth (L_{i}), we multiply the flux by Earth's cross-sectional area ($\pi {R_\oplus}^2$):

$$L_{\mathrm{i}} = \pi {R_\oplus}^2 \frac{\sigma {R_\odot}^2 {T_\odot}^4}{{d_\oplus}^2} \, .$$

The reflected power is proportional to the albedo of the Earth, according to $L_{\mathrm{r}} = A_\oplus L_{\mathrm{i}}$. The difference between the incident and reflected power is equal to the power (P) absorbed by the Earth:

$$P = L_{\mathrm{i}} - L_{\mathrm{r}} = L_{\mathrm{i}}(1 - A_\oplus) = \pi {R_\oplus}^2 \frac{\sigma {R_\odot}^2 {T_\odot}^4}{{d_\oplus}^2} (1 - A_\oplus) \, .$$

In a stationary regime (at constant temperature) the energy absorbed is equal to the energy emitted, according to the black body radiation formula:

$$P = 4\pi\sigma {R_\oplus}^2 {T_\oplus}^4 \, .$$

Solving for ${T_\oplus}^4$, we find:

$$T_\oplus = T_\odot \sqrt{\frac{R_\odot}{2d_\oplus}} (1 - A_\oplus)^{1/4} \, . \qquad (15.6)$$

The temperature is independent on the radius of the planet. Inserting the numerical values and assuming Earth has an albedo of $A_\oplus = 0.31$:

$$T_\oplus = 254\,^\circ\mathrm{K} = -19\,^\circ\mathrm{C}\,.$$

This is lower by about 30° C, compared to the average temperature on Earth. In fact, we neglected the role of the Earth's atmosphere.

8.3 Earth's temperature, another model

For the Sun, the wavelength of maximum emission is $\lambda_{\max,s} = 501.5\,\mathrm{nm}$, which corresponds to visible light. The Earth, as shown in Pr. 8.2, has an average temperature of around 250 K, hence its wavelength of maximum emission should be around $\lambda_{\max,t} = 11500\,\mathrm{nm}$, which corresponds to the infra-red part of the electromagnetic spectrum. Since $e_i < e_v$, this phenomenon raises the Earth's temperature.
Let us model the Earth as surrounded by a thin sheet of atmosphere, with a radius of about $R_\oplus$, and constant temperature T_{atm}. The radiation that enters the atmosphere is:

$$L_{\mathrm{eff}} = (1 - A_{\mathrm{atm}})k\pi R_\oplus^2 = (1 - A_{\mathrm{atm}})\frac{\sigma R_\odot^2 T_\odot^4}{d_\oplus^2}\pi R_\oplus^2\,,$$

where k is the solar constant. The atmosphere directly absorbs a luminosity of $L_{\mathrm{eff}}(1 - e_v)$, and transmits the remaining $L_{\mathrm{eff}}\, e_v$ to the surface. The energy arriving on the surface and then reflected is $L_{\mathrm{eff}}\, e_v A_s$, while the energy absorbed is $L_{\mathrm{eff}}\, e_v(1 - A_s)$. The energy reflected by the surface and absorbed by the atmosphere is $L_{\mathrm{eff}}\, e_v A_s(1 - e_v)$, the remaining $L_{\mathrm{eff}}\, {e_v}^2 A_s$ being lost in space. Energy is not only emitted by the Sun, but also by the atmosphere and the surface of the Earth. The atmosphere absorbs a fraction equal to $(1 - e_i)$ of the thermal energy emitted by the Earth, which is $4\pi R_\oplus^2 \sigma T_\oplus^4$. On the other hand, the Earth absorbs half of the energy emitted by the atmosphere, which is $8\pi R_\oplus^2 T_{\mathrm{atm}}^4$ (taking into account both inner and outer surfaces). Therefore, the conditions for the atmosphere and the Earth to be in thermal equilibrium are:

$$\begin{cases} L_{\mathrm{eff}}[(1 - e_v) + A_s e_v(1 - e_v)] + 4\pi\sigma R_\oplus^2 T_\oplus^4(1 - e_i) = 8\pi\sigma R_\oplus^2 T_{\mathrm{atm}}^4 \\ L_{\mathrm{eff}}\, e_v(1 - A_s) + 4\pi\sigma R_\oplus^2 T_{\mathrm{atm}}^4 = 4\pi\sigma R_\oplus^2 T_\oplus^4\,. \end{cases}$$

Isolating $4\pi\sigma R_\oplus^2 T_\oplus^4$ in the second equation, and substituting into the first:

$$L_{\mathrm{eff}}[(1 - e_v) + A_s e_v(1 - e_v) + e_v(1 - A_s)(1 - e_i)] = 4\pi\sigma R_\oplus^2 T_{\mathrm{atm}}^4[1 + e_i]\,.$$

Isolating $4\pi R_\oplus^2 \sigma T_{\mathrm{atm}}^4$:

$$4\pi R_\oplus^2 \sigma T_{\mathrm{atm}}^4 = \frac{L_{\mathrm{eff}}[1 - A_s {e_v}^2 - e_v e_i + A_s e_v e_i]}{1 + e_i}\,.$$

Substituting $4\pi R_\oplus^2 \sigma T_{\mathrm{atm}}^4$ into the second equation of the system, and isolating $T_\oplus^4$:

$$T_\oplus^4 = \frac{L_{\mathrm{eff}}}{4\pi R_\oplus^2 \sigma}\frac{1 + e_v - A_s e_v(1 + e_v)}{1 + e_i}\,.$$

Inserting the expression for L_{eff} and taking the fourth root of both sides:

$$T_\oplus = T_\odot\sqrt{\frac{R_\odot}{2d_\oplus}}\left[(1 - A_{\mathrm{atm}})(1 - A_s e_v)\cdot\frac{1 + e_v}{1 + e_i}\right]^{1/4}\,.$$

Inserting numerical values, we find $T_\oplus \approx 283\,\mathrm{K} = 10°\mathrm{C}$. In reality, the Sun and the Earth do not emit at only one wavelength, hence the effective e_v, e_i could be slightly different. At the same time, we have assumed that the Earth and atmosphere only reflect visible light. This model predicts that the temperature of the Earth increases with decreasing e_i (*greenhouse effect*), to a maximum temperature of $T_{\max} \approx 300\,\mathrm{K} = 27°\mathrm{C}$, when e_i goes to zero and the albedo of the Earth is reduced, due to the melting of the ice, to $A_s = 0.06$. The change in A_{atm} is, instead, harder to estimate. In any case, it wouldn't be such a friendly world!

9. Cosmic Distance Ladder

9.1 The Astronomical Unit

As explained in Sec. 11.1, the distance of Venus from the Sun can be obtained from its maximum elongation:

$$\sin\epsilon_{\rm max} = \frac{d_{♀}}{d_{\oplus}} \Rightarrow d_{♀} = 0.723\, d_{\oplus}\,.$$

The angle between the two images of Venus is $\Delta\phi = 23'' \pm 6''$, while the parallax baseline is $b = 2R_{\oplus}\sin 60^{\circ} = \sqrt{3}R_{\oplus}$. Hence, the distance ($d$) of Venus from the Earth is:

$$\Delta\phi_{\rm rad} = \frac{b}{d} \Rightarrow d = (9.91 \pm 2.5)\cdot 10^{10}\,\mathrm{m}\,.$$

We can determine the astronomical unit by applying the Pythagorean theorem:

$$\begin{aligned} d^2 &= d_{\oplus}^2 - d_{♀}^2 = (1 - 0.723^2)d_{\oplus}^2 \\ d_{\oplus} &\approx d\cdot 1.4475 = (1.43 \pm 0.35)\cdot 10^{11}\,\mathrm{m}\,. \end{aligned} \tag{15.7}$$

9.2 Luminosity-radius relationship

Differentiating Stefan-Boltzmann law ($L = 4\pi R^2\sigma T^4$), we find:

$$\begin{aligned} dL &= 8\pi R\sigma T^4\, dR + 16\pi R^2\sigma T^3\, dT \\ &= 4\pi R^2\sigma T^4\Big(2\frac{dR}{R} + 4\frac{dT}{T}\Big)\,. \end{aligned}$$

Dividing by L:

$$\frac{dL}{L} = 2\frac{dR}{R} + 4\frac{dT}{T}\,.$$

Writing $V = 4/3\pi R^3$ in the adiabatic relationship, we find $TR^{3(\gamma-1)}$ = const, which differentiated gives:

$$\begin{aligned} dTR^{3(\gamma-1)} &+ 3(\gamma-1)TR^{3(\gamma-1)-1}dR = 0 \\ \Rightarrow \frac{dT}{T} &= -3(\gamma-1)\frac{dR}{R}\,. \end{aligned}$$

Substituting this expression for dT/T in the equation for dL/L, we find:

$$\begin{aligned} \frac{dL}{L} &= 2\frac{dR}{R} + 4\Big[-3(\gamma-1)\frac{dR}{R}\Big] \\ &= 2\frac{dR}{R} - 12(\gamma-1)\frac{dR}{R}\,. \end{aligned}$$

Hence, using $\gamma = 5/3$:

$$\frac{dL}{L} = -6\frac{dR}{R}\,.$$

When the star expands, dR is positive, thus the above equation tells us that dL must be negative, i.e. the luminosity decreases.

9.3 Period-density relationship

The pressure at a distance r from the centre of the star can be obtained using the equation of hydrostatic equilibrium (Pr. 7.8):

$$\frac{\mathrm{d}P}{\mathrm{d}r} = -\frac{Gm(r)\rho}{r} = -\frac{G\rho}{r^2}\cdot\Big(\frac{4}{3}\pi r^3\rho\Big) = -\frac{4}{3}\pi G\rho^2 r\,,$$

which can be easily integrated assuming that the pressure on the surface of the star is zero:

$$\int_{P(r)}^{0} dP = \int_{r}^{R} -\frac{4}{3}\pi G\rho^2 r$$
$$-P(r) = -\frac{2}{3}\pi G\rho^2(R^2 - r^2)$$
$$\Rightarrow P(r) = \frac{2}{3}\pi G\rho^2(R^2 - r^2)\,.$$

The velocity of sound is therefore:

$$v = \sqrt{\frac{\gamma P}{\rho}} = \sqrt{\frac{2}{3}\gamma\pi G\rho(R^2 - r^2)}\,.$$

The period of pulsation is:

$$T = 2\int_0^R \frac{dr}{v} \approx 2\int_0^R \frac{dr}{\sqrt{\frac{2}{3}\gamma\pi G\rho(R^2 - r^2)}} = \frac{2}{\sqrt{\frac{2}{3}\gamma\pi G\rho}}\int_0^R \frac{dr}{\sqrt{(R^2 - r^2)}}\,.$$

The last integral can be solved by substituting $r = R\sin\theta$ and $dr = R\cos\theta\, d\theta$. The lower and upper limits become $\theta_i = \arcsin(0/R) = 0$ and $\theta_f = \arcsin(R/R) = \pi/2$, so we have:

$$\int_0^{\pi/2} \frac{R\cos\theta\, d\theta}{\sqrt{R^2(1 - \sin^2\theta)}} = \int_0^{\pi/2} d\theta = \frac{\pi}{2}\,.$$

Hence, substituting in the equation for the period:

$$T = \frac{2}{\sqrt{\frac{2}{3}\gamma\pi G\rho}} \cdot \frac{\pi}{2} \Rightarrow T = \sqrt{\frac{3\pi}{2\gamma G\rho}}\,.$$

This shows that the period of pulsation is inversely proportional to the square root of the density. Indeed, the period of pulsation is greater for supergiants than for white dwarfs. Taking $M = 5M_\odot$ and $R = 50R_\odot$ for a typical Cepheid, we find $T = 10\,\text{days}$, which is close to the observed period.

10. Gravitation and Kepler's Laws

10.1 Planets as point masses

In Sec. 10.2 we stated that a sphere behaves in the same way as a point mass, with the same mass, placed at its centre. The easiest way to prove this statement is by computing the potential of a spherical shell, since the force can be obtained by differentiating the potential, according to $F = -dU(r)/dr$. It is sufficient to show that a spherical shell obeys this statement, since a solid sphere can be regarded as a collection of shells.
Let R be the radius of the spherical shell, r the distance between the centre of the shell and a point P at its exterior, d the distance between P and a circular annulus constituting the shell, and θ the angle between d and r (Fig. 15.12). The distance d can be written as a function of r, R and θ, by applying the law of cosines (Eq. A.14) to triangle APO:

$$d = \sqrt{R^2 + r^2 - 2rR\cos\theta}\,.$$

The area of the circular annulus between θ and $\theta + d\theta$ is equal to the circumference times the height: $(2\pi R\sin\theta)\cdot(Rd\theta)$. Let $\sigma = M/(4\pi R^2)$ be the surface density. The potential energy at point P due to the circular annulus between θ and $\theta + d\theta$ is:

$$dU = -\frac{Gm\sigma(Rd\theta)(2\pi R\sin\theta)}{d}\,.$$

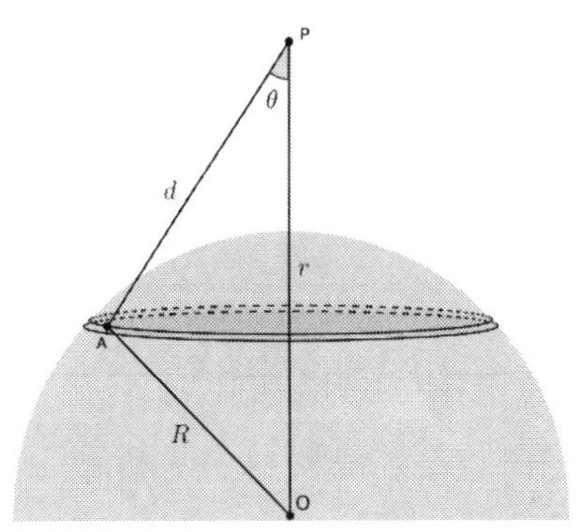

Figure 15.12: The potential at a point P, at a distance r from the centre of a spherical shell of radius R, can be calculated by sectioning the shell into infinitesimal circular annuli and summing (integrating) all of these contributions to the potential.

To obtain the potential energy of the spherical shell, we need to sum (integrate) the contribution of all circular annuli:

$$\begin{aligned} U(r) &= -\int_0^{\pi} \frac{2\pi\sigma G R^2 m \sin\theta}{\sqrt{R^2 + r^2 - 2rR\cos\theta}} d\theta \\ &= -\frac{2\pi\sigma G R m}{r} \left[\sqrt{R^2 + r^2 - 2rR\cos\theta}\right]_0^{\pi} . \end{aligned} \tag{15.8}$$

We now consider two cases. If $r \geqslant R$:

$$U(r) = -\frac{2\pi\sigma G R m}{r}\big[(r+R) - (r-R)\big] = -\frac{4\pi R^2 \sigma G m}{r} = -\frac{GMm}{r} ,$$

which is the potential due to a point mass, with the same mass of the shell, and placed at its centre. If $r < R$:

$$U(r) = -\frac{2\pi\sigma G R m}{r}\big[(r+R) - (R-r)\big] = -\frac{4\pi R^2 \sigma G m}{R} = -\frac{GMm}{R} ,$$

which is independent of r.
By differentiating the potential $U(r)$ with respect to r, we find $F(r)$ in both cases:

$$F(r) = \begin{cases} -\dfrac{GMm}{r^2}, & \text{if} \quad r \geqslant R \\ 0, & \text{if} \quad r < R . \end{cases}$$

By symmetry, $F(r)$ is entirely radial. Since a solid sphere is made up of many spherical shells, and since each spherical shell can be treated as a material point, it follows that a solid sphere produces a gravitational field, on its exterior, equal to the gravitational field produced by a material point, with the same mass, and placed at its centre. This result is valid as long as the density is spherically symmetry.
The force experienced inside a spherical shell is zero. Therefore, when calculating the force inside a solid sphere, at a distance r from its centre, it is sufficient to consider the mass inside a spherical surface with radius r (the mass outside does not contribute to the total gravitational field).

10.2 Using Gauss' theorem

As explained in Sec. 10.2, for $r \geqslant R_{\oplus}$, we can treat the Earth as a point mass, and use Eq. 10.1 in conjunction with Newton's second law to obtain:

$$g(r) = G\frac{M_{\oplus}}{r^2} .$$

For the Earth, $M_\oplus = 5.972 \cdot 10^{24}$ kg and $R_\oplus = 6.378 \cdot 10^6$ m, hence:

$$g(R_\oplus) = 6.674 \cdot 10^{-11} \frac{\mathrm{Nm}^2}{\mathrm{kg}^2} \cdot \frac{5.972 \cdot 10^{24}\,\mathrm{kg}}{6.371 \cdot 10^6\,\mathrm{m}^2} = 9.81\,\mathrm{m/s}^2\,,$$

which is the well-known value for the gravitational acceleration on the surface of the Earth. To compute the gravitational acceleration inside the Earth, at a distance r from its centre, we only need to take into account the mass $m(r)$ inside a sphere of radius r:

$$m(r) = \frac{4}{3}\pi\rho r^3\,, \qquad M_\oplus = \frac{4}{3}\pi\rho R_\oplus^3\,.$$

Dividing the two equations:

$$m(r) = M_\oplus \frac{r^3}{R_\oplus^3}\,.$$

This is the mass responsible for the gravitational acceleration at a distance r. Using Eq. 10.1 in conjunction with Newton's second law, we find, for $r < R_\oplus$:

$$g(r) = \frac{GM_\oplus}{R_\oplus^3} r\,,$$

where the gravitational field is directed towards the centre of the body. To sum up:

$$g(r) = \begin{cases} \dfrac{GM_\oplus}{r^2}\,, & \text{if} \quad r \geqslant R_\oplus \\ \dfrac{GM_\oplus}{R_\oplus^3} r\,, & \text{if} \quad r < R_\oplus\,. \end{cases}$$

We note that the gravitational field at the centre of the Earth is zero, therefore an object would be in (stable) equilibrium at that point.

10.3 Deep tunnels

A body being transported on a circular orbit to the antipode, must cover half the orbit. Hence, the time taken is:

$$t_1 = \pi\sqrt{\frac{R_☾^3}{GM_☾}}\,.$$

We now calculate the time taken to reach the antipode through a tunnel. In Pr. 10.2, we found that the gravitational field inside a spherical body is:

$$\vec{g(r)} = -\frac{GM_☾}{R_☾^3}\vec{r}\,,$$

where the minus sign has been introduced because the gravitational field points towards the centre of the body, in direction opposite the position vector $\vec{r}$. Consider, by analogy, the motion of a body of mass m, given by Hooke's law:

$$\vec{F} = -k\vec{r} \Rightarrow \vec{a(r)} = -\frac{k}{m} r\,.$$

where k is the *spring constant*. The above equation can be written as:

$$\vec{a(r)} = -\omega^2 r\,,$$

where ω is the angular frequency. In this case, the period of oscillation is:

$$T = \frac{2\pi}{\omega}\,.$$

The equation of motion of a body falling through the tunnel and one attached to a spring are the same, if we set $\omega^2 = GM_{☾}/R_{☾}^3$. The time taken to reach the antipode, i.e. to cover half a full oscillation, is:

$$t_2 = \frac{\pi}{\omega} = \pi\sqrt{\frac{R_{☾}^3}{GM_{☾}}}\,.$$

We note that $t_1 = t_2$, hence it takes the same time to transfer a load using a circular orbit or a tunnel. For a more accurate calculation, we would need to consider the centrifugal force due to the rotation of the Moon about its axis, which slightly increases the transfer period. You can verify that the transfer time is the same for any straight tunnel connecting two points on the surface of the Moon (try it!).

10.4 Potential energy of a spherical body

Imagine we assemble a spherical body by successively compounding spherical shell. The potential energy of the whole body is then the sum of the potential energy of each of these shells. The potential energy of a shell of radius r, thickness dr and mass $dm = (4\pi r^2 dr)\rho$, which is placed on the surface of a body of mass $m(r)$ and radius r, is:

$$dU = -\frac{Gm(r)\,dm}{r}\,.$$

Assuming the density is constant, $m(r) = Mr^3/R^3$. We then obtain:

$$dU = -\frac{4\pi GM\rho\, r^4}{R^3}\,dr\,.$$

Integrating both sides to find the total potential energy:

$$\int_0^U dU = -\int_0^R \frac{4\pi GM\rho r^4\,dr}{R^3}$$

$$\Rightarrow U = -\frac{4\pi GM\rho R^2}{5}\,.$$

Finally, substituting $\rho = 3M/(4\pi R^3)$:

$$\boxed{U = -\frac{3GM^2}{5R}}\,. \tag{15.9}$$

Eq. 15.9 gives the potential energy of a body of mass M, radius R and constant density.

10.5 Escape velocity

Without loss of generality, let us assume that m_1 is initially at rest and that m_2 has an initial velocity of $\vec{v}$. We want to find the minimum velocity needed for both masses to travel indefinitely far away from each other. In Sec. 10.5, we assumed that m_1 remains stationary while m_2 is escaping. However, it is clear that m_1 will move in the same direction as m_2, because of the reciprocal gravitational attraction. Looking at Fig. 15.13 a), at the point of maximum distance, the relative velocity between the two bodies is zero, therefore they will both travel with the same velocity $\vec{v'}$.
The key idea is to consider the reference system of the centre of mass of m_1 and m_2. In this system, the velocity of the centre of mass if zero, by definition, and the total momentum is always zero, hence the bodies will move in opposite directions, as shown in Fig. 15.13 b). The maximum distance is attained when both bodies are stationary, i.e. when their final kinetic energy is zero in the CM frame.
In the lab frame, the centre of mass moves with velocity:

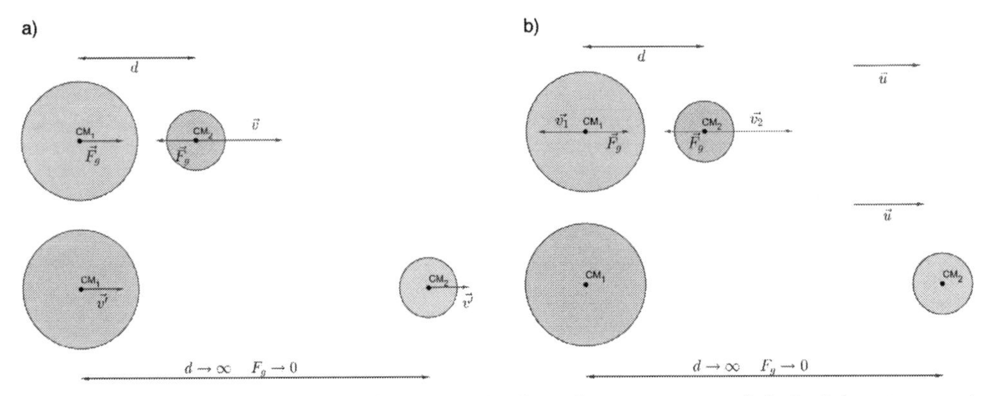

Figure 15.13: Initial and final configurations in the reference system of a) the laboratory and b) the centre of mass.

$$\vec{u} = \frac{m_1 \cdot 0 + m_2\vec{v}}{m_1 + m_2} = \frac{m_2}{m_1 + m_2}\vec{v}\,.$$

Hence, in the CM frame, m_2 and m_1 move with velocities:

$$\vec{v_2} = \vec{v} - \vec{u} = \frac{m_1}{m_1 + m_2}\vec{v}\,,$$
$$\vec{v_1} = -\vec{u} = -\frac{m_2}{m_1 + m_2}\vec{v}\,.$$

The initial kinetic energy in the frame of the CM is thus:

$$K_{\mathrm{CM},i} = \frac{1}{2}m_1 v_1{}^2 + \frac{1}{2}m_1 v_1{}^2 = \frac{1}{2}\frac{m_1 m_2}{m_1 + m_2}v^2\,.$$

At the end, both the kinetic and potential energy are zero in the CM frame, hence $E_{f,\mathrm{CM}} = 0$. Applying conservation of energy, $E_{i,\mathrm{CM}} = E_{f,\mathrm{CM}}$, which implies $K_{\mathrm{CM},i} + U_i = 0$, thus:

$$\frac{1}{2}\frac{m_1 m_2}{m_1 + m_2}v_\mathrm{e}{}^2 = \frac{G m_1 m_2}{d}$$

$$\Rightarrow \boxed{v_\mathrm{e} = \sqrt{\frac{2G(m_1 + m_2)}{d}}}\,.$$

In the limit $m_1 \gg m_2$, the above equation reduces to Eq. 10.27.

10.6 Roche limit

Let us consider a central body with mass M and radius R, and a secondary body with mass m and radius r, which is at a distance x from the central body (Fig. 15.14). Let A and B be the points of the secondary body which are farther and closer to the central body, respectively. The gravitational field is the vector sum of the gravitational fields of the central body g_c and of the secondary body g_s. Taking the magnitudes:

$$g_\mathrm{A} = g_c + g_s = \frac{GM}{(x+r)^2} + \frac{Gm}{r^2}\,,$$
$$g_\mathrm{B} = g_c - g_s = \frac{GM}{(x-r)^2} - \frac{Gm}{r^2}\,.$$

We require the relative acceleration between A and B to be zero, since the body is rigid:

$$g_\mathrm{A} - g_\mathrm{B} = 0$$
$$\Rightarrow \frac{GM}{(x-r)^2} - \frac{GM}{(x+r)^2} = 2\frac{Gm}{r^2}\,.$$

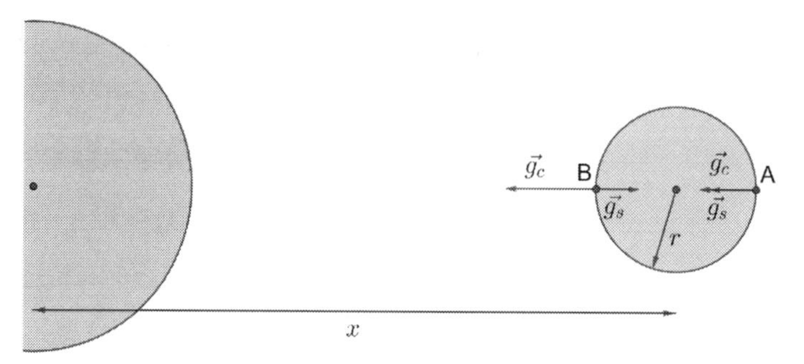

Figure 15.14: The rigid body-condition is satisfied if the gravitational accelerations in A and B are the same.

Multiplying both sides by $(r+x)^2\,(r-x)^2$:

$$2rx = \alpha\left(r^2 + \frac{x^4}{r^2} - 2x^2\right),$$

where $\alpha = m/M$. When $M \gg m$, we expect the body to disintegrate at a distance x much greater than its radius r. Hence, on the RHS of the above equation, we only retain the term x^4/r^2:

$$2rx = \alpha\frac{x^4}{r^2} \Rightarrow x = r\sqrt[3]{\frac{2M}{m}}\,.$$

The ratio of the masses can also be written as a function of the densities and radii of the two bodies:

$$\frac{M}{m} = \frac{\frac{4}{3}\pi\rho_c R^3}{\frac{4}{3}\pi\rho_s r^3} = \frac{\rho_c}{\rho_s}\left(\frac{R}{r}\right)^3.$$

Hence:

$$\boxed{x = R\sqrt[3]{\frac{2\rho_c}{\rho_s}}}\,. \tag{15.10}$$

This is the Roche limit for a non-rotating body. If the rotation of the body about its axis cannot be neglected, we need to add the contribution of the centrifugal force:

$$g_{\mathrm{A}} = g_c + g_s - a_{cf} = \frac{GM}{(x+r)^2} + \frac{Gm}{r^2} - \omega^2 r\,,$$

$$g_{\mathrm{B}} = g_c - g_s + a_{cf} = \frac{GM}{(x-r)^2} - \frac{Gm}{r^2} + \omega^2 r\,.$$

Equating the accelerations and using the approximation $x \gg r$, we find, following a similar reasoning as before:

$$\boxed{x_{\mathrm{rot}} = R\sqrt[3]{\frac{2\rho_c}{\rho_s - \frac{3\omega^2}{4\pi G}}}}\,. \tag{15.11}$$

As expected, the centrifugal force increases the distance at which the body disintegrates. In the case $\omega = 0$, Eq. 15.11 reduces to Eq. 15.10. Denoting with x_s and x_r the Roche limits for a stationary and a rotating body, respectively, in the case of the Earth-Sun system we find:

$$x_{e,s} \approx 556330\,\mathrm{km} \quad x_{e,r} \approx 556980\,\mathrm{km}\,.$$

While, for the Earth-Moon system:

$$x_{m,s} \approx 9485\,\mathrm{km} \quad x_{m,r} \approx 9485\,\mathrm{km}\,.$$

We see that rotation can be neglected in both cases (remember that the approximation $x \gg r$ likely introduces an error comparable to the correction due to rotation). If a secondary body revolves about a central body at a distance equal to its Roche limit, and its rotation is synchronous with its revolution:

$$\omega^2 = \frac{GM}{x^3} = \frac{4}{3}\pi R^3 \rho_c \cdot \frac{G}{x^3}.$$

Substituting ω in 15.11, we find:

$$x = R\sqrt[3]{\frac{2\rho_c}{\rho_s - \rho_c \frac{R^3}{x^3}}},$$

which, after some manipulation, leads to the synchronous Roche limit:

$$\boxed{x_{\text{syn}} = R\sqrt[3]{\frac{3\rho_c}{\rho_s}}}. \tag{15.12}$$

10.7 Jeans Limit

The force of gravity tends to compress the nebula, while the gas pressure tends to expand it. In equilibrium, these two forces are equal. Instead of working with forces however, it is more convenient to think in terms of energy. We can equal the difference in potential energy of the nebula, following an infinitesimal contraction dr, with the work done by the gas during this contraction. Equating the differences in energy and work is just another way of equating the forces. As we found in Pr. 10.4, the potential energy of the nebula is:

$$U = -\frac{3GM^2}{5r}.$$

Differentiating the above equation:

$$dU = \frac{3GM^2}{5r^2} dr.$$

The work done by the gas during the contraction is:

$$dW = PdV.$$

Assuming the gas is ideal, we know that $P = nRT/V$, where $V = 4/3\,\pi r^3$. Substituting $dV = 4\pi r^2\, dr$, we find:

$$dW = \frac{3nRT}{r} dr.$$

Equating dU and dW:

$$\frac{GM^2}{5r} = nRT.$$

Let μ be the molar mass of the gas, we then have $n = M/\mu$. Furthermore, $M = 4/3\pi\rho r^3$, where ρ is the density of the nebula, hence $r = [3M/(4\pi\rho)]^{1/3}$. Substituting the expressions for n and r in the last equation, isolating M:

$$\boxed{M_J = \sqrt{\frac{3}{4\pi\rho}}\left(5\frac{RT}{G\mu}\right)^{3/2}}. \tag{15.13}$$

10.8 Hill sphere and Lagrangian points L_1, L_2

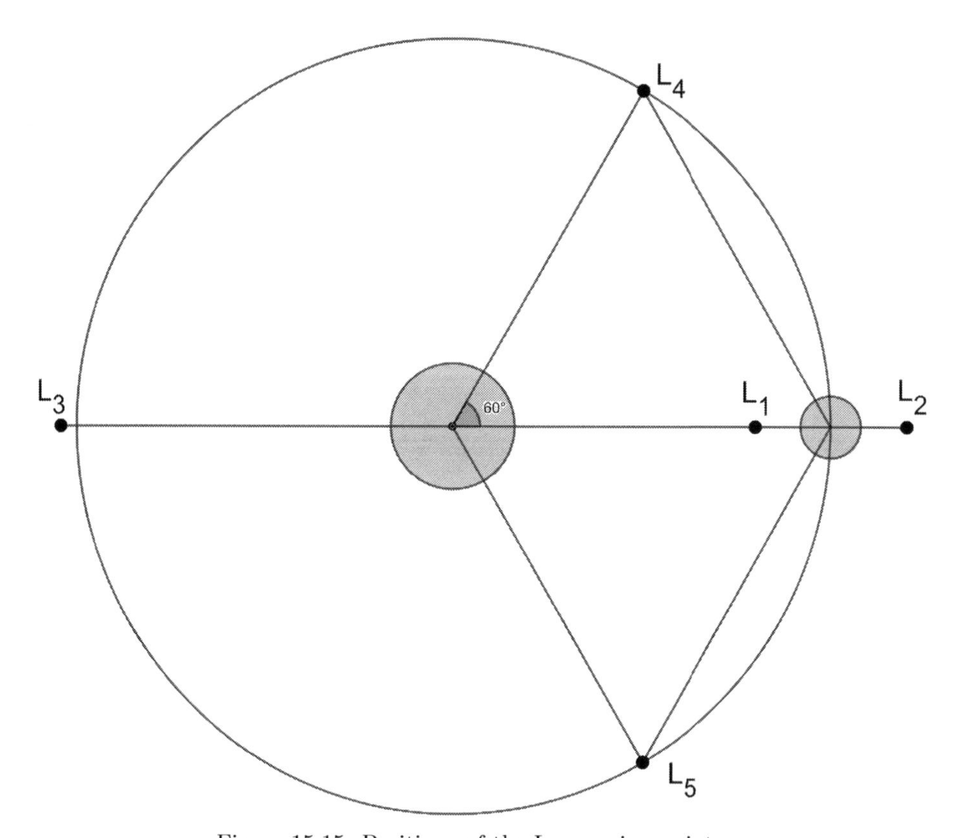

Figure 15.15: Positions of the Lagrangian points.

The Lagrangian points are five points in the Solar System whose positions relative to the Earth and Sun are fixed (Fig. 15.15). It can be shown that only L_4 and L_5 are stable points, while L_1, L_2 and L_3 are all unstable. Nonetheless, L_1 is very convenient for observing the Sun, and many probes, such as the SOHO (*Solar and Heliospheric Observatory*) and the ACE (*Advanced Composition Explorer*), have taken advantage of it. Instead, L_2, placed on the opposite side of the Earth relative to the Sun and shielded from the solar radiation, is used for studying deep space, especially in the infra-red (which requires low temperatures). Among the best known satellites in L_2 we find Gaia, and the future James Webb telescope. Now, let us calculate the position of L_1 and L_2. Let m be the mass of the Earth, M the mass of the Sun, r the distance of the Earth from the Sun and x the distance of the points L_1 and L_2 from the Earth. For now, we don't make any assumptions about the m/M ratio. The two bodies orbit around the common centre of mass, which is at a distance $[m/(M+m)]r$ from the centre of the Sun. The distance r_{L_1} between L_1 and the centre of mass corresponds to the radius of its orbit:

$$r_{L_1} = r - x - \frac{m}{M+m}r = \frac{M}{M+m}r - x\,.$$

Equating the gravitational and centrifugal forces in L_1:

$$\omega^2 r_{L_1} = \frac{GM}{(r-x)^2} - \frac{Gm}{x^2}\,,$$

where $\omega^2 = G(M+m)/r^3$ is the square of the angular velocity of revolution of the Earth. Multiplying by $(r-x)^2/GM$:

$$(1+\alpha)\left(\frac{1}{1+\alpha}-\frac{x}{r}\right)\left(1-\frac{x}{r}\right)^2 = 1-\alpha\left(1-\frac{x}{r}\right)^2$$

$$\Rightarrow \alpha\left(1-2\frac{x}{r}+\frac{x^2}{r^2}+\frac{x^3}{r^3}-2\frac{x^4}{r^4}+\frac{x^5}{r^5}\right) = 3\frac{x^3}{r^3}-3\frac{x^4}{r^4}+\frac{x^5}{r^5},$$

where $\alpha = m/M \approx 3 \cdot 10^{-6}$. Since $r \gg x$ (this hypothesis can be verified at the end, but appears reasonable given that $m/M \ll 1$), in approximating the above equation we only retain terms of zero order in x/r on the LHS (ignoring x/r, x^2/r^2, x^3/r^3 and so on), and only the third order term on the RHS. Thus:

$$3x^3/r^3 = \alpha\,.$$

It follows that:

$$x = r\sqrt[3]{\frac{m}{3M}}\,. \tag{15.14}$$

For the Earth-Sun system, we obtain $x = 149.6 \cdot 10^7\,\text{m} = 0.01\,\text{ua}$. Therefore, L_1 is at a distance from the Earth of one hundredth of an astronomical unit, in the direction of the Sun. Hence, the assumption $r \gg x$ was indeed valid. With a similar reasoning, we can show that the distance of L_2 from Earth is also equal to x (in the direction opposite the Sun). If, instead, we numerically solve the fifth order equation, we find

$$x_{L_1} = 1.4914 \cdot 10^9\,\text{m}\,,$$
$$x_{L_2} = 1.5014 \cdot 10^9\,\text{m}\,.$$

As we expect, the error is on the order of 1/100 (i.e. the size of x/r). The points L_1 and L_2 define a sphere around the Earth such that all the bodies inside it gravitationally belong to the Earth, while the bodies outside it belong to the Sun. If we take into account the eccentricity of the Earth's orbit, we need to replace $r \to a(1-e)$ in Eq. 15.14. Therefore, a more general formula for the radius of the Hill sphere is:

$$\boxed{r_\text{H} = (1-e)a\sqrt[3]{\frac{m}{3M}}}\,. \tag{15.15}$$

For the above formula to hold, the mass of the secondary body should be much smaller than the mass of the primary body, so that the approximations $m/M, x/r \ll 1$ are valid. While this is certainly true for the Earth-Sun system, it is not a very good approximation for the Earth-Moon system (indeed the Hill sphere of the Moon is about 1/6 of the Earth-Moon distance). In reality, there are other phenomena (such as radiation pressure and the *Yarkovsky effect*), that make a body at the edge of the Hill sphere unstable. Therefore, bodies are stable, in the long term, only if their distance is less than half of the Hill sphere.

10.9 Position of Lagrangian point L_3

Let m be the mass of the Earth, M the mass of the Sun, r the Earth-Sun distance and x the distance of L_3 from the point on Earth's orbit opposite the Earth. Looking at Fig. 15.16 (overleaf), we find the distances L_3-centre of mass, L_3-Sun and L_3-Earth, respect-

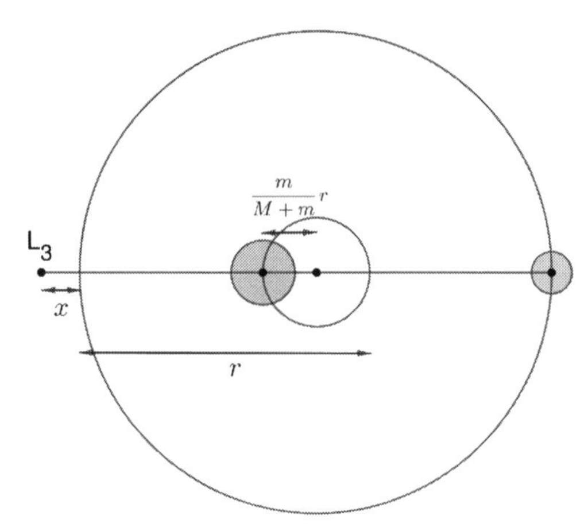

Figure 15.16: Schematics showing the orbits of the Sun and the Earth (small and large circles, respectively).

ively:

$$d_{L_3,c} = r - \frac{m}{M+m}r + x = \frac{M}{M+m}r + x\,,$$
$$d_{L_3,s} = r - 2\frac{m}{M+m}r + x = \frac{M-m}{M+m}r + x\,,$$
$$d_{L_3,e} = d_{L_3,s} + r = \frac{2M}{M+m}r + x\,.$$

Equating the centrifugal and gravitational forces at L_3, we find:

$$\omega^2 d_{L_3,c} = \frac{GM}{d^2_{L_3,s}} + \frac{Gm}{d^2_{L_3,e}}\,,$$

where $\omega^2 = G(M+m)/r^3$ is the square of the angular velocity of revolution of the Earth. Denoting by $\alpha = m/M \approx 3\cdot 10^{-6}$, similarly to the previous problem, it is possible to neglect the terms of order greater than one in x/r and α. After some manipulation, we get:

$$\frac{17}{4}\alpha = \left(3 + \frac{53}{4}\alpha\right)\frac{x}{r}\,.$$

But $\alpha x/r$ is an infinitesimal of second order, hence, to be consistent with the previous approximation, we neglect it. We finally obtain:

$$x = \frac{17}{12}\alpha r\,. \tag{15.16}$$

Therefore, the point L_3 is at a distance $x \approx 636\,\text{km}$ away from the point on Earth's orbit opposite the Earth. It can be shown that L_3 is a point of unstable equilibrium, therefore there aren't any bodies that orbit the Sun at that point. We can also write down the position of L_3 with respect to the centre of mass of the system:

$$d_{L_3,c} = \frac{M}{M+m}r + x \approx (1-\alpha)r + \frac{17}{12}\alpha r = \left(1 + \frac{5}{12}\alpha\right)r\,. \tag{15.17}$$

10.10 The three-body problem and Lagrangian points L_4, L_5

Let us place the origin of our coordinate system in the centre of mass of m_1, m_2 and m_3. Then:

$$m_1\vec{r_1} + m_2\vec{r_2} + m_3\vec{r_3} = 0\,.$$

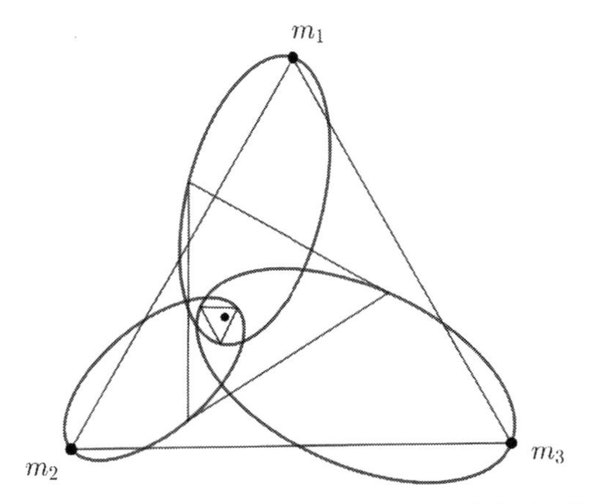

Figure 15.17: The three-body problem has a simple solution if the bodies form an equilateral triangle.

The forces acting on the first body are:

$$\vec{F_{2,1}} = -\frac{Gm_1m_2}{a_{1,2}^3}(\vec{r_1} - \vec{r_2}),$$

$$\vec{F_{3,1}} = -\frac{Gm_1m_3}{a_{1,3}^3}(\vec{r_1} - \vec{r_3}).$$

The centripetal force is:

$$\vec{F_c} = m_1\omega^2 \cdot \vec{r_1}.$$

Equating the centripetal and gravitational forces, we find:

$$\frac{Gm_1m_2}{a_{1,2}^3}(\vec{r_1} - \vec{r_2}) + \frac{Gm_1m_3}{a_{1,3}^3}(\vec{r_1} - \vec{r_3}) - m_1\omega^2 \cdot \vec{r_1} = 0.$$

Solving for $\vec{r_3}$ in the CM equation, we find $\vec{r_3} = -(m_1/m_3)\vec{r_1} - (m_2/m_3)\vec{r_2}$. Substituting in the above equation:

$$\frac{Gm_1m_2}{a_{1,2}^3}(\vec{r_1} - \vec{r_2}) + \frac{Gm_1m_3}{a_{1,3}^3}\left[\vec{r_1}\left(1 + \frac{m_1}{m_3}\right) + \frac{m_2}{m_3}\vec{r_2}\right] - m_1\omega^2 \cdot \vec{r_1} = 0$$

$$\Rightarrow \vec{r_1}\left[\frac{m_2}{a_{1,2}{}^3} + \frac{m_3}{a_{1,3}{}^3}\left(1 + \frac{m_1}{m_3}\right) - \frac{\omega^2}{G}\right] = \vec{r_2}\left[\frac{m_2}{a_{1,3}{}^3} - \frac{m_2}{a_{1,2}{}^3}\right].$$

The three bodies cannot be aligned, hence $\vec{r_1}$ and $\vec{r_2}$ cannot be on the same line. For the above equation to be satisfied, the coefficients of $\vec{r_1}$ and $\vec{r_2}$ should be both zero. We therefore obtain two equations:

$$\frac{m_2}{a_{1,3}{}^3} - \frac{m_2}{a_{1,2}{}^3} = 0,$$

$$\frac{m_2}{a_{1,2}{}^3} + \frac{m_3}{a_{1,3}{}^3}\left(1 + \frac{m_1}{m_3}\right) - \frac{\omega^2}{G} = 0.$$

The first equation tells us that $a_{1,2} = a_{1,3}$ but, as we could have chosen any of the three bodies at the beginning, by symmetry we must also have $a_{1,2} = a_{1,3} = a_{2,3} = a$. Now, the second equation gives:

$$\omega^2 = \frac{G}{a^3}(m_1 + m_2 + m_3) = \frac{GM}{a^3},$$

where M is the total mass of the system. We have shown that the three bodies all move with the same orbital period and that they are placed on the vertices of an equilateral triangle (Fig. 15.17). In the Solar System, the points L_5 and L_4 are at the same distance from the Sun as the Earth, and are placed 60° before and after the Earth, respectively.
Analysing the energy of the system in L_4 and L_5, we discover that it is a local maximum. Therefore, even a small disturbance would move the body away from the Lagrangian points and, consequently, there would be no stable orbits around L_4 and L_5. However, in the Solar System, we observe a large amount of so-called Trojans asteroids distributed in two oblong regions around the points L_4 and L_5 of Jupiter. The reason why these points can host stable orbits is due to the Coriolis force in the rotating frame of the Sun. This force is large enough to sufficiently curve the trajectory of an asteroid that moves at moderate speeds, forcing it to stay close to L_4 or L_5.

10.11 The fate of abandoned satellites

Part 1 Let r be the radius of the orbit of the satellite. Its total energy is:

$$E_{\text{tot}} = -\frac{GmM}{2r}\,.$$

Since the orbit is circular, $\langle K\rangle = K$ and $\langle U\rangle = U$. According to the Virial theorem:

$$K = -\frac{1}{2}U = \frac{GmM}{2r}\,.$$

Friction with air reduces the total energy of the satellite. The above equations shows that, as r decreases, K increases: it seems like friction increases the velocity of the satellite. However, we have not taken into account the gravitational force. In a perfectly circular orbit, the gravitational force is perpendicular to the instantaneous velocity, hence it does no work. In this case, the orbit is not perfectly circular, and the force of gravity has a small but non-zero tangential component. In the next part, we will show that the value of this component is exactly two times the friction force, so that the total force acting on the body is equal to the friction force, in magnitude, but opposite in direction.

Part 2 The work done by the friction force is $dW_a = \vec{F_a}\cdot\vec{ds}$, where $\vec{ds} = \vec{v}\,dt$ and v is the velocity of the satellite. The satellite looses energy due to friction, according to $dE_{\text{tot}} = dW_a$. Hence:

$$dE_{\text{tot}} = \vec{F_a}\cdot\vec{ds}\,.$$

Note that $\vec{F_a}$ is directed opposite to $\vec{ds}$, hence the scalar product is negative, and the energy decreases. The kinetic energy is $K = \frac{1}{2}mv^2$. Differentiating, we find:

$$dK = mv\,dv = mv\frac{\text{d}v}{\text{d}t}\frac{\text{d}t}{\text{d}s}\,ds = ma\,ds = \vec{F_{\text{tot}}}\cdot\vec{ds} = (\vec{F_g}+\vec{F_a})\cdot\vec{ds}$$
$$\Rightarrow dK = (\vec{F_g}+\vec{F_a})\cdot\vec{ds}\,.$$

Applying the Virial theorem $dE_{\text{tot}} = -dK$. Substituting the expressions obtained previously for dE_{tot} and dK, we have:

$$\vec{F_a} = -(\vec{F_g}+\vec{F_a}) \Rightarrow \vec{F_g} = -2\vec{F_a}\,.$$

The tangential component of the gravitational force is therefore equal to two times the friction force, and has opposite direction. Hence, the total force acting on the body in the tangential direction is $\vec{F_{\text{tot}}} = \vec{F_g} + \vec{F_a} = 2\vec{F_a} - \vec{F_a} = \vec{F_a}$, equal to the friction force, but opposite in direction. That is why the satellite accelerates!

Part 3 Differentiating the kinetic energy with respect to the radius (r) of the orbit:

$$\frac{dK}{dr} = -\frac{GMm}{2r^2}\,.$$

If the radius of the orbit decreases, dr is negative and dK is positive, i.e. the velocity increases. After every revolution, the radius of the orbit decreases by $dr = 100\,\text{m}$. The work done by the force acting on the satellite is:

$$dW = \oint \vec{F_{\text{tot}}} \cdot d\vec{s} = F_a \oint ds = F_a(2\pi r)\,.$$

The friction force is $F_a = c\rho v^2$. The velocity in the circular orbit is $v^2 = GM/r$, hence:

$$dW = \left(c\rho\frac{GM}{r}\right) \cdot (2\pi r) \Rightarrow dW = 2\pi c\rho GM\,.$$

This work is equal to the increase in kinetic energy after one revolution. Therefore:

$$2c\pi\rho GM = -\frac{GMm}{2r^2}dr\,.$$

Taking the magnitude of dr and isolating ρ, we find an expression for the density:

$$\rho = \frac{m}{4c\pi r^2}|dr| \approx 4 \cdot 10^{-10}\,\text{kg/m}^3\,.$$

10.12 Kepler's equation

In the usual notation, the equation of an ellipse is:

$$\frac{x^2}{a^2} + \frac{y^2}{b^2} = 1\,.$$

Placing the origin at the rightmost focus, we perform the translation $x \to x + c$:

$$\frac{(x+c)^2}{a^2} + \frac{y^2}{b^2} = 1$$

Consider a plane polar coordinate system, with $x = r\cos\theta$ and $y = r\sin\theta$. Substituting in the previous equation:

$$\frac{(r\cos\theta + c)^2}{a^2} + \frac{r^2\sin^2\theta}{b^2} = 1$$

Since $c = \sqrt{a^2 - b^2}$ and $c = a\,e$, we have $b^2 = a^2(1 - e^2)$:

$$(r\cos\theta + a\,e)^2 + \frac{r^2\sin^2\theta}{1-e^2} = a^2$$

$$r^2\left(\cos^2\theta + \frac{1-\cos^2\theta}{1-e^2}\right) + a^2e^2 + 2a\,er\cos\theta = a^2$$

$$r^2\frac{1-e^2\cos^2\theta}{1-e^2} + 2ae\,r\cos\theta - a^2(1-e^2) = 0\,.$$

Solving the quadratic equation, we find:

$$\begin{aligned}
r &= \frac{1-e^2}{1-e^2\cos^2\theta}\Big[-a\,e\cos\theta \pm \sqrt{a^2e^2r^2\cos^2\theta + a^2(1-e^2\cos^2\theta)}\Big] \\
&= \frac{1-e^2}{1-e^2\cos^2\theta}\Big[-a\,e\cos\theta \pm a\Big] \\
&= \frac{a(1-e^2)}{(1-e\cos\theta)(1+e\cos\theta)}\Big[1 \pm e\cos\theta\Big] \\
&= \frac{a(1-e^2)}{1 \pm e\cos\theta}\,.
\end{aligned}$$

Since the origin is in the rightmost focus, $r(0°) = a(1-e)$, we chose the positive sign in the previous equation. We then find:

$$r(\theta) = \frac{a(1-e^2)}{1+e\cos\theta},$$

as we wanted to show.
Let H be the intersection of the line passing through EP and the major axis. Looking at Fig. 10.8, we have:

$$\overline{\mathrm{FP}}^2 = \overline{\mathrm{PH}}^2 + \overline{\mathrm{FH}}^2.$$

Since an ellipse is just the scaled version of a circle, by factors a and b along the major and minor axis, respectively:

$$\begin{aligned} r^2 &= (b\sin E)^2 + (a\cos E - ea)^2 \\ &= a^2(1-e^2)\sin^2 E + a^2\cos^2 E + e^2a^2 - 2ea^2\cos E \\ &= a^2 + e^2a^2(1-\sin^2 E) - 2ea^2\cos E \\ &= a^2 + e^2a^2\cos^2 E - 2ea^2\cos E \\ &= a^2(1-e\cos E)^2. \end{aligned}$$

Hence:

$$r = a(1-e\cos E),$$

as we wanted to show.
To derive Kepler's equation, we differentiate the above formula with respect to time:

$$\dot{r} = ae\sin E\,\dot{E}.$$

On the other hand, r is related to the true anomaly by:

$$r(\theta) = \frac{a(1-e^2)}{1+e\cos\theta} \Rightarrow \frac{1}{r} = \frac{1+e\cos\theta}{a(1-e^2)}.$$

Differentiating the above equation with respect to time:

$$\frac{\dot{r}}{r^2} = \frac{e\sin\theta\,\dot{\theta}}{a(1-e^2)}.$$

From Kepler's second law, we know that the radius vector sweeps out equal areas in equal times. Multiplying Eq. 5.9 by r^2, setting $\theta = 0°$:

$$r^2\dot{\theta} = \omega a^2\sqrt{1-e^2},$$

where $\omega = \sqrt{GM/a^3}$ is the mean angular velocity of revolution. Substituting $\dot{\theta}$ in the previous equation:

$$\dot{r} = \omega\frac{ae\sin\theta}{\sqrt{1-e^2}}.$$

Comparing with $\dot{r} = ae\sin E\,\dot{E}$:

$$\dot{E} = \omega\frac{\sin\theta}{\sqrt{1-e^2}\sin E}.$$

Since an ellipse is just the scaled version of a circle, by factors a and b along the major and minor axis, respectively; $\overline{\mathrm{PH}} = b\,\overline{\mathrm{EH}}$, which can be written as:

$$r\sin\theta = a\sqrt{1-e^2}\sin E.$$

Comparing with the equation for $\dot{E}$, we then have:

$$r\dot{E} = \omega\, a\,.$$

Since $r = a(1 - e\cos E)$, this becomes:

$$(1 - e\cos E)\dot{E} = \omega\,.$$

Integration over time yields:

$$E - e\sin E = \omega(t - t_0)\,.$$

If t_0 is interpreted as the time elapsed since the body passed through perihelion, then $M = \omega(t - t_0)$. Hence:

$$M = E - e\sin E\,,$$

which is Kepler's equation.

10.13 Relaxation time

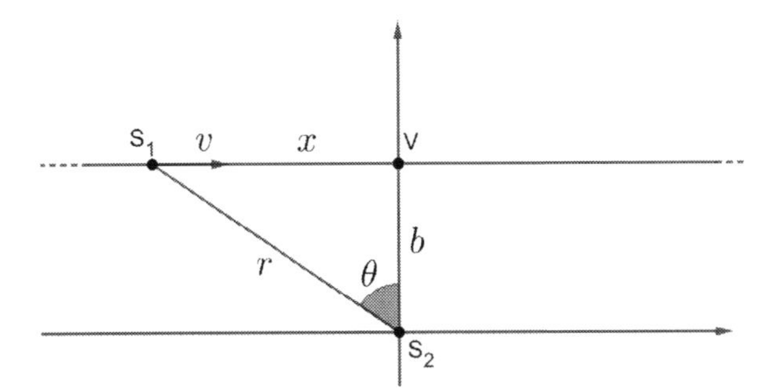

Figure 15.18: If two stars interact weakly, their motion is approximately a straight line, and only the component of the velocity perpendicular to the direction of motion is changed significantly.

If the impact parameter is large enough, the incoming star travels approximately on a straight line with velocity v, as shown in Fig. 15.18. The distance between the two stars is $r = b/\cos\theta$, hence the gravitational force along the y direction is:

$$F_y = F_g\cos\theta = \frac{Gm^2}{b^2}\cos^3\theta\,.$$

After an infinitesimal time dt, the variation of linear momentum in the y direction is $dp_y = F_y\,dt = (F_y/v)\,dx$. But $\tan\theta = x/b$, which differentiated is $1/\cos^2\theta\,d\theta = dx/b$. We then find $dp_y = Gm^2/(vb)\cos\theta\,d\theta$. Integrating between $-\pi/2$ and $\pi/2$:

$$\int_{p_{y,i}}^{p_{y,f}} dp_y = \int_{-\pi/2}^{\pi/2} \frac{Gm^2}{vb}\cos\theta\,d\theta$$

$$\Rightarrow \Delta p = \frac{2Gm^2}{vb}\,.$$

The variation of the velocity along the y direction is $\Delta v_y = \Delta p_y/m$. Clearly, this approximation is valid only in the case $\Delta v_y \ll v$, while for $\Delta v_y \gtrsim v$ the interaction is substantial and should be computed using the concepts presented in Appendix B.
We want to understand if stars for which $\Delta v_y \gtrsim v$ give rise to a greater perturbation compared to stars for which $\Delta v_y \ll v$. In the first case, the interaction is significant, but, at the same time, it is rare that two stars come very close to each other. In the second case, the

interaction is weak, but the stars can be far away from each other.
Significant interactions happen when the impact parameter is smaller than the critical value b_c:

$$\frac{2Gm}{vb} \gtrsim v \Rightarrow b \lesssim b_c \approx \frac{Gm}{v^2},$$

where we ignore numerical factors, since this is just an estimate. Therefore, a significant interaction takes place only if the incoming star passes through an area of $\sigma \approx b_c^2 \approx G^2m^2/v^4$. Let n be the number density of stars. In a volume $\sigma\, dx$ there are $n\sigma\, dx$ stars, hence, every second, a number $n\sigma\, dx/dt = n\sigma v$ of stars pass through this volume. On average, to register the passage of a star, we need to wait a time of $1 = t_{\rm gc}\, n\sigma v \Rightarrow t_{\rm gc} = 1/(n\sigma v)$. Let R be the dimension of the system and N the total number of stars. Then:

$$t_{\rm gc} \approx \frac{1}{n\sigma v} \approx \frac{R^3v^3}{N(G^2m^2)} \approx \frac{NR^3v^3}{G^2M^2}.$$

where we have used $n \approx N/R^3$, and $M = Nm$ is the total mass of the system. If the system is in gravitational equilibrium, the velocity (v) can be found from the Virial theorem, i.e. $v = \sqrt{GM/R}$. It follows that:

$$t_{\rm gc} \approx N\left(\frac{R}{v}\right)\left(\frac{R^2v^4}{G^2M^2}\right) \approx N\frac{R}{v}.$$

This is the time scale in the case of significant interactions between stars.
If the interaction is weak, $b \gg b_c$. Following a collision, the velocity of the star changes by $dv_y \approx Gm/(vb)$. Since stars are equally likely to come from any direction, we expect the mean velocity to be constant. The star therefore traces a chaotic path about the orbit it would have in the case of a continuous mass distribution. The mean square speed produced by these interaction in an interval $[b, b+db]$ and time Δt is:

$$\langle dv_y^2\rangle = (2\pi b\, db)(v\Delta t)n\left(\frac{Gm}{vb}\right)^2.$$

Integrating over all possible impact parameters, we find the total mean square speed:

$$\langle dv_y^2\rangle_{tot} \approx \Delta t\int_{b_1}^{b_2}(2\pi b\, db)(vn)\left(\frac{G^2m^2}{b^2v^2}\right) = \frac{2\pi nG^2m^2}{v}\Delta t\ln\left(\frac{b_2}{b_1}\right).$$

We can take $b_2 \approx R$ and $b_1 \approx b_c \approx Gm/v^2$. Therefore $b_2/b_1 \approx Rv^2/Gm = N(Rv^2/GM) \approx N$ if the system is in Virial equilibrium. The system significantly deviates from the model of continuous mass distribution on the time scale for which $\langle dv_y^2\rangle_{\rm tot} \approx v^2$. Isolating Δt from the above equation, it follows that:

$$\Delta t_{\rm gc} \approx \frac{v^3}{2\pi G^2m^2n\ln N} \approx \left(\frac{N}{\ln N}\right)\left(\frac{R}{v}\right) \approx \frac{t_{\rm gc}}{\ln N},$$

which is shorter than the scale time t_{gc}. This is the relaxation time.
For a galaxy, $N \approx 10^{11}$, $R \approx 10^{20}$ m and $m \approx 10^{30}$ kg, hence $R/v \approx \sqrt{R^3/GNm} \approx 10^{15}$ s $\approx 10^9$ years, therefore $t_{\rm gc} \approx 10^{17}$ years. This is much larger than the age of the universe, therefore galaxies are considered to be "relaxed", and can be approximated with a continuous mass distribution.
For a star cluster, $N \approx 10^6$, $R \approx 10^{17}$ m and $m \approx 10^{30}$ kg, thus $R/v \approx \sqrt{R^3/GNm} \approx 10^{13}$ s $\approx 10^5$ years, and $t_{\rm gc} \approx 5\cdot 10^9$ years. The age of the universe is on the same order of magnitude, therefore stellar interactions are important, and effectively randomize the velocities of all stars in the cluster. It is for this reason that star clusters are spherical.

11. Motion of the Planets

11.1 Great opposition

We neglect the precession of Mars' apse line. The sidereal period of Mars is $P_♂ = 1.88082$ years, while the synodic period can be calculated using Eq. 11.3:

$$P_{\text{syn,yr}} = \frac{P_{♂,\text{yr}}}{P_{♂,\text{yr}} - 1} = 2.1353\,\text{years}\,.$$

At the end of a synodic period, Mars will have covered an angle corresponding to $P_{\text{syn}}/P_{\text{sid}} = 1.13531$ of a full circle, and will be displaced by 0.13531 of a full circle, relative to the starting position. The time taken to return to the same point with respect to its apse line is $1/0.13531 = 7.39067$ synodic periods, which corresponds to $7.39067 \cdot 2.1353 = 15.78$ years. Therefore, great oppositions happen approximately once every $15 - 17$ years. A more precise cycle is five times longer than the previous one, with a periodicity of $78.906 \approx 79$ years.

12. Orbital Manoeuvres

12.1 Circular orbits

The most effective way to put a satellite in an equatorial orbit around the Earth is to launch it from a location on the equator, in the same direction as the rotation of the Earth, from west to est. Indeed, the Earth completes a rotation about its axis every 86164 seconds, thus its angular velocity of rotation is $\omega = 2\pi/T = 7.29 \cdot 10^5\,\text{rad/s}$. The equatorial radius is $R_\oplus = 6378\,\text{km}$, therefore a satellite launched in this way has an initial velocity of $\omega R_\oplus = 465.1\,\text{m/s}$. The velocity of a circular orbit at sea level is $v = \sqrt{GM/R_\oplus} \approx 7.9\,\text{km/s}$, hence the minimum Δv required to put a satellite in circular orbit is approximately 7.4 km/s.
To insert a satellite in a polar orbit, perpendicular to the equator, it is best to launch it from a location near the poles, so that the component of the velocity along the equator (which will have to be cancelled) is the smallest possible. The minimum Δv required is simply equal to the velocity of a circular orbit at sea level, which is 7.9 km/s.
To escape Earth's gravitational field, the velocity must be equal to the escape velocity from the Earth, i.e $v = \sqrt{2GM/R_\oplus} \approx 11.2\,\text{km/s}$. If the satellite is launched from the equator, in the same direction as the rotation of the Earth, the Δv is reduced to 10.7 km/s.
In reality, the Δv should be a bit larger than the previously stated limits. Of course, it is not possible to put a body in circular orbit at sea level, but we must first reach a height of a few hundred kilometres above the surface. This requires an additional Δv to contrast the gravitational force (which adds about 1 km/s), while the velocity of the new circular orbit is essentially the same, since $v = \sqrt{GM/(R_\oplus + h)}$ and $h \ll R_\oplus$. In addition, the friction force with the atmosphere (1.5 km/s) also tends to slow down the satellite. Therefore, it is reasonable to assume we need a further 3 km/s above the previously stated limits.

12.2 Changing the plane of the orbit

The manoeuvre should be carried out in the ascending or descending node, depending on whether the desired final orbit is above or below the starting orbit. Let v_0 be the velocity of the circular orbit. Then, the projections of $\vec{v_0}$ on the plane of the final orbit and on the perpendicular to it are:

$$v_{i_\parallel} = v_0 \cos i\,, \qquad v_{i_\perp} = v_0 \sin i\,.$$

The final velocity must be:

$$v_{f_\parallel} = v_0\,, \qquad v_{f_\perp} = 0\,.$$

Hence, the necessary Δv is:

$$\Delta v_{\parallel} = v_{f_{\parallel}} - v_{i_{\parallel}} = v_0(1-\cos i)\,,$$
$$\Delta v_{\perp} = v_{f_{\perp}} - v_{i_{\perp}} = -v_0 \sin i\,.$$

Adding the two components in quadrature:

$$\Delta v_{\text{tot}} = \sqrt{\Delta v_{\parallel}^2 + \Delta v_{\perp}^2} = \sqrt{v_0^2(1-\cos i)^2 + v_0^2 \sin^2 i}$$
$$= v_0\sqrt{1+\cos^2 i - 2\cos i + \sin^2 i} = \sqrt{2}v_0\sqrt{1-\cos i}\,.$$

Using the half-angle formula in Eq. A.22, we find:

$$\Delta v_{\text{tot}} = 2v_0 \sin\left(\frac{i}{2}\right).$$

Usually, satellites are directly launched with the desired orbital inclination, in order to avoid wasting fuel with this orbital manoeuvre. A satellite in a circular orbit, at a height of $h = 200\,\text{km}$ above the surface, would require $\Delta v = 270\,\text{m/s}$ for a change of 1° in its orbital inclination. If the expulsion speed is $4.5\,\text{km/s}$ and the mass of the satellite is $1000\,\text{kg}$, we need an additional $\Delta m = 60\,\text{kg}$ of fuel for this manoeuvre. However, to bring additional fuel in orbit, we also require more fuel to lift the satellite. Suppose $\Delta v_{\text{orb}} \approx 9\,\text{km/s}$ is needed to place the satellite into orbit (taking into account friction with air and the effect of gravity). Then:

$$\Delta v_{\text{orb}} = v_e \ln\frac{m_i}{m_f}\,.$$

If an additional mass of fuel Δm must be left at the end, the initial mass must be greater by an amount ΔM:

$$\Delta v_{\text{orb}} = v_e \ln\frac{m_i + \Delta M}{m_f + \Delta m}\,.$$

Dividing the above equations, we find:

$$\ln\frac{m_i}{m_f} = \ln\frac{m_i+\Delta M}{m_f+\Delta m} \Rightarrow \Delta M = \Delta m\frac{m_i}{m_f}\,,$$

where $m_i/m_f = e^{v_{\text{orb}}/v_e} \approx 7$. Hence, the additional mass of fuel needed to launch the satellite is:

$$\Delta M = \Delta m\frac{m_i}{m_f} = 420\,\text{kg}\,,$$

approximately $40\,\%$ of the mass of the satellite itself.

12.3 Third cosmic velocity

The trajectory of the probe escaping the Solar System can be divided into two parts.
First, the satellite moves far away from Earth, to enter the sphere of influence of the Sun. The energy needed to achieve this is approximately equal to the energy required to escape the gravitational field of the Earth entirely. Indeed, the gravitational potential energy is inversely proportional to the distance, thus the error made in assuming that the final distance is infinite rather than the Hill sphere of the Earth (about 1.5 million kilometres) is not significant. Denoting with v_i the initial speed of the probe and with v_f its speed upon entering the sphere of influence of the Sun, applying energy conservation:

$$\frac{1}{2}mv_i^2 - \frac{GmM_\oplus}{R_\oplus} = \frac{1}{2}mv_f^2 \Rightarrow v_f = \sqrt{v_i^2 - \frac{2GM_\oplus}{R_\oplus}}\,.$$

In the second part, the gravitational field of the Earth can be neglected. Now, the probe is at a distance $d_\oplus$ from the Sun and is moving with a velocity of $v_f + v_\oplus$ in the frame of the Sun. To find the minimum velocity required to escape the Solar System, we require the speed of the probe to be zero at infinity. Therefore, the probe must travel in a parabolic orbit, with energy $E = 0$:

$$\frac{1}{2}m(v_f + v_\oplus)^2 - \frac{GmM_\odot}{d_\oplus} = 0\,.$$

Solving for v_f:

$$v_f = (\sqrt{2} - 1)\sqrt{\frac{GM_\odot}{d_\oplus}}\,.$$

Substituting v_f in the first equation and solving for v_i, we find:

$$v_i = \sqrt{\frac{2GM_\oplus}{R_\oplus} + (3 - 2\sqrt{2})\frac{M_\odot}{d_\oplus}} \approx 16.6\,\mathrm{km/s}\,.$$

If the satellite is launched from the equator, the required velocity is smaller by about $0.5\,\mathrm{km/s}$, therefore the third cosmic velocity is approximately $v \approx 16\,\mathrm{km/s}$. Notice that the energy scales with the square of the velocity. Since the escape velocity from Earth is $11.2\,\mathrm{km/s}$, and from the Sun is $12.3\,\mathrm{km/s}$, a quicker way to find v_i is by adding in quadrature the two contributions: $v_i = \sqrt{11.2^2 + 12.3^2} - 0.5 \approx 16\,\mathrm{km/s}$.
To collide with the Sun, the probe first needs to escape Earth's gravitational field, which requires a velocity of $11.2\,\mathrm{km/s}$. After this, the probe would orbit the Sun, at the same distance and with the same velocity as the Earth. To reach the centre, the angular momentum of the probe needs to be zero (see Pr. 12.5), hence it is necessary to cancel all of its tangential velocity, which requires a Δv equal to Earth's orbital velocity. To minimize this Δv, we chose to launch the probe when the Earth is at aphelion, its velocity being:

$$-\frac{GmM}{2a} = -\frac{GmM}{a} + \frac{1}{2}mv_a^2 \Rightarrow v_a = \sqrt{\frac{1-e}{1+e}\frac{GM}{a}} = 29.3\,\mathrm{km/s}\,.$$

Since energy scales with the square of the velocity, the speed needed to collide with the Sun is $v_c = \sqrt{11.2^2 + 29.3^2} - 0.5 \approx 31\,\mathrm{km/s}$. Hence, escaping the Sun requires less energy than colliding with it!
Notice that, in both cases, we must deliver at take off all the Δv the probe needs for both parts of its motion. By doing so, we take advantage of the fact that the energy is proportional to the square of the velocity (Oberth effect). In fact, if we only gave the probe enough energy to escape the Earth first, and then fired the engine again to escape the Sun, the total Δv needed would have been $11.2 + 12.3 - 0.5 = 23\,\mathrm{km/s}$, instead of $\sqrt{11.2^2 + 12.3^2} - 0.5 \approx 16\,\mathrm{km/s}$.

12.4 Rocket in a gravitational field

While the linear momentum is conserved for a rocket moving in free space, this is not true when the rocket is in a gravitational field of strength g (for instance, it is escaping the Earth), since its momentum changes by $dp = -mg\,dt$ after a time dt. Looking at Fig. 12.1:

$$dp = (m + dm)(v + dv) - (-dm)\cdot(v_e - v) - mv = -mg\,dt\,.$$

Simplifying, we obtain:

$$m\,dv + v_e\,dm = -mg\,dt\,.$$

Denoting with $r = -dm/dt$ the rate of exhaustion, it follows that $dt = -dm/r$ (for $r \neq 0$), and substituting in the above equation:

$$dv = -\left(\frac{v_e}{m} - \frac{g}{r}\right)dm\,.$$

If r is independent of time, we can easily integrate both sides, to get:

$$\Delta v = v_e \ln\frac{m_i}{m_f} - \frac{g}{r}(m_i - m_f)\,. \tag{15.18}$$

If the gravitational field goes to zero, or the rate of fuel consumption goes to infinity, the previous equation reduces to Eq.12.1, which describes a rocket moving in empty space. Therefore, the gravitational field reduces the effective Δv a rocket can deliver. This is the effect of gravity mentioned in Pr. 12.1. If $v_e/m = g/r$, then $dv = 0$, therefore the rocket hovers at a constant height. If $r = 0$, we must revert back to the equation expressing the conservation of linear momentum to deduce that $dv/dt = -g$, as expected.

12.5 Reaching the centre

We assume the potential has the form:

$$U = -\frac{\alpha}{r^k}\,.$$

We want to find out what k needs to be for the body to reach the centre, with non-zero angular momentum ($L \geqslant 0$). The body reaches the centre if the radial velocity, for $r \to 0$, is positive, i.e. if the radial kinetic energy is positive in this limit:

$$\frac{1}{2}m\dot{r}^2 = E - U(r) - \frac{L^2}{2mr^2} > 0\,.$$

It follows that r can take any arbitrary small value only if:

$$[r^2 U(r)]_{r\to 0} < -\frac{L^2}{2m} \Rightarrow [r^{-k+2}]_{r\to 0} > \frac{L^2}{2m\alpha}\,.$$

Let us distinguish three cases:

- if $k > 2$, then $\infty > L^2/2m\alpha$ is always true, therefore the body can reach the centre;
- if $k = 2$, then $1 > L^2/2m\alpha$ and the body reaches the centre only if $L < \sqrt{2m\alpha}$;
- if $k < 2$, the body never reaches the centre if $L \neq 0$.

The last case corresponds to the gravitational potential ($k = 1$). This is the reason why is necessary to cancel the tangential speed of the probe (so that $L = 0$) in order to reach the centre of the Sun in Pr. 12.3.

12.6 Gravitational slingshot

Let v_0 be the velocity of the probe with respect to the Sun once it has left the gravitational field of the Earth. Let $d_\oplus$ be the radius of the Earth's orbit and $d_p = xd_\oplus$ the radius of the orbit of the planet which is used as a gravitational slingshot. Let v_t and v_r be the tangential and radial velocities of the probe when it is at a distance d_p from the Sun, with mass M (Fig. 15.19). Applying conservation of energy and angular momentum:

$$\frac{1}{2}m{v_0}^2 - \frac{GMm}{d_\oplus} = \frac{1}{2}m({v_t}^2 + {v_r}^2) - \frac{GMm}{xd_\oplus}$$

$$md_\oplus v_0 = mxd_\oplus v_t \Rightarrow v_t = \frac{v_0}{x}\,.$$

Substituting v_t, obtained from the second equation into the first, and isolating v_r:

$$v_r = \sqrt{{v_0}^2\Big(1 - \frac{1}{x^2}\Big) - 2\frac{GM}{d_\oplus}\Big(1 - \frac{1}{x}\Big)}\,.$$

The velocities of the Earth and of the planet are, respectively:

$$\frac{GmM}{d_\oplus^2} = m\frac{{v_t}^2}{d_\oplus} \Rightarrow {v_\oplus}^2 = \frac{GM}{d_\oplus}\,,$$

$$\frac{GmM}{d_\oplus^2 x^2} = m\frac{{v_p}^2}{xd_\oplus} \Rightarrow {v_p}^2 = \frac{1}{x}{v_\oplus}^2\,.$$

Hence, v_r can be written as:

$$v_r = \sqrt{{v_0}^2\left(1 - \frac{1}{x^2}\right) - 2v_\oplus^2\left(1 - \frac{1}{x}\right)}\,.$$

Requiring that the term under the square root is positive, we have a condition on v_0:

$$\frac{v_0}{v_\oplus} \geqslant g(x) = \sqrt{\frac{2x}{1+x}}\,.$$

If $v_0/v_\oplus$ were smaller than $g(x)$, the probe wouldn't have enough energy to reach the planet. Let us now consider the reference system of the planet. We assume that the gravitational field of the Sun can be neglected close to the planet. Let v_{rel} be the relative velocity between the probe and the planet when the probe is just entering its sphere of influence:

$$v_{\text{rel}} = \sqrt{(v_t - v_p)^2 + {v_r}^2}\,.$$

The probe will travel on a hyperbolic orbit, with speed at infinity v_{rel} and some angle of deviation (Fig. 15.20). In the reference system of the planet, it appears not much has happened; however, in the frame of the Sun, the velocity of the probe has increased. In the best case, the probe is deviated in such a way that its velocity, upon exiting the sphere of influence of the planet, is directed along the velocity of the planet. Therefore, in the frame of the Sun, the final velocity of the probe is $v_f = v_{\text{rel}} + v_p$. If this velocity is greater than the escape velocity from the Sun at that distance, which is $\sqrt{2}v_p$, then the probe will leave the Solar System. The condition is therefore $v_{\text{rel}} \geqslant (\sqrt{2}-1)v_p$. Substituting v_{rel} as a function of v_t, v_p and v_r, we obtain:

$$\frac{v_0}{v_\oplus} \geqslant \frac{1}{\sqrt{x^3}} + \sqrt{\frac{1}{x^3} + 2 - \frac{\sqrt{8}}{x}} \equiv f(x)\,.$$

We therefore have two constraints on the velocity v_0. The first ensures the probe does indeed reach the planet, while the second is needed for the planet to escape the Solar System.

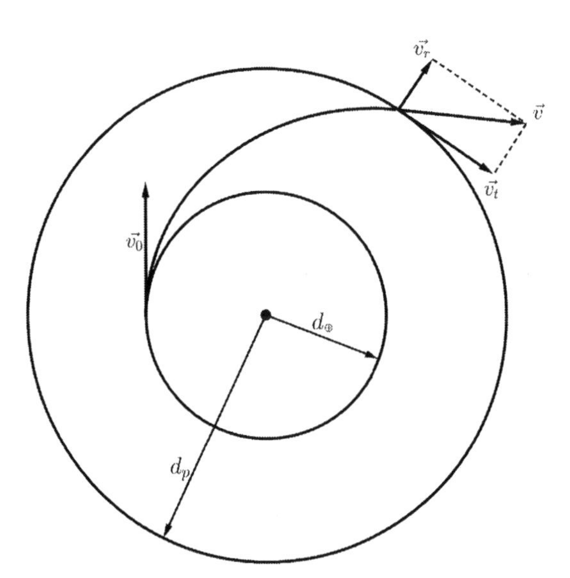

Figure 15.19: Applying conservation of energy and angular momentum, it is possible to find the components of the velocity of the probe when it reaches the other planet.

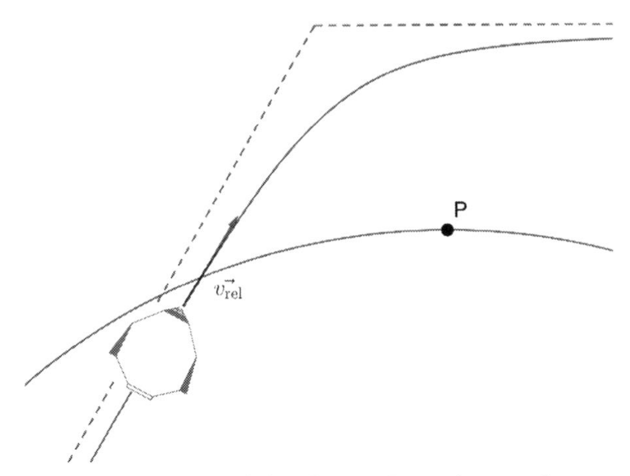

Figure 15.20: In the reference system of the planet, the probe travels on an hyperbolic orbit, with velocity $\vec{v_{\rm rel}}$ at infinite distance. The planet deflects the path of the probe, therefore changing its velocity in the reference system of the Sun.

Plotting both functions on the same graph, we see that $f(x)$ is always greater than $g(x)$ for $x < 2 + \sqrt{8}$ (Fig. 15.21). At this point, the two functions have the same value, and, for $x > 2 + \sqrt{8}$, $g(x)$ is greater than $f(x)$. The minimum velocity is therefore:

$$\frac{v_0}{v_\oplus} = \begin{cases} \dfrac{1}{\sqrt{x^3}} + \sqrt{\dfrac{1}{x^3} + 2 - \dfrac{\sqrt{8}}{x}} & \text{for} \quad 1 \leqslant x \leqslant 2 + \sqrt{8} \\ \sqrt{\dfrac{2x}{1+x}} & \text{for} \quad x > 2 + \sqrt{8}\,. \end{cases}$$

Solving $f(x)' = 0$, we find the planet that gives the lowest possible velocity v_0:

$$x = \left(9 + \sqrt{81 - 24\sqrt{8}}\right)/8 = 1.58\,.$$

Surprisingly, this value is very close to the Sun-Mars distance (1.52 au). Therefore, Mars is the best planet to use as a gravitational slingshot.

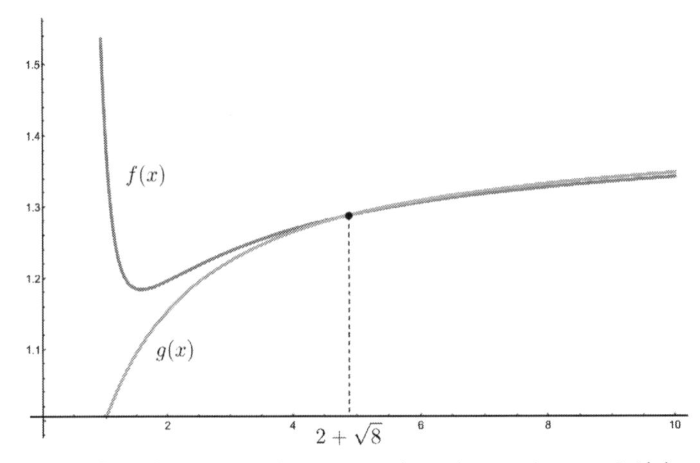

Figure 15.21: The velocity must be greater than the maximum of $f(x)$ and $g(x)$.

Appendix

A

Mathematics

A.1 Vectors

The simplest physical quantity one can imagine is the *scalar*, specified only by its magnitude, a number complete with units in which it is measured. Examples of scalar quantities are temperature, energy, time and density.
A vector is a quantity that requires both a *magnitude*, and a *direction* in space to specify it completely; we may think of it as an arrow in space. Examples of vectors are velocity, acceleration, force, linear and angular momentum.

Addition and Subtraction

There are two ways to sum vectors. With the parallelogram method, we translate the first vector in such a way that its tail coincides with the tail of the second vector. As you can see in Fig. A.1 a), the vector sum is the diagonal of the parallelogram formed by these two vectors. With the head-to-tail method, we translate the first vector in such a way that its tail coincides with the head of the second vector. In Fig. A.1 b), the resultant vector is the vector whose head and tail are the head and tail of the first and second vector, respectively. The sum of two vectors is associative, i.e. $\vec{a}+\vec{b}=\vec{b}+\vec{a}$. Subtraction is similar to addition, since $\vec{a}-\vec{b}=\vec{a}+(-\vec{b})$, where $-\vec{b}$ is the vector with the same magnitude as $\vec{b}$, but opposite direction.

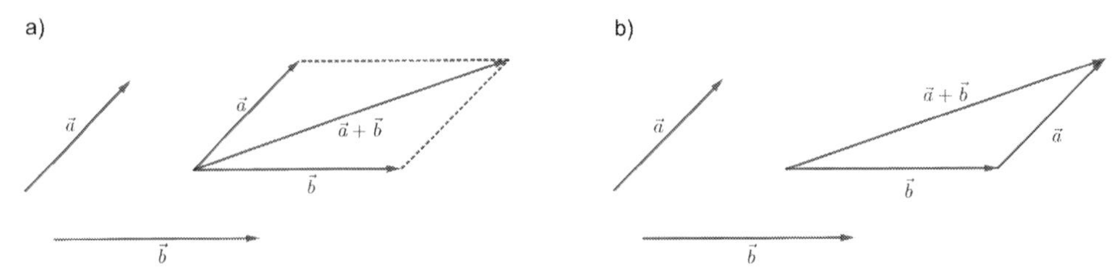

Figure A.1: Vector addition using: a) parallelogram and b) head-to-tail methods.

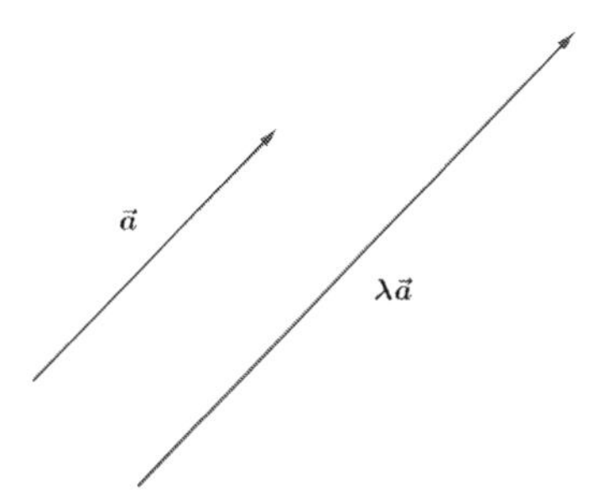

Figure A.2: Multiplication of a vector $\vec{a}$ by a scalar λ.

Multiplication by a scalar

Not to be confused with the scalar product, the multiplication of a vector $\vec{a}$ by a scalar λ returns a vector with magnitude $|\lambda||\vec{a}|$ (where $|\vec{a}|$ denotes the magnitude of $\vec{a}$), and the same direction as $\vec{a}$ if $\lambda > 0$, otherwise opposite direction (Fig. A.2). The following properties hold:

$$(\lambda\mu)\vec{a} = \lambda(\mu\vec{a}) = \mu(\lambda\vec{a}),$$
$$\lambda(\vec{a} + \vec{b}) = \lambda\vec{a} + \lambda\vec{b},$$
$$(\lambda + \mu)\vec{a} = \lambda\vec{a} + \mu\vec{a}.$$

Basis vectors and components

Given three distinct and non-coplanar vectors $\vec{u_1}, \vec{u_2}, \vec{u_3}$, it is possible to write any other vector $\vec{a}$ in space as:

$$\vec{a} = a_1\vec{u_1} + a_2\vec{u_2} + a_3\vec{u_3}.$$

The three vectors $\vec{u_1}, \vec{u_2}, \vec{u_3}$ are said to form a *basis*, and the scalars a_1, a_2, a_3 are called the components of the vector $\vec{a}$ with respect to the basis. We say that the vector has been resolved into components. The basis is called *orthogonal* if $\vec{u_1}, \vec{u_2}, \vec{u_3}$ are all perpendicular to each other; *orthonormal* if, additionally, all three vectors have unit modulus. An example of an orthonormal basis is the Cartesian basis, specified by three *unit vectors* $\hat{\imath}, \hat{\jmath}, \hat{k}$ which point along the x, y, z axes, respectively. We can therefore write $\vec{a}$ as:

$$\vec{a} = a_x\hat{\imath} + a_y\hat{\jmath} + a_z\hat{k}, \tag{A.1}$$

where a_x, a_y, a_z are the projections of $\vec{a}$ on the axes x, y, z, respectively. Using a Cartesian basis, it is easy to write down vector addition:

$$\begin{aligned}\vec{a} + \vec{b} &= (a_x\hat{\imath} + a_y\hat{\jmath} + a_z\hat{k}) + (b_x\hat{\imath} + b_y\hat{\jmath} + b_z\hat{k}) \\ &= (a_x + b_x)\hat{\imath} + (a_y + b_y)\hat{\jmath} + (a_z + b_z)\hat{k}.\end{aligned}$$

Multiplication of $\vec{a}$ by a scalar can be written as:

$$\lambda\vec{a} = \lambda(a_x\hat{\imath} + a_y\hat{\jmath} + a_z\hat{k}) = (\lambda a_x)\hat{\imath} + (\lambda a_y)\hat{\jmath} + (\lambda a_z)\hat{k}.$$

Magnitude of a vector

The magnitude of a vector $\vec{a}$, denoted by $|\vec{a}|$, can be written in terms of its Cartesian components as:

$$|\vec{a}| = \sqrt{a_x^2 + a_y^2 + a_z^2}\,, \tag{A.2}$$

which can be viewed as the Pythagorean theorem in three dimensions. A vector whose magnitude equals unity is called a *unit vector*. The unit vector in the direction $\vec{a}$ is usually denoted by $\hat{a}$, and may be evaluated as:

$$\hat{a} = \frac{\vec{a}}{|\vec{a}|}\,. \tag{A.3}$$

Scalar product

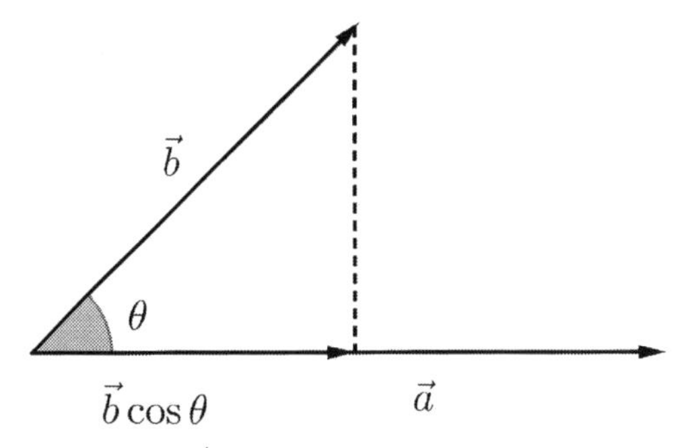

Figure A.3: The scalar product $\vec{a}\cdot\vec{b}$ equals the magnitude of $\vec{a}$ multiplied by the projection of $\vec{b}$ onto $\vec{a}$, or vice versa.

The scalar product of two vectors $\vec{a}$ and $\vec{b}$, denoted by $\vec{a}\cdot\vec{b}$, is the scalar:

$$\vec{a}\cdot\vec{b} = |\vec{a}||\vec{b}|\cos\theta\,, \tag{A.4}$$

where θ is the angle between the two vectors, placed tail to tail or head to head. Thus, the scalar product $\vec{a}\cdot\vec{b}$ equals the magnitude of $\vec{a}$ multiplied by the projection of $\vec{b}$ onto $\vec{a}$, or vice versa (Fig. A.3). If $\vec{a}$ is perpendicular to $\vec{b}$, their scalar product is zero. If $\vec{a}$ and $\vec{b}$ are parallel to each other and point in the same direction, their scalar product is $|\vec{a}||\vec{b}|$, otherwise it is $-|\vec{a}||\vec{b}|$, if their direction is opposite. An example of scalar product is the work $W = \vec{F}\cdot\vec{r}$.

Vector product

The vector product of $\vec{a}$ and $\vec{b}$, denoted by $\vec{a}\times\vec{b}$, is the vector perpendicular to both $\vec{a}$ and $\vec{b}$, with magnitude:

$$\vec{a}\times\vec{b} = |\vec{a}||\vec{b}|\sin\theta\,, \tag{A.5}$$

where θ is the angle between the two vectors, placed tail to tail or head to head (Fig. A.4). The direction can be found with the *right hand rule*: if the thumb is pointed in the direction of the first vector, the index in the direction of the second vector, then the middle finger gives the direction of their vector product. If we exchange the order of the two vectors, their vector product changes sign, as you can verify with the right hand rule. Therefore, the vector product is *anti-commutative*: $\vec{a} \times \vec{b} = -\vec{b} \times \vec{a}$. From its definition, we see that the vector product has the very useful property that if $\vec{a} \times \vec{b} = 0$, then either $\vec{a}$ is parallel or anti-parallel to $\vec{b}$ (unless either of them is zero). We also note that $\vec{a} \times \vec{a} = 0$. An example of vector product is the torque $\vec{\tau} = \vec{F} \times \vec{r}$.

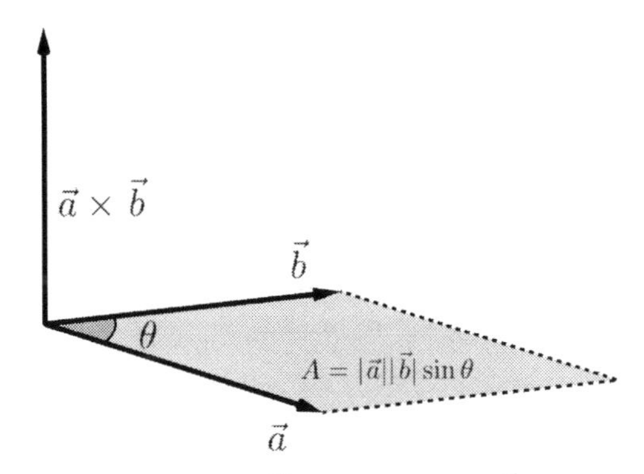

Figure A.4: The vector product of $\vec{a}$ and $\vec{b}$, denoted by $\vec{a} \times \vec{b}$, is the vector perpendicular to both $\vec{a}$ and $\vec{b}$, with magnitude equal to the area A of the parallelogram formed by the two vectors. The direction can be found with the right hand rule.

A.2 Conic Sections

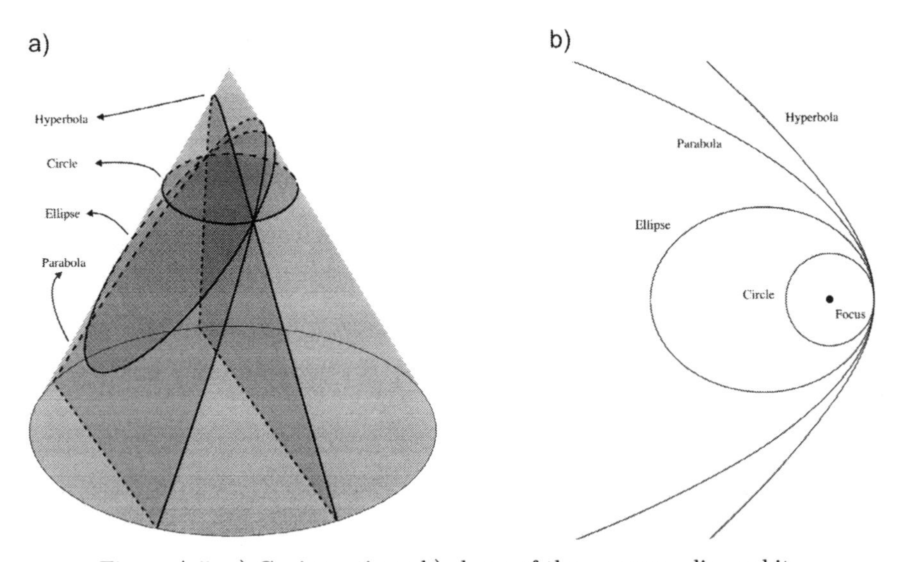

Figure A.5: a) Conic sections, b) shape of the corresponding orbits.

As the name suggests, conic sections are curves obtained from the intersection of a circular cone with a plane (Fig. A.5).

Circle

The circle is the locus of points in the plane that are equidistant from a fixed point, called the *centre*. Let r be the radius of the circle, i.e. the constant distance of any point P= (x, y) on the circle to the centre C= $(0, 0)$. Then:

$$\boxed{x^2 + y^2 = r^2}\,.$$

Ellipse

The ellipse is the locus of points in the plane such that the sum of the distances from two fixed points, called *foci*, is constant. Consider a Cartesian coordinate system and let the position of the two foci be $\mathrm{F}_1 = (-c, 0)$ and $\mathrm{F}_2 = (c, 0)$. Take a point P= (x, y) on the ellipse, and let $2a$ be the sum of the distances of P from the two foci. Then:

$$2a = \sqrt{(x+c)^2 + y^2} + \sqrt{(x-c)^2 + y^2}\,.$$

Moving the second square root to the LHS and squaring both sides:

$$\begin{aligned}
\left(2a - \sqrt{(x-c)^2 + y^2}\,\right)^2 &= (x+c)^2 + y^2 \\
4a^2 + (x-c)^2 + y^2 - 4a\sqrt{(x-c)^2 + y^2} &= (x+c)^2 + y^2 \\
4a^2 - 4xc &= 4a\sqrt{(x-c)^2 + y^2} \\
a - x\frac{c}{a} &= \sqrt{(x-c)^2 + y^2}\,.
\end{aligned}$$

Squaring again both sides:

$$\begin{aligned}
a^2 + x^2\left(\frac{c}{a}\right)^2 - 2xc &= x^2 + c^2 - 2xc + y^2 \\
a^2 - c^2 &= -x^2\left(\frac{c}{a}\right)^2 + x^2 + y^2 \\
a^2 - c^2 &= x^2\left[1 - \left(\frac{c}{a}\right)^2\right] + y^2\,.
\end{aligned}$$

For simplicity, let us define $b^2 = a^2 - c^2$. The above equation becomes:

$$x^2\frac{b^2}{a^2} + y^2 = b^2\,.$$

Dividing both sides by b^2, we obtain the canonical equation of an ellipse:

$$\boxed{\frac{x^2}{a^2} + \frac{y^2}{b^2} = 1}\,. \qquad \text{(A.6)}$$

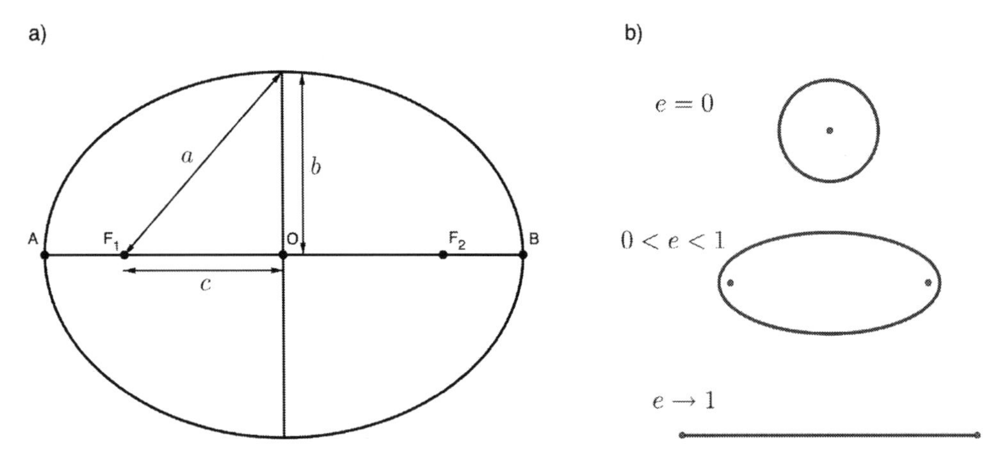

Figure A.6: a) Schematics of an ellipse, b) the degree of flattening along the semi-major axis increases with increasing eccentricity.

Here, a is the *semi-major axis*, b the *semi-minor axis* and c the *semi-focal distance*, as shown in Fig. A.6 a). The ratio between the semi-focal distance and the semi-major axis is called the *eccentricity* of the ellipse, and is denoted by e:

$$e = \frac{c}{a} \quad \text{with} \quad 0 < e < 1\,.$$

The eccentricity is always contained in the range 0 to 1, and measures the degree of flattening along the major axis. As you can see in Fig. A.6 b), if $a = b$, then $e = 0$, and the ellipse reduces to a circle. With increasing eccentricity, the ellipse is more and more flattened along the major axis until, in the degenerate case $e \to 1$, it becomes a segment connecting the two foci.
The area of a sector on an ellipse between the major axis and a segment forming an angle θ with it is:

$$A(\theta) = \frac{1}{2} a\, b \arctan\left(\frac{a}{b}\tan\theta\right). \tag{A.7}$$

Which is easy to prove by thinking of an ellipse as a circle scaled by factors a and b along the x and y axes, respectively. The total area of an ellipse is:

$$A = \pi a\, b\,. \tag{A.8}$$

Parabola

The parabola is the locus of points in the plane that are equidistant from a fixed line, called *directrix*, and a fixed point, called *focus* (Fig. A.7, overleaf). The line passing through the focus and perpendicular to the directrix is the *axis of symmetry* of the parabola.
Consider a Cartesian coordinate system and let F$= (p, 0)$ be the position of the

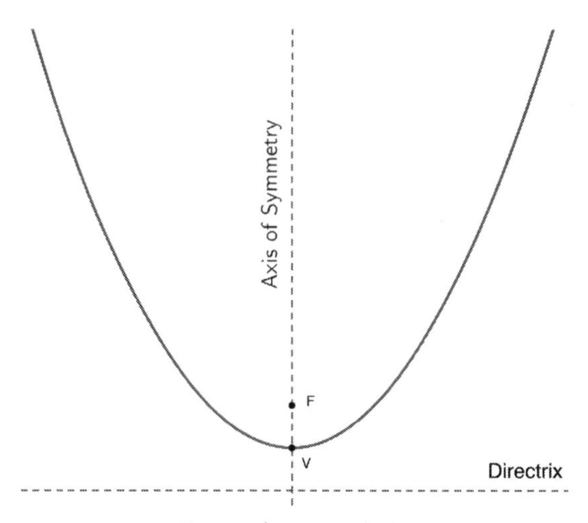

Figure A.7: Parabola

focus and $x = -p$ the equation defining the directrix. Taking a point P$= (x, y)$ on the parabola, we have:

$$\sqrt{(x-p)^2 + y^2} = x + p\,.$$

Squaring both sides:

$$x^2 + p^2 - 2xp + y^2 = x^2 + p^2 + 2px$$

Hence, the canonical equation for a parabola, with focus in F$= (p, 0)$ and directrix $x = -p$, is:

$$\boxed{y^2 = 4px}\,. \tag{A.9}$$

An important property of the parabola is that any incident ray parallel to its axis of symmetry is reflected in its focus.

Hyperbola

The hyperbola is the locus of points in the plane such that the difference in the distances from two fixed points, called foci, is constant (Fig. A.8). Consider a Cartesian coordinate system and let $F_1 = (-c, 0)$ and $F_2 = (c, 0)$ be the position of the two foci. Take a point P$=(x, y)$ on the hyperbola, and let $2a$ be the difference of the distances from the two foci. Then:

$$2a = \left|\,\sqrt{(x+c)^2 + y^2} - \sqrt{(x-c)^2 + y^2}\,\right|,$$

where we have taken the modulus of the difference since we don't know, a priori, which distance is greater. Moving the second square root to the LHS,

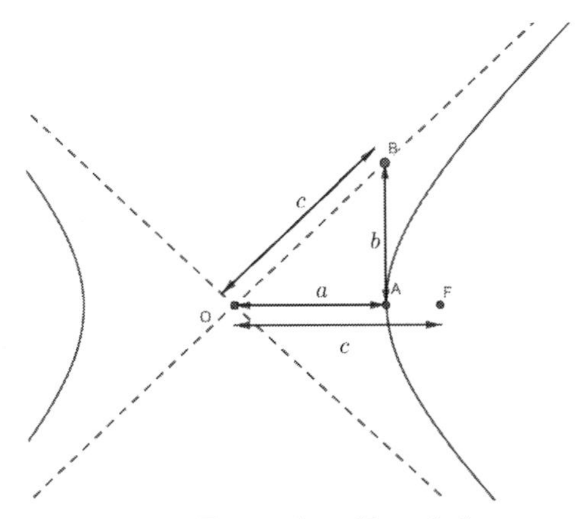

Figure A.8: Hyperbola

squaring both sides, and considering both positive and negative cases:

$$\begin{aligned}
(2a \pm \sqrt{(x-c)^2+y^2})^2 &= (x+c)^2+y^2 \\
4a^2+(x-c)^2+y^2 \pm 4a\sqrt{(x-c)^2+y^2} &= (x+c)^2+y^2 \\
\pm 4a\sqrt{(x-c)^2+y^2} &= -4a^2+4xc \\
\pm\sqrt{(x-c)^2+y^2} &= x\,\frac{c}{a}-a\,.
\end{aligned}$$

Squaring both sides again:

$$\begin{aligned}
a^2+x^2\left(\frac{c}{a}\right)^2-2xc &= x^2+c^2-2xc+y^2 \\
x^2\left(\frac{c}{a}\right)^2-x^2-y^2 &= c^2-a^2 \\
x^2\left[\left(\frac{c}{a}\right)^2-1\right]-y^2 &= c^2-a^2\,.
\end{aligned}$$

For simplicity, define $b^2 = c^2 - a^2$ (note that c and a are swapped compared to the ellipse, since $c > a$). The above equation becomes:

$$x^2\frac{b^2}{a^2}-y^2=b^2\,.$$

Dividing both sides by b^2, we finally obtain the canonical equation for the hyperbola:

$$\boxed{\frac{x^2}{a^2}-\frac{y^2}{b^2}=1}\,. \tag{A.10}$$

The hyperbola has eccentricity $e > 1$, since $c > a$. The asymptotes are described by:

$$y=\pm\frac{b}{a}\,.$$

A.3 Plane Trigonometry

Measuring Angles

The *radian* is the SI unit for measuring angles. Given a circle of radius r, the radian is defined as the angle at the centre subtended by an arc of length r. Therefore, an arc of length s subtends an angle of:

$$\alpha_{\text{rad}} = \frac{s}{r}\,.$$

A full circle, measured in radians, is equal to the circumference divided by the radius, i.e. $\alpha_{\text{c, rad}} = 2\pi r/r = 2\pi$. On the other hand, a full circle measured in degrees is $\alpha_{\text{c, deg}} = 360°$, hence the conversion from degrees to radians is:

$$\frac{\alpha_{\text{rad}}}{2\pi} = \frac{\alpha_{\text{deg}}}{360\,°} \Rightarrow \boxed{\alpha_{\text{rad}} = \frac{\pi}{180}\alpha_{\text{deg}}}\,. \tag{A.11}$$

Similarly, the *steradian* is the SI unit for measuring solid angles. Given a sphere of radius r, the steradian is numerically equal to the solid angle at the centre subtended by an area of r^2 on the sphere. The solid angle subtended by an element of area A, at a distance r, is thus:

$$\omega_{\text{sr}} = \frac{A}{r^2}\,.$$

Since the area of a sphere is $4\pi r^2$, a full sphere measures 4π steradians.

Sine, cosine and tangent

Trigonometry is useful in physics to describe periodic motion. Given a right triangle (Fig. A.9), we define three trigonometric functions:

- the *sine* of an angle, denoted by "sin", is the ratio of the leg opposite the angle to the hypotenuse;
- the *cosine* of an angle, denoted by "cos", is the ratio of the leg adjacent the angle to the hypotenuse;
- the *tangent* of an angle, denoted by "tan", is the ratio of the leg opposite the angle to the adjacent leg.

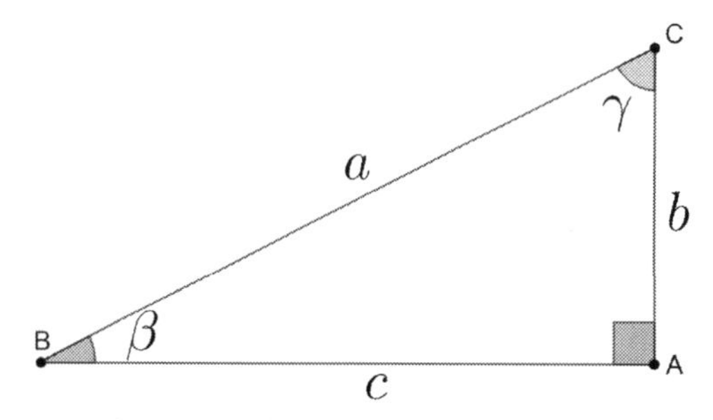

Figure A.9: Definition of sine, cosine and tangent.

Hence, we can write:

$$b = a\,\sin\beta\,, \qquad c = a\,\sin\gamma\,;$$
$$b = a\,\cos\gamma\,, \qquad c = a\,\cos\beta\,;$$
$$b = c\,\tan\beta\,, \qquad c = b\,\tan\gamma\,.$$

Pythagorean identity

The sum of the sine squared and cosine squared of any angle is equal to unity. Indeed, looking at Fig. A.9, the lengths of the opposite and adjacent legs to β are equal to $b = a\sin\beta$ and $c = a\cos\beta$. According to the Pythagorean theorem, $a^2 = b^2 + c^2$. Hence, substituting for b and c, dividing both sides by a^2, we find:

$$\boxed{\sin^2\beta + \cos^2\beta = 1}\,. \tag{A.12}$$

Law of Sines

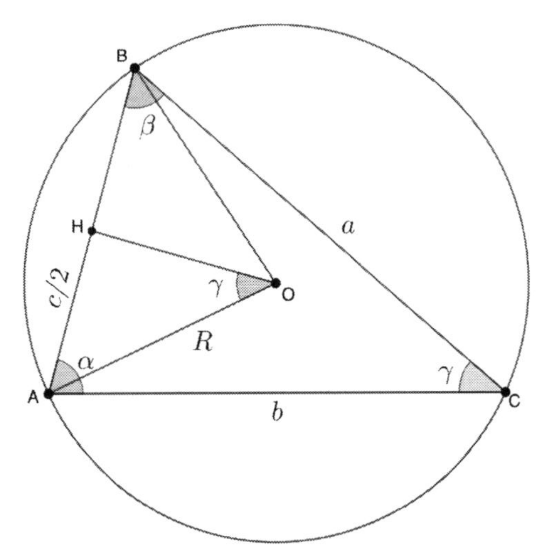

Figure A.10: Proof of the law of sines.

Consider a generic triangle with sides of length a, b, c and angle opposite those sides of α, β, γ, respectively. Let R be the radius of the circumference that circumscribes the triangle. The law of sines states that:

$$\boxed{\frac{a}{\sin\alpha} = \frac{b}{\sin\beta} = \frac{c}{\sin\gamma} = 2R}\,. \tag{A.13}$$

Let us prove that $c/\sin\gamma = 2R$; the other equations follow by symmetry. Looking at Fig. A.10, let H be the height of triangle ABO relative to the base AB. Since the angle at the circumference is half the angle at the centre subtended by the same arc, we have $\angle\text{AOB} = 2\gamma$. Because triangles AHO and HBO are both right triangles in H, $\overline{\text{AO}} = \overline{\text{BO}} = R$, and since they share OH, it follows that they are congruent, hence $\angle\text{AOH} = \gamma$. Then, $\overline{\text{AH}} = c/2$, but we also know that $\overline{\text{AH}} = R\sin\gamma$, hence $c/\sin\gamma = 2R$, as we wanted to show.

Law of Cosines

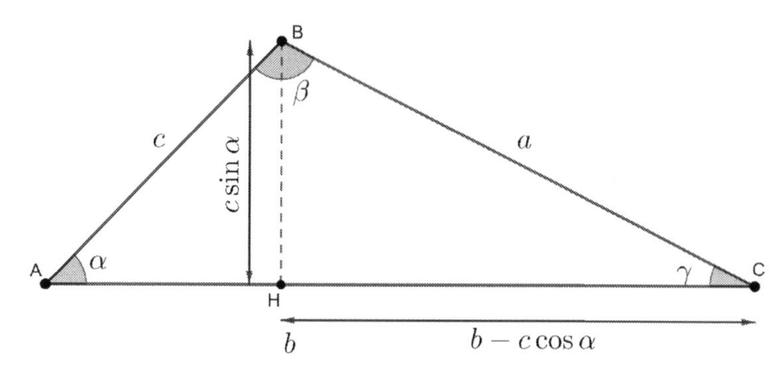

Figure A.11: Proof of the law of cosines.

The law of cosines allows us to compute the length of one side of a triangle, knowing the length of the other two and the angle between them. In particular:

$$\boxed{a^2 = b^2 + c^2 - 2bc\,\cos\alpha}\,. \tag{A.14}$$

By symmetry, we can substitute $a \to b$, $b \to c$, $c \to a$ and $\alpha \to \beta$ to obtain $b^2 = c^2 + a^2 - 2ca\,\cos\beta$ and, with a similar transformation, $c^2 = a^2 + b^2 - 2ab\,\cos\gamma$. Swapping every index with the next (imagine they are placed on a circle) is called performing a *cyclic permutation*. We only prove Eq. A.14; the others follow by cyclic permutation.

Looking at Fig. A.11, we draw the height relative to the base AC and we call H the point of intersection. We have:

$$\overline{\text{AH}} = c\,\cos\alpha\,, \qquad \overline{\text{BH}} = c\,\sin\alpha\,.$$

It follows that $\overline{\text{HC}} = b - c\,\cos\alpha$. Applying the Pythagorean theorem to triangle HCB:

$$\begin{aligned} a^2 &= \overline{\text{BH}}^2 + \overline{\text{HC}}^2 = (c\,\sin\alpha)^2 + (b - c\,\cos\alpha)^2 \\ &= c^2(\sin^2\beta + \cos^2\beta) + b^2 - 2bc\,\cos\alpha\,. \end{aligned}$$

Using the Pythagorean identity:

$$a^2 = b^2 + c^2 - 2bc\,\cos\alpha\,,$$

which is what we wanted to prove.

Reduction Formulae

In this section we look for a simple way to memorize the value of trigonometric functions for some common angles. Since the sine and cosine have a period of 2π (i.e. a full circle), we only need to know the value of trigonometric functions for angles in the interval $[0, 2\pi]$. We can do even better: from the trigonometric function of an angle α in $[0, \pi/4]$, we would like to find the trigonometric functions for:

$$\frac{\pi}{2} \pm \alpha\,, \quad \frac{3}{2}\pi \pm \alpha\,, \quad \pi \pm \alpha\,, \quad -\alpha\,.$$

The strategy is to visualize the angles on the trigonometric circle (Fig. A.12), applying rotations or reflections in order to reduce the problem to that of an angle in the interval $[0, \pi/4]$. We then find:

- for $\frac{\pi}{2} \pm \alpha$, $\sin\left(\frac{\pi}{2} \pm \alpha\right) = \cos\alpha\,,$ $\cos\left(\frac{\pi}{2} \pm \alpha\right) = \mp\sin\alpha\,;$
- for $\pi \pm \alpha$, $\sin\left(\pi \pm \alpha\right) = \mp\sin\alpha\,,$ $\cos\left(\pi \pm \alpha\right) = -\cos\alpha\,;$
- for $\frac{3}{2}\pi \pm \alpha$, $\sin\left(\frac{3}{2}\pi \pm \alpha\right) = -\cos\alpha\,,$ $\cos\left(\frac{3}{2}\pi \pm \alpha\right) = \mp\sin\alpha\,;$
- for $-\alpha$, $\sin(-\alpha) = -\sin\alpha\,,$ $\cos(-\alpha) = \cos\alpha\,.$

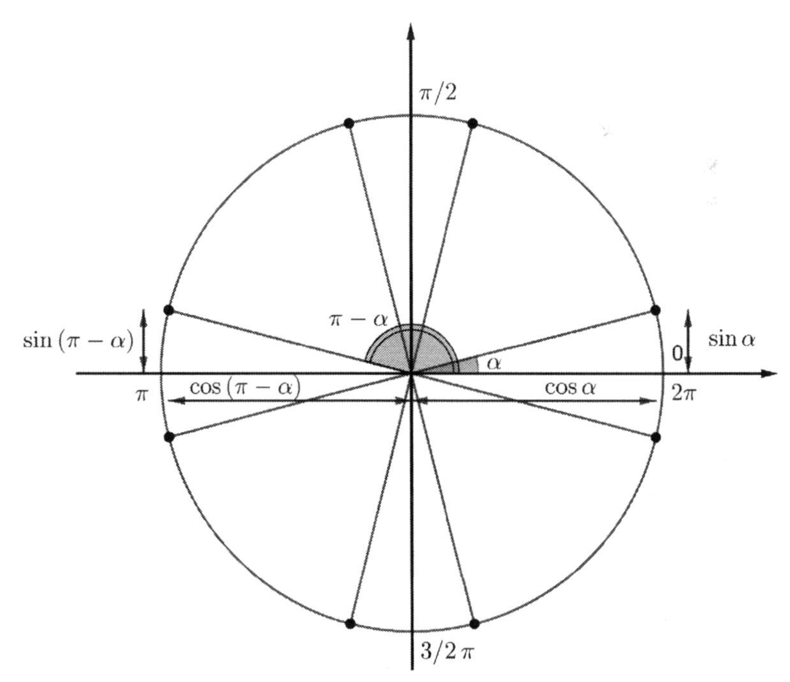

Figure A.12: By visualizing the angles on the trigonometric circle, and applying rotations or reflections, it is possible to reduce the problem of finding the trigonometric function of an angle in $[0, 2\pi]$, to that of an angle in $[0, \pi/4]$.

Looking at Fig. A.12, let us compute, for instance, the sine, cosine and tangent of $\pi - \alpha$. Clearly, the two angles have the same sine, hence $\sin(\pi - \alpha) = \sin\alpha$. Instead, the cosine has opposite sign (because the corresponding triangles are in the first and second quadrant, respectively), hence $\cos(\pi - \alpha) = -\cos\alpha$. The tangent is equal to the ratio of sine and cosine, therefore $\tan(\pi - \alpha) = -\tan\alpha$. With a similar reasoning, it is possible to obtain all the above relations.
For completeness, we give the trigonometric values of some common angles in the interval $[0, \pi/4]$:

$$\cos 30^\circ = \frac{\sqrt{3}}{2}, \qquad \sin 30^\circ = \frac{1}{2}, \qquad \tan 30^\circ = \frac{1}{\sqrt{3}};$$
$$\cos 45^\circ = \frac{1}{\sqrt{2}}, \qquad \sin 45^\circ = \frac{1}{\sqrt{2}}, \qquad \tan 45^\circ = 1.$$

Using the reduction formula for $\pi/2 - \alpha$, and taking $\alpha = 30^\circ$, we can then compute the trigonometric functions for 60°:

$$\cos 60^\circ = \frac{1}{2}, \qquad \sin 60^\circ = \frac{\sqrt{3}}{2}, \qquad \tan 60^\circ = \sqrt{3}.$$

Trigonometric Identities

Below, we give some of the most useful trigonometric identities, which should be remembered.

Addition Formulae:

$$\sin(\alpha \pm \beta) = \sin\alpha\cos\beta \pm \cos\alpha\cos\beta, \tag{A.15}$$

$$\cos(\alpha \pm \beta) = \cos\alpha\cos\beta \mp \sin\alpha\sin\beta, \tag{A.16}$$

$$\tan(\alpha \pm \beta) = \frac{\tan\alpha \pm \tan\beta}{1 \mp \tan\alpha\tan\beta}. \tag{A.17}$$

Double Angle Formulae: Setting $\alpha = \beta$ in the addition formulae:

$$\sin 2\alpha = 2\sin\alpha\cos\alpha, \tag{A.18}$$

$$\cos 2\alpha = \cos^2\alpha - \sin^2\alpha, \tag{A.19}$$

$$\tan 2\alpha = \frac{2\tan\alpha}{1 - \tan^2\alpha}. \tag{A.20}$$

Half-angle Formulae: Using the Pythagorean identity in Eq. A.19, setting $\alpha \to \alpha/2$:

$$\sin\frac{\alpha}{2} = \sqrt{\frac{1-\cos\alpha}{2}}, \tag{A.21}$$

$$\cos\frac{\alpha}{2} = \sqrt{\frac{1+\cos\alpha}{2}}. \tag{A.22}$$

A.4 Spherical Trigonometry

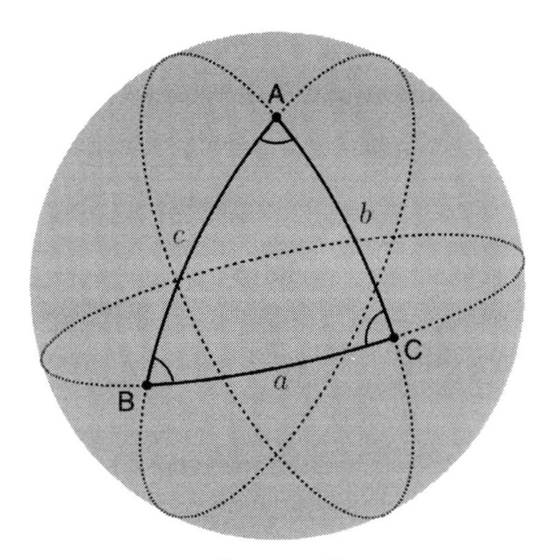

Figure A.13

Let ABC be a triangle on a sphere of unit radius, with sides of length a, b, c and angles opposite the respective sides of A, B and C (Fig. A.13).

Spherical Law of Cosines

The spherical law of cosines states that:

$$\boxed{\cos c = \cos a \cos b + \sin a \sin b \cos C}, \tag{A.23}$$

and the corresponding cyclic permutations. In the case $C = 90\,^\circ$, we obtain the Pythagorean theorem on a sphere:

$$\cos c = \cos a \cos b\,. \tag{A.24}$$

If the triangle is small, i.e. $a, b, c \ll 1$, we can use the approximation $\cos x \approx 1 - x^2/2$ for $x \ll 1$ (see Sec. A.6), to obtain the Pythagorean theorem on a plane:

$$1-\frac{c^2}{2}\approx\left(1-\frac{a^2}{2}\right)\left(1-\frac{b^2}{2}\right)$$
$$1-\frac{c^2}{2}\approx 1-\frac{a^2}{2}-\frac{b^2}{2}$$
$$\Rightarrow c^2\approx a^2+b^2\,,$$

where we have neglected the term $a^2b^2/4$, since it is an infinitesimal of higher order.

Spherical Law of Sines

The spherical law of sines states that:

$$\boxed{\frac{\sin a}{\sin A}=\frac{\sin b}{\sin B}=\frac{\sin c}{\sin C}}\,. \tag{A.25}$$

If $a, b, c \ll 1$, we can use the approximation $\sin x \approx x$ for $x \ll 1$, to obtain the law of sines on a plane.
If the sphere has radius r, it is sufficient to substitute $a \to a/r$, $b \to b/r$ and $c \to c/r$ in the previous formulae.

Spherical Excess

The sum of the angles of a spherical triangle is always greater than 180°. The quantity:

$$E = A + B + C - 180^\circ\,, \tag{A.26}$$

is called the *spherical excess.* It is not a constant, but depends on the triangle. The area of a spherical triangle is related to the spherical excess in a very simple way:

$$\boxed{A = Er^2} \tag{A.27}$$

A.5 Special Functions

Exponentials

Let us consider a function of the form $f(x) = a^x$, where $a > 0$, called the *exponential* function. We can distinguish three cases:

- $a = 1$, then $f(x) = 1$ for every x;
- $a > 1$, as in Fig. A.14 a);
- $0 < a < 1$, as in Fig. A.14 b).

The exponential function has some important properties:

- $f(0) = 1$ for every a. Indeed $a^0 = 1$ for every a;
- $f(x) > 0$ for every x. Indeed $a > 0$ implies $a^x > 0$.

In the case $a = e$, where e is *Euler's number*, we obtain the *natural* exponential function. The functions $f(x) = e^x$ and $f(x) = e^{-x}$ are shown in Fig. A.14 c).

Logarithms

The logarithmic function has the form $f(x) = \log_a x$, where $a > 0$ and $a \neq 1$. The meaning of this notation is the following:

$$f(x) = \log_a x \qquad \text{if and only if} \qquad a^{f(x)} = x\,. \tag{A.28}$$

Clearly, the case $a = 1$ is not valid because $1^{f(x)}$ is always equal to unity. Since $a > 0$, it follows that $a^{f(x)} > 0$ for every x, i.e. the function is only defined for a positive argument (the domain is $x > 0$).
The graphs of the logarithmic function for $a > 1$ and $0 < a < 1$ are shown in Figs. A.14 d) and e), respectively. As shown in Fig. A.14 f), the logarithmic function is simply a reflection of the exponential function about the line $f(x) = x$. In other words, the two functions are related by an exchange of axes: x becomes $f(x)$, and $f(x)$ becomes x.
The logarithmic function has some important properties, which can be proven by using A.28 and reducing the problem to one involving exponentials:

- logarithm of a product: $\log_a (x \cdot y) = \log_a x + \log_a y$;
- exponent rule: $\log_a(x^y) = y \log_a x$;
- logarithm of a fraction: $\log_a (x/y) = \log_a x - \log_a y$;
- change of basis: $\log_a x = \log_y x / \log_y a$.

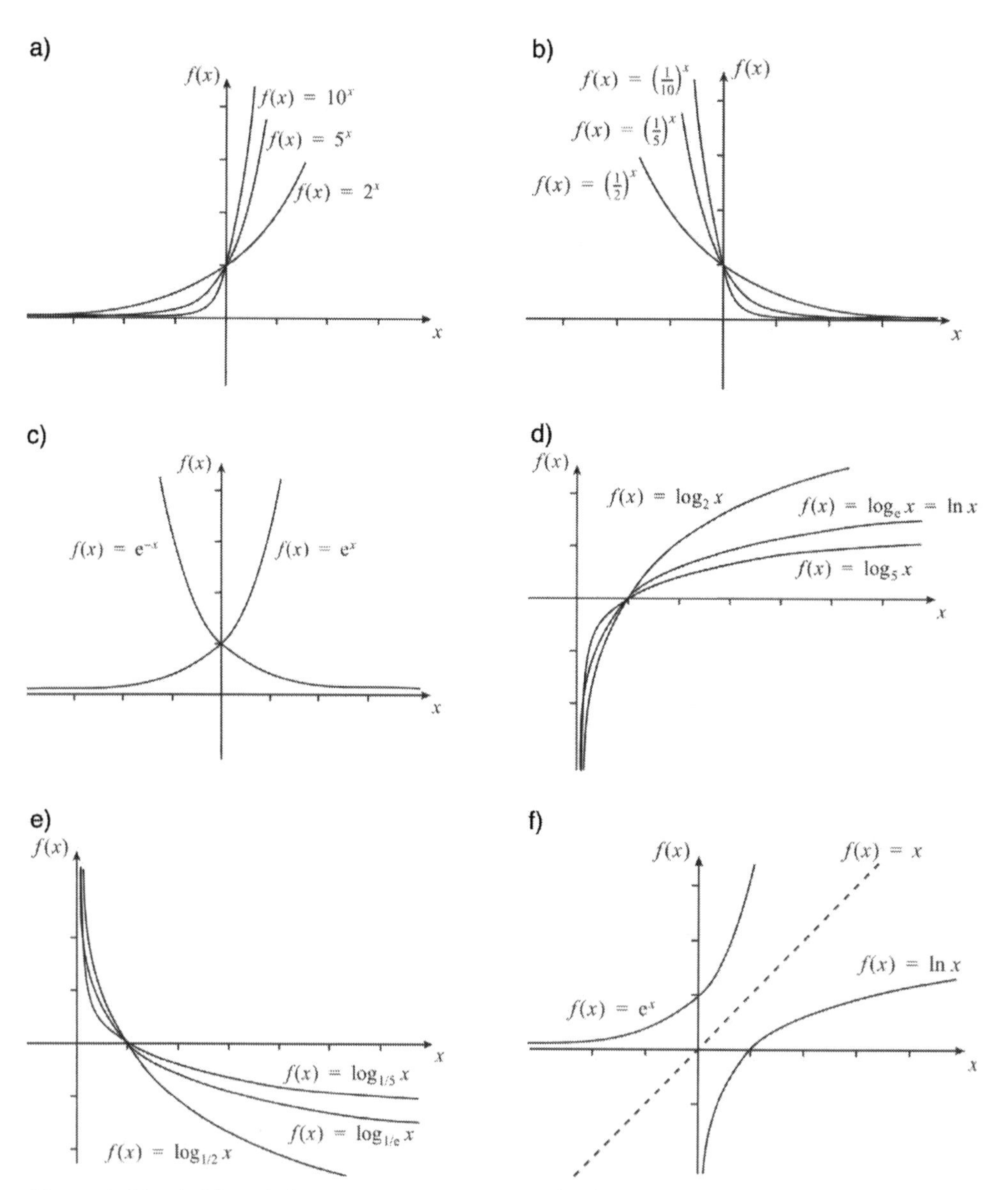

Figure A.14: a), b) and c) exponential functions; d), e) logarithmic functions. In f) we see that the logarithmic function is simply a reflection of the exponential function about the line $f(x) = x$.

A.6 Derivatives

In physics and astronomy, differential calculus is of fundamental importance. In this section we introduce the concept of a derivative in a straightforward and intuitive manner.
Let $f(x)$ be a function with real argument. Consider two points $P_1 = (x_1, f(x_1))$ and $P_2 = (x_1 + \Delta x, f(x_1 + \Delta x))$ on $f(x)$. The line passing through P_1 and P_2 is a secant of $f(x)$ but, as we make Δx smaller, it becomes closer to the line tangent to $f(x)$ in P_1. The angular coefficient of this line is $\Delta f/\Delta x$, with $\Delta f = f(x + \Delta x) - f(x)$. Thus, for Δx infinitesimally small, $\Delta f/\Delta x$ tends to the angular coefficient of the line tangent to $f(x)$ in P_1. Instead of using the notation Δ for a generic variation, we will use the letter d to denote an infinitesimal variation. Hence, Δx becomes dx and Δf becomes $df = f(x+dx) - f(x)$. In this limit, the angular coefficient of the line tangent to $f(x)$ in x, which we denote with $f'(x)$, is called the first derivative of $f(x)$ in x, and is given by:

$$\boxed{f'(x) = \lim_{\Delta x \to 0} \frac{f(x + \Delta x) - f(x)}{\Delta x} = \frac{\mathrm{d}f}{\mathrm{d}x}}\,. \tag{A.29}$$

Qualitatively, the derivative measures the rate of variation of a function, as x is changed. The greater the derivative, the stronger the growth is; the smaller the derivative, the faster the decrease.
From the definition of the derivative, it is clear that the value of Δf for a small (but non-infinitesimal) displacement Δx is:

$$\Delta f \approx \frac{\mathrm{d}f}{\mathrm{d}x} \Delta x\,.$$

The accuracy of the above relation increases as Δx becomes smaller. For Δx that goes to zero, we can substitute Δx and Δf with dx and df, respectively:

$$\boxed{df = \frac{\mathrm{d}f}{\mathrm{d}x} dx}\,. \tag{A.30}$$

In the same way, we can define the second derivative of a function $f(x)$ with respect to x, which we denote by $f''(x)$, as the angular coefficient of the line tangent to $f'(x)$ in x:

$$f''(x) = \lim_{\Delta x \to 0} \frac{f'(x + \Delta x) - f'(x)}{\Delta x} = \frac{\mathrm{d}f'(x)}{\mathrm{d}x}\,. \tag{A.31}$$

A physical example of a second derivative is the acceleration, equal to the derivative of velocity w.r.t. time, with velocity being the derivative of distance w.r.t time. Hence, acceleration is the second derivative of distance w.r.t. time. Since we often consider derivatives w.r.t. time, it is convenient to define $\dot{x}$, $\ddot{x}$

as a short-hand notation for dx/dt and d^2x/dt^2, respectively.
We can then define the n-th derivative of a function as:

$$f^{(n)}(x) = \lim_{\Delta x \to 0} \frac{f^{(n-1)}(x+\Delta x) - f^{(n-1)}(x)}{\Delta x}. \quad (A.32)$$

In this notation, we have $f'(x) = f^{(1)}(x)$, $f''(x) = f^{(2)}(x)$, etc. and, formally, $f^{(0)}(x) = f(x)$.

Basic derivatives

All derivatives can be obtained by solving the limit A.29. In most cases, the limit can be solved by simply thinking what quantities can be neglected as Δx goes to zero. Let us calculate the derivative with respect to x of $f(x) = x^2$:

$$\begin{aligned} f'(x) = \frac{df(x)}{dx} &= \lim_{\Delta x \to 0} \frac{f(x+\Delta x) - f(x)}{\Delta x} \\ &= \lim_{\Delta x \to 0} \frac{(x+\Delta x)^2 - x^2}{\Delta x} \\ &= \lim_{\Delta x \to 0} \frac{2x\Delta x + (\Delta x)^2}{\Delta x} \\ &= \lim_{\Delta x \to 0} (2x + \Delta x). \end{aligned}$$

As Δx goes to zero, $2x + \Delta x$ is just $2x$:

$$f'(x) = 2x.$$

It is useful to memorize the basic derivatives in Tab. A.1. Using the techniques explained in the next section, they can be used to find the derivatives of more complex functions.

Function	Derivative
k	0
x^α	$\alpha x^{\alpha-1}$
e^x	e^x
$\log x$	$1/x$
$\sin x$	$\cos x$
$\cos x$	$-\sin x$
$\tan x$	$1/\cos^2 x$
$\arcsin x$	$1/\sqrt{1-x^2}$
$\arccos x$	$-1/\sqrt{1-x^2}$
$\arctan x$	$1/(1+x^2)$

Table A.1

Product Rule

We want to calculate the derivative of a function $f(x)$ which can be written as the product of two or more functions: $f(x) = u(x)\,v(x)$. For example, if $f(x) = x^3 \sin x$, we see that $u(x) = x^3$ and $v(x) = \sin x$. Of course, the separation is not unique: we could have chosen $u(x) = x^2, v(x) = x \sin x$ or even $u(x) = x^4 \tan x, v(x) = \cos x/x$. However, the point of separating $f(x)$ is to find two functions, $u(x)$ and $v(x)$, whose derivatives are simpler to calculate. Ultimately, we would like to trace back the calculation of the derivative of any function to the basic derivatives. Since $f(x) = u(x)\,v(x)$, it follows that:

$$\begin{aligned} f(x+\Delta x) - f(x) &= u(x+\Delta x)v(x+\Delta x) - u(x)v(x) \\ &= u(x+\Delta x)[v(x+\Delta x) - v(x)] + [u(x+\Delta x) - u(x)]v(x). \end{aligned}$$

From the definition A.29:

$$\frac{\mathrm{d}f}{\mathrm{d}x} = \lim_{\Delta x \to 0} \frac{f(x+\Delta x) - f(x)}{\Delta x}$$
$$= \lim_{\Delta x \to 0} \left\{ u(x+\Delta x)\Big[\frac{v(x+\Delta x) - v(x)}{\Delta x}\Big] + \Big[\frac{u(x+\Delta x) - u(x)}{\Delta x}\Big] v(x) \right\}.$$

As Δx becomes infinitesimally small, the terms in the square brackets become du/dx and dv/dx, respectively, while $u(x+\Delta x)$ becomes $u(x)$. Therefore:

$$\frac{\mathrm{d}f}{\mathrm{d}x} = \frac{d}{dx}[u(x)v(x)] = u(x)\frac{dv(x)}{dx} + \frac{du(x)}{dx}v(x),$$

which can be written as:

$$\boxed{f' = (uv)' = uv' + u'v} \qquad \text{(A.33)}$$

This equation is known as the *product rule*. Hence, the initial problem is reduced to:

$$\frac{d}{dx}(x^3 \sin x) = x^3 \frac{d}{dx}(\sin x) + \frac{d}{dx}(x^3) \sin x = x^3 \cos x + 3x^2 \sin x .$$

We can extend the product rule to three or more separations:

$$f(x) = u(x)v(x)w(x) \Rightarrow f'(x) = u\frac{d}{dx}(vw) + \frac{du}{dx}vu,$$

where we have used Eq. A.33, applied to u and vw. Using again Eq. A.33 to expand $(vw)'$:

$$\frac{d}{dx}(uvw) = uv\frac{dw}{dx} + u\frac{dv}{dx}w + \frac{du}{dx}vw,$$

which can be written as:

$$(uvw)' = uvw' + uv'w + u'vw . \qquad \text{(A.34)}$$

It is clear that the same reasoning can be extended to the product of any number of terms.

Chain Rule

Consider a function of a function $f(u(x))$. For example, if $f(x) = (3 + x^2)^3$, we can take $u(x) = 3 + x^2$, so that $f(x) = u(x)^3$. If Δf, Δu and Δx are finite, we can write:

$$\frac{\Delta f}{\Delta x} = \frac{\Delta f(u)}{\Delta u(x)} \frac{\Delta u(x)}{\Delta x} .$$

When the above quantities become infinitesimally small, we can write:

$$\boxed{\frac{\mathrm{d}f}{\mathrm{d}x} = \frac{\mathrm{d}f(u)}{\mathrm{d}u(x)}\frac{\mathrm{d}u(x)}{\mathrm{d}x}}\,. \tag{A.35}$$

The above equation is known as the *chain rule*, and can be used to compute the derivative of a function of a function. We can now solve the example given at the beginning of the section:

$$\frac{\mathrm{d}f}{\mathrm{d}x} = \frac{\mathrm{d}f(u)}{\mathrm{d}u}\frac{\mathrm{d}u}{\mathrm{d}x} = \left[\frac{d}{du}(u^3)\right]\cdot\left[\frac{d}{dx}(3+x^2)\right] = 3u^2\cdot 2x = 6x(3+x^2)^2\,.$$

In a similar way, we can use the chain rule to compute the derivative of the reciprocal of a function. Let $f(x) = 1/v(x) = v(x)^{-1}$. Applying Eq. A.35:

$$\frac{df}{dx} = -\frac{1}{v^2}\frac{dv}{dx}\,. \tag{A.36}$$

Quotient Rule

Using the product rule and the chain rule (Eqs. A.33, A.35), it is easy to find the derivative of the quotient of two functions:

$$f' = \left(\frac{u}{v}\right)' = u\left(\frac{1}{v}\right)' + u'\left(\frac{1}{v}\right) = u\left(-\frac{v'}{v^2}\right) + \frac{u'}{v}\,.$$

The last equation can be written as:

$$\boxed{f' = \left(\frac{u}{v}\right)' = \frac{vu' - uv'}{v^2}}\,, \tag{A.37}$$

which is often referred to as the *quotient rule*. For example, the derivative of $f(x) = \sin x/x$ can be found by taking $u = \sin x$ and $v = x$, from which $u' = \cos x$ and $v' = 1$. Applying A.37:

$$f'(x) = \frac{x\cos x - \sin x}{x^2} = \frac{\cos x}{x} - \frac{\sin x}{x^2}\,.$$

Special Points

We have seen that the derivative measures the rate of change of a function $f(x)$ in the neighbourhood of a point. If the increment of $f(x)$ in the neighbourhood of x_0 is zero, we call this a *stationary point*. Stationary points can be divided into three categories.

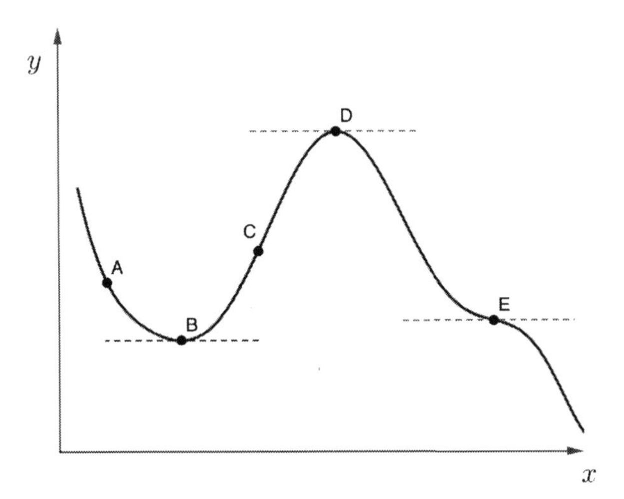

Figure A.15: In B, D and E the first derivative is zero: these are called *stationary points*. In particular, B and D are points of minimum and maximum, respectively (collectively known as *turning points*), while E is a *point of inflection*.

Referring to Fig. A.15, B is a point of *minimum*, since the function increases in both directions moving away from that point; D is a point of *maximum*, since the function decreases in both directions away from that point; finally, E is an *inflection* point, as the function increases on one side of E and decreases on the other. Note that B and D are not absolute maxima or minima, but, rather, local maxima and minima. From a mathematical point of view, we can distinguish maxima, minima and inflection points by studying the second derivative of the function at those points. By definition, all stationary points are such that $df/dx = 0$. Looking at the minimum B, we see that $df(x)/dx$ is negative in A and positive in C, passing through zero in B. Therefore, in the neighbourhood of B, the first derivative is increasing, hence the second derivative (rate of change of the first derivative) is positive. Similarly, the second derivative must be negative in D. It is less obvious that in E the second derivative is zero. Intuitively, we note that, on the left of E, the function is concave upwards, therefore the first derivative is increasing and the second derivative is positive. Similarly, on the right of E, the function is concave downwards, therefore the second derivative is negative. Since the second derivative varies continuously, it must be zero in E.
To sum up, stationary points are characterized by $df/dx = 0$ and:

- $d^2f/dx^2 > 0$, for a minimum;
- $d^2f/dx^2 < 0$, for a maximum;
- $d^2f/dx^2 = 0$ for an inflection point, and d^2f/dx^2 changes sign through the point.

Taylor Series

The Taylor series allows us to approximate a function, in the neighbourhood of a point, with a polynomial of degree n. The general form of the Taylor series is:

$$f(x+x_0) = f(x_0) + f'(x_0)x + \frac{f''(x_0)}{2!}x^2 + \frac{f'''(x_0)}{3!}x^3 + .. \tag{A.38}$$

The validity of the above formula can be checked with the following reasoning. Taking the first derivative with respect to x and setting $x = 0$, we obtain $f'(x_0)$ on both sides. Indeed, the derivative of the LHS is $f'(x+x_0)$, which, in $x = 0$, is equal to $f'(x_0)$. The derivative of the RHS is $f'(x_0) + f''(x_0)x + ..$, hence, in $x = 0$, the only non-zero term is $f'(x_0)$. Similarly, taking the second derivative and setting $x = 0$, we obtain $f''(x)$ on both sides, and so on. Since both sides are equal in $x = 0$, and their n-th derivatives in $x = 0$ are the same for every n, it follows that they are the same function. Some examples of Taylor series and their first order approximations are given in Tab. A.2.

Function	Taylor series	For $x \ll 1$
$\frac{1}{1-x}$	$1 + x + x^2 + x^3 + ...$	$1 + x$
$\ln(1+x)$	$x - x^2/2 + x^3/3 - ...$	x
e^x	$1 + x + x^2/2! + x^3/3! + ...$	$1 + x$
$\sin x$	$x - x^3/3! + x^5/5! - ...$	x
$\cos x$	$1 - x^2/2! + x^4/4! - ...$	1
$(1+x)^n$	$1 + nx + \binom{n}{2}x^2 + \binom{n}{3}x^3 + ...$	$1 + nx$

Table A.2

To obtain the first formula in the table, we need to compute the $n-$th derivatives of $1/(1-x)$ at $x_0 = 0$. Using the chain rule with $u(x) = 1 - x$:

$$\begin{aligned}
f(x) &= \frac{1}{1-x} \Rightarrow f(0) = 1\,, \\
f'(x) &= \frac{\mathrm{d}f}{\mathrm{d}u}\frac{\mathrm{d}u}{\mathrm{d}x} = \frac{1}{(1-x)^2} \Rightarrow f'(0) = 1\,, \\
f''(x) &= \frac{\mathrm{d}f'(x)}{\mathrm{d}x} = -2\frac{1}{(1-x)^3}\cdot(-1) = \frac{2}{(1-x)^3} \Rightarrow f''(0) = 2\,, \\
f'''(x) &= \frac{6}{(1-x)^3} \Rightarrow f'''(0) = 6\,, \\
&\dots, \\
f^n(x) &= \frac{n!}{(1-x)^n} \Rightarrow f'(0) = n!\,.
\end{aligned}$$

Substituting the derivatives in Eq. A.38, with $x_0 = 0$:

$$
\begin{aligned}
f(x) &= f(0) + f'(0)x + \frac{f''(0)}{2!} + ... + \frac{f^n(0)}{n!}x^n + ... \\
&= 1 + x + \frac{2}{2!} + \frac{6}{3!} + ... + \frac{n!}{n!}x^n + ... \\
\Rightarrow \frac{1}{1-x} &= 1 + x + x^2 + x^3 + ...,
\end{aligned}
$$

as we wanted to show.
We define the *order of a power series* as the minimum degree of the independent variable. For instance, the order of x^2 is 2, while the order of $x^2 + 3x$ is 1. The Taylor series allows us to approximate a function with a power series. For $x \ll 1$, we can often truncate the Taylor series to first order, since $x^n \ll x$ for $n > 1$. For example, the first order approximation of $1/(1-x)$ is obtained by neglecting all terms of degree greater than 1, hence $1/(1-x) \approx 1 + x$.

A.7 Integrals

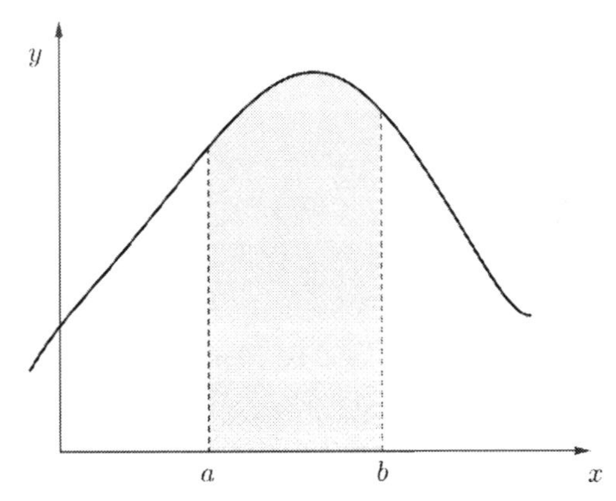

Figure A.16: The shaded area under the curve is the integral of $f(x)$ between the extremes a and b, and is denoted by $\int_a^b f(x)\,dx$.

The notion of an integral as the area under a curve will already be familiar to many. In Fig. A.16, where the function $f(x)$ is drawn as a solid line, the shaded area represents the quantity denoted by:

$$
I = \int_a^b f(x)\,dx\,. \tag{A.39}
$$

This expression is known as the *definite integral* of $f(x)$ between the lower and upper limits $x = a$ and $x = b$, respectively. Here, $f(x)$ is called the *integrand*.

Integration from basic principles

According to Riemann's definition, the integral of a function $f(x)$ between the extremes a and b is the limit of the sum of the areas of the rectangles with height $f(x_i)$ and width $(x_i - x_{i-1})$, for $a = x_0 < x_1 ... < x_i < ... x_n = b$, as n goes to infinity. Hence:

$$S = \lim_{n \to \infty} \sum_{i=1}^{n} f(x_i)(x_i - x_{i-1}) . \qquad \text{(A.40)}$$

In this limit, with the constraint that the length of the interval $[x_{i-1}, x_i]$ goes to zero (i.e. to dx), S converges, for a continuous function in a finite interval, to the definite integral of $f(x)$. The symbol representing the integral (see Eq. A.39) is a reminder of the S, which stands for summation.

Properties of the integral

Let $a < b < c$ be three real numbers. The following properties hold:

- reversing the limits of integration changes the sign of the definite integral:

$$\int_a^b f(x)\, dx = - \int_b^a f(x)\, dx ;$$

- if the upper and lower limits are the same, the integral is zero:

$$\int_a^a f(x)\, dx = 0 ;$$

- the definite integral of $f(x)$ over the interval $[a, c]$ is equal to the sum of the integrals over the intervals $[a, b]$ and $[b, c]$:

$$\int_a^c f(x) dx = \int_a^b f(x)\, dx + \int_b^c f(x)\, dx ;$$

- the definite integral of the sum of two functions is equal to the sum of the integrals of these functions:

$$\int_a^b (f(x) + g(x))\, dx = \int_a^b f(x)\, dx + \int_a^b g(x)\, dx ;$$

- a constant factor can be moved outside the integral sign:

$$\int_a^b c \cdot f(x)\, dx = c \int_a^b f(x)\, dx .$$

How to solve integrals

Let us define the *indefinite integral* $F(x)$ of $f(x)$ as:

$$\boxed{F(x) = \int_{x_0}^{x} f(x')\,dx'}\,, \tag{A.41}$$

where x' has been used to distinguish the independent variable from the extreme of integration. Using the properties above, we find:

$$\boxed{\int_a^b f(x)\,dx = F(a) - F(b)}\,. \tag{A.42}$$

If we know $F(x)$, it is then possible to compute the definite integral of $f(x)$. But how do we find $F(x)$?

Fundamental theorem of calculus

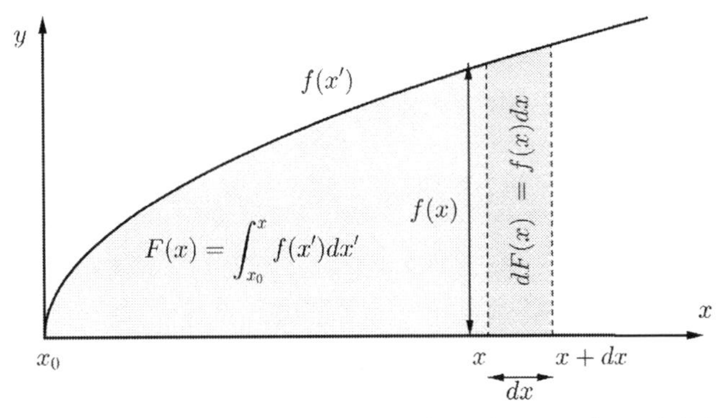

Figure A.17: $F(x)$ is the indefinite integral, equal to the area bounded by $f(x)$ between $[x_0, x]$. Considering a small additional stripe between x and $x + dx$, with width dx and area $f(x)\,dx$, the change in $F(x)$ is $f(x)\,dx$. Therefore, $dF(x) = f(x)\,dx$, i.e. $F'(x) = f(x)$. For this reason, $F(x)$ is also called the anti-derivative of $f(x)$.

The fundamental theorem of calculus states that $F'(x) = f(x)$, i.e. the derivative of $F(x)$ is equal to $f(x)$. Since the derivative of a constant factor is zero, if follows that $F(x)$ is defined up to an additive constant. Integration and differentiation are therefore inverse operations. This theorem is easy to visualize graphically: for an infinitesimal variation dx, the increment $dF(x)$ of $F(x)$ is equal to the area of the rectangle between x and $x + dx$, with base dx and height $f(x)$, i.e. $dF(x) = f(x)dx$ (Fig. A.17). By the definition of the derivative, $F'(x) = dF(x)/dx$. Hence $dF(x) = f(x)\,dx$ implies $F'(x) = f(x)$, as we wanted to show. Now that we know how to find $F(x)$, it is sufficient to apply Eq. A.42 to find the definite integral of $f(x)$, as we will see in the following examples.

Let us calculate, for instance, the integral $\int_0^a x\,dx$. We want to find the function $F(x)$ whose derivative is $f(x) = x$. It is easy to see that $F(x) = x^2/2 + c$ satisfies the required condition. We then use Eq. A.42 and calculate the difference of $F(x)$, evaluated at the extremes:

$$\int_0^a x\,dx = F(a) - F(0) = \frac{a^2}{2}\,.$$

As another example, let us compute the integral $\int_0^a 2xe^{-x^2}$. It is easy to see that $F(x) = -e^{-x^2} + c$ is the anti-derivative of $2xe^{-x^2}$. Hence:

$$\int_0^a 2xe^{-x^2} = F(a) - F(0) = -e^{-a^2} + 1\,.$$

At times, we may want to compute an integral without specifying the extremes of integration. This can be useful, for instance, to examine how the integral itself varies as the extremes are changed. In this case, it suffices to find $F(x)$ and write down Eq. A.41. Notice that the solution will carry an arbitrary constant, which we call *constant of integration.*
For the previous examples:

$$\int x\,dx = x^2/2 + c\,,$$

$$\int 2xe^{-x^2}\,dx = -e^{-x^2} + c\,.$$

Here, the extremes of integration are implicitly x_0 and x, where x_0 is an arbitrary constant related to c by $F(x_0) = c$. In Tab. A.3, we present some common indefinite integrals, where the constant of integration has been omitted for clarity. These are the basis for computing more complex integrals, using the rules explained in the next section.

$f(x)$	$F(x)$
k	$kx + c$
x^n, $n \neq -1$	$x^{n+1}/(n+1) + c$
$1/x$	$\ln x + c$
e^x	$e^x + c$
$\sin x$	$-\cos x + c$
$\cos x$	$\sin x + c$
$1/\sqrt{1-x^2}$	$\arcsin x + c$
$1/\sqrt{1-x^2}$	$-\arccos x + c$
$1/1+x^2$	$\arctan x + c$

Table A.3

Integration by substitution

The indefinite integral of simple integrands can be found immediately by inspection, as shown in the previous section. For more complex functions, we can use some strategies to simplify the calculation.
If the integral does not appear to be elementary, the first strategy we can try is *integration by substitution.* The idea is to define a new variable that simplifies the integral, possibly reducing it to an elementary one. For example, if we are given $\int f(x)\,dx$, we may want to define a new variable $u = u(x)$, which

transforms the starting integral into one of the form $\int g(u)\, du$. To find du in terms of dx, we use Eq. A.30. It is important to note that once the variable has been changed and the new integral has been obtained, no reference to x or dx should remain. Choosing the new variable is not always easy: below we see some examples in order of difficulty.

$$\int \cos(2x + 7)\, dx\,.$$

The choice $u(x) = 2x + 7$ looks promising. Using Eq. A.30 to compute du:

$$du = \frac{\mathrm{d}u}{\mathrm{d}x}\, dx \Rightarrow du = 2\, dx\,.$$

Substituting in the initial integral:

$$\int \cos(2x + 7)\, dx = \frac{1}{2} \int \cos u\, du\,,$$

which is a basic integral:

$$\frac{1}{2} \int \cos u\, du = \frac{1}{2} \sin u + c\,.$$

Reverting back to the expression for $u(x)$, we then obtain:

$$\int \cos(2x + 7)\, dx = \frac{1}{2} \sin\,(2x + 7) + c\,.$$

Let us now calculate:

$$\int \frac{3}{x(\ln x)^2}\, dx\,.$$

After some trials, we see that the substitution $u(x) = \ln x$ will simplify the integral. Then, $du = dx/x$, and substituting in the integral:

$$\int \frac{3}{x(\ln x)^2}\, dx = 3 \int \frac{1}{(\ln x)^2} \frac{dx}{x} = 3 \int \frac{1}{u^2}\, du\,,$$

which is an elementary integral:

$$3 \int \frac{1}{u^2}\, du = 3 \int u^{-2}\, du = 3 \frac{u^{-1}}{-1} + c = -\frac{3}{u} + c\,.$$

We can now substitute u in terms of x:

$$\int \frac{3}{x(\ln x)^2}\, dx = -\frac{3}{\ln x} + c\,.$$

Integration by parts

Another strategy to simplify integrals is *integration by parts.* Using Eq. A.33 for the derivative of a product:

$$\frac{d}{dx}(uv) = u\frac{dv}{dx} + \frac{du}{dx}v\,.$$

Isolating the first term on the RHS:

$$u\frac{dv}{dx} = \frac{d}{dx}(uv) - \frac{du}{dx}v\,.$$

Integrating both sides w.r.t. x:

$$\int u\frac{dv}{dx}\,dx = uv - \int \frac{du}{dx}v\,dx\,,$$

which can be written as:

$$\boxed{\int uv'\,dx = uv - \int u'v\,dx}\,. \tag{A.43}$$

The above equation is the rule of *integration by parts.* To integrate by parts, we write the integrand as a product of two functions u and v', so that the integral of $u'v$ is easier to calculate compared to the initial integral. As an example, let us calculate:

$$\int x\cos x\,dx\,.$$

Substitution fails here, so we try integration by parts. Taking $dv/dx = x$ and $u = \cos x$, we find $v = x^2/2$ and $du/dx = -\sin x$. Applying Eq. A.43:

$$\int x\cos x\,dx = \frac{x^2}{2}\cos x + \frac{1}{2}\int x^2\sin x\,dx\,.$$

This integral looks more complicated than the starting integral, so we try another choice for u and v. If we take $u = x$ and $dv/dx = \cos x$, we have $du/dx = 1$ and $v = \sin x$, and it follows that:

$$\int x\cos x\,dx = x\sin x - \int \sin x\,dx\,.$$

This is an elementary integral:

$$\int x\cos x\,dx = x\sin x + \cos x + c\,.$$

B

Kepler's Laws

B.1 Solving the equation of motion

The velocity of a body orbiting around a centre of attraction can be decomposed into radial and tangential components:

$$v_r = \frac{\mathrm{d}r}{\mathrm{d}t} = \dot{r}\,,$$
$$v_t = r\omega = r\frac{\mathrm{d}\theta}{\mathrm{d}t} = r\dot{\theta}\,.$$

The energy can be written as:

$$E = \frac{1}{2}m\dot{r}^2 + \frac{1}{2}mr^2\dot{\theta}^2 - \frac{GmM}{r}\,. \tag{B.1}$$

Angular momentum is also conserved:

$$L = mr^2\dot{\theta}\,. \tag{B.2}$$

We want an expression in which the only unknown is r. Hence, we isolate $\dot{\theta}$ in Eq. B.2, and substitute it in Eq. B.1:

$$E = \frac{1}{2}m\dot{r}^2 + \frac{L^2}{2mr^2} - \frac{GmM}{r}\,. \tag{B.3}$$

Isolating $\dot{r}$:

$$\dot{r} = \sqrt{\frac{2}{m}}\sqrt{E - \frac{L^2}{2mr^2} + \frac{GmM}{r}}\,. \tag{B.4}$$

From the above equation we can obtain $r(t)$. Since we are interested in the shape of the orbit, we want to find $r(\theta)$. Using the chain rule (Eq. A.35):

$$\dot{r} = \frac{\mathrm{d}r}{\mathrm{d}t} = \frac{\mathrm{d}r}{\mathrm{d}\theta}\frac{\mathrm{d}\theta}{\mathrm{d}t} = \dot{\theta}\frac{\mathrm{d}r}{\mathrm{d}\theta} = \frac{L}{mr^2}\frac{\mathrm{d}r}{\mathrm{d}\theta}\,.$$

Hence, substituting in B.4 and reordering:

$$\frac{dr}{r^2\sqrt{E-\frac{L^2}{2mr^2}+\frac{GmM}{r}}}=\frac{\sqrt{2m}}{L}\,d\theta\,. \tag{B.5}$$

Since this equation is valid for any r and θ, we can imagine summing all such equations separated by infinitesimal increments $d\theta$. This sum is just an integral:

$$\int\frac{dr}{r^2\sqrt{E-\frac{L^2}{2mr^2}+\frac{GmM}{r}}}=\int\frac{\sqrt{2m}}{L}\,d\theta\,. \tag{B.6}$$

Now, it comes down to solve two integrals. The one on the RHS is straightforward:

$$\int\frac{\sqrt{2m}}{L}\,d\theta=\frac{\sqrt{2m}}{L}\theta+c_1\,.$$

To solve the integral on the LHS, we change variable to $x=1/r$, so that $dx=-1/r^2\,dr$:

$$-\frac{1}{\sqrt{E}}\int\frac{dx}{\sqrt{1-\frac{L^2}{2Em}x^2+\frac{GmM}{E}x}}\,.$$

Completing the square at the denominator:

$$-\frac{1}{\sqrt{E}}\int\frac{dx}{\sqrt{(1+\frac{G^2m^3M^2}{2L^2E})-(\frac{L}{\sqrt{2Em}}x-\frac{GMm}{L}\sqrt{\frac{m}{2E}})^2}}\,.$$

Using the substitution:

$$\sqrt{1+\frac{G^2m^3M^2}{2L^2E}}\cos y=\frac{L}{\sqrt{2Em}}x-\frac{GMm}{L}\sqrt{\frac{m}{2E}}\,,$$

$$\sqrt{1+\frac{G^2m^3M^2}{2L^2E}}\sin y\,dy=-\frac{L}{\sqrt{2Em}}\,dx\,,$$

the previous integral simplifies to:

$$\frac{\sqrt{2m}}{L}\int dy=\frac{\sqrt{2m}}{L}y+c_2=\frac{\sqrt{2m}}{L}\arccos\left[\frac{\frac{L}{\sqrt{2Em}}x-\frac{GMm}{L}\sqrt{\frac{m}{2E}}}{\sqrt{1+\frac{G^2m^3M^2}{2L^2E}}}\right]+c_2\,.$$

The axes are often chosen so that the integration constant is zero. Then, Eq. B.6 becomes:

$$\frac{\frac{L}{\sqrt{2Em}}\frac{1}{r}-\frac{GMm}{L}\sqrt{\frac{m}{2E}}}{\sqrt{1+\frac{G^2m^3M^2}{2L^2E}}}=\cos\theta\,.$$

Isolating r:

$$r = \frac{\frac{L^2}{GMm^2}}{1 + \sqrt{\frac{2L^2E}{G^2m^3M^2} + 1}\cos\theta}.$$

If we denote with:

$$r_0 = \frac{L^2}{GMm^2}, \tag{B.7}$$

$$e = \sqrt{1 + \frac{2E}{GmM}r_0}. \tag{B.8}$$

The equation for $r(\theta)$ can be written in a simpler form:

$$\boxed{r = \frac{r_0}{1 + e\cos\theta}}. \tag{B.9}$$

B.2 The orbits

Starting from Eq. B.9, it is possible to determine the type of orbits described by the planets. In any case, Eq. B.9 represents a conic section. Setting $dr/d\theta = 0$ in Eq. B.9, it is possible to obtain the maximum and minimum distances of the body from the centre of attraction:

$$r_{\max} = \begin{cases} \frac{r_0}{(1-e)} & \text{if} \quad e < 1 \\ \infty & \text{if} \quad e \geqslant 1. \end{cases} \tag{B.10}$$

$$r_{\min} = \frac{r_0}{1+e}. \tag{B.11}$$

We can write Eq. B.9 in Cartesian coordinates, using $r = \sqrt{x^2 + y^2}$ and $\cos\theta = x/(\sqrt{x^2 + y^2})$:

$$x^2 + y^2 = r_0^2 - 2r_0\, e\, x + e^2x^2. \tag{B.12}$$

Circle $(e = 0)$

In the case $e = 0$, Eq. B.9 becomes $r = r_0$, hence the planet describes a circular orbit with radius r_0. From Eq. B.8 it follows that:

$$E = -\frac{GmM}{2r_0}.$$

We see that the energy is negative.

Ellipse $(0 < e < 1)$

From Eq. B.8, it follows that:

$$-\frac{GmM}{2r_0} < E < 0\,.$$

Hence, the energy is negative, but it is greater than the energy of a circular orbit with radius r_0. Completing the square for x in Eq. B.12, we obtain:

$$\frac{(x + \frac{r_0\, e}{1-e^2})^2}{a^2} + \frac{y^2}{b^2} = 1 \quad \text{with} \quad a = \frac{r_0}{1-e^2} \quad \text{and} \quad b = \frac{r_0}{\sqrt{1-e^2}}\,. \tag{B.13}$$

This is the equation of an ellipse with semi-major axis a, semi-minor axis b and centre in $(-r_0\, e/(1-e^2), 0)$. The semi-focal distance is $c = \sqrt{a^2 - b^2} = r_0\, e/(1-e^2)$, hence one focus is in the origin. Now that we know the semi-major axis a, we can write the eccentricity as $e = \sqrt{1 - r_0/a}$. Substituting e in Eq. B.8, we find:

$$\sqrt{1 - \frac{r_0}{a}} = \sqrt{1 + \frac{2E}{GmM} r_0}$$
$$\Rightarrow E = -\frac{GmM}{2a}\,,$$

which is the well-known expression for the energy of an elliptical orbit.

Parabola $(e = 1)$

From Eq. B.8, it follows that $E = 0$. Hence, the velocity of the body tends to zero when its distance from the centre of attraction tends to infinity. The minimum and maximum distances are $r_{\rm min} = r_0/(1+e)$ and $r_{\rm max} = \infty$, respectively. In this case, Eq. B.12 becomes:

$$y^2 = r_0^2 - 2r_0\, x = -2r_0(x - r_0/2)\,. \tag{B.14}$$

This is the equation of a parabola with vertex in $(r_0/2, 0)$ and focal length $r_0/2$. Hence, the focal point is in the origin.

Hyperbola $(e > 1)$

From Eq. B.8, it follows that $E > 0$. The body can escape from the central potential with a non-zero velocity at infinity, given by $v = \sqrt{2E/m}$. Again, $r_{\rm min} = r_0/(1+e)$ and $r_{\rm max} = \infty$. Completing the square for x, Eq. B.12 becomes:

$$\frac{(x - \frac{r_0\, e}{e^2-1})^2}{a^2} - \frac{y^2}{b^2} = 1 \quad \text{with} \quad a = \frac{r_0}{e^2-1} \quad \text{and} \quad b = \frac{r_0}{\sqrt{e^2-1}}\,. \tag{B.15}$$

This is the equation for a hyperbola with centre (point of intersection of the asymptotes) at $(r_0\, e/(e^2 - 1), 0)$. The semi-focal length is $c = \sqrt{a^2 + b^2} = r_0\, e/(e^2 - 1)$. The impact parameter is the minimum distance from the centre of the potential that the body would reach if it were to travel on a straight line from infinity. It can be shown that the impact parameter is equal to b.

B.3 Proof of Kepler's third law

In Ch. 10 we proved Kepler's third law only in the case of a circular orbit. We are now in a position to prove Kepler's third law for elliptical orbits. As shown in Sec. "The Second Law":

$$\frac{\mathrm{d}A}{\mathrm{d}t} = \frac{L}{2m} .$$

Integrating over a period:

$$A_{\text{tot}} = \frac{LT}{2m} ,$$

where T is the orbital period. The area of an ellipse is $A_{\text{tot}} = \pi ab$, where $b = a\sqrt{1 - e^2}$. Substituting and squaring both sides:

$$\pi^2 a^4 = \left[\frac{L^2}{m(1 - e^2)}\right]\frac{T^2}{4m} . \tag{B.16}$$

Using $L^2 = GMm^2 r_0$, the term in square brackets becomes $GmMr_0/(1 - e^2)$. Since $a = r_0/(1 - e^2)$, this can also be written as $GmMa$:

$$\pi^2 a^4 = (GmMa)\frac{T^2}{4m}$$

$$\Rightarrow T^2 = \frac{4\pi^2}{GM} a^3 ,$$

which is what we wanted to prove.

C

Virial Theorem

To prove the Viral theorem, consider the quantity:

$$G = \sum_i \vec{p_i} \cdot \vec{r_i}\,,$$

where the summation extends over all the particles in the system. The total time derivative of this quantity is:

$$\frac{\mathrm{d}G}{\mathrm{d}t} = \sum_i \dot{\vec{r_i}} \cdot \vec{p_i} + \sum_i \dot{\vec{p_i}} \cdot \vec{r_i}\,. \tag{C.1}$$

The first term can be transformed into:

$$\sum_i \dot{\vec{r_i}} \cdot \vec{p_i} = \sum_i m_i \dot{\vec{r_i}} \cdot \dot{\vec{r_i}} = \sum_i m_i \cdot v_i^2 = 2K\,.$$

Since force equals rate of change of momentum, the second term is:

$$\sum_i \dot{\vec{p_i}} \cdot \vec{r_i} = \sum_i \vec{F_i} \cdot \vec{r_i}\,.$$

Then, Eq. C.1 reduces to:

$$\frac{\mathrm{d}G}{\mathrm{d}t} = 2K + \sum_i \vec{F_i} \cdot \vec{r_i}\,.$$

The time average of the last equation, over an interval τ, is obtained by integrating both sides from 0 to τ, and dividing by τ:

$$\frac{1}{\tau}\int_0^\tau \frac{\mathrm{d}G}{\mathrm{d}t} = \overline{\frac{\mathrm{d}G}{\mathrm{d}t}} = \overline{2K} + \overline{\sum_i \vec{F_i} \cdot \vec{r_i}}\,,$$

or:

$$\overline{2K} + \overline{\sum_i \vec{F_i} \cdot \vec{r_i}} = \frac{1}{\tau}\Big[G(\tau) - G(0)\Big]\,. \tag{C.2}$$

If the motion is periodic, i.e. all coordinates repeat after a certain time, the term $G(\tau) - G(0)$ is bound, while τ can be taken arbitrarily large. Therefore, the right hand side of Eq. C.2 vanishes. It then follows that:

$$\overline{K} = -\frac{1}{2}\overline{\sum_i \vec{F_i} \cdot \vec{r_i}}\,. \tag{C.3}$$

Now, for a potential of the form $U = -\alpha r^{-n}$, the force is:

$$F = -\frac{\mathrm{d}U}{\mathrm{d}r} = -n\alpha r^{-n-1} = n\frac{U}{r}\,.$$

Hence, the right hand side of Eq. C.3 is just $-n\overline{U}/2$, and we obtain the Virial equation for a general potential $U = -\alpha r^{-n}$:

$$\boxed{\overline{K} = -\frac{n}{2}\overline{U}}\,.$$

For the gravitational potential, $n = 1$, hence $\overline{K} = -\overline{U}/2$.

D

Tables and constants

Sun

Mean radius	695475 km
Mass	$1.99 \cdot 10^{30}$ kg
Surface temperature	5778 K
Absolute magnitude	+4.83
Apparent magnitude	−26.74
Distance to the galactic centre	2700 ly
Spectral class	G2V

Moon

Mean radius	1738 km
Mass	$7.35 \cdot 10^{22}$ kg
Semi-major axis	$3.844 \cdot 10^{6}$ km
Sidereal period	27.322^{d}
Orbital eccentricity	0.0549
Orbital inclination	5.145°
Apparent magnitude (full)	−12.74

Symbols of celestial bodies

Name	Symbol
Sun	☉
Mercury	☿
Venus	♀
Earth	⊕
Moon	☾
Mars	♂
Jupiter	♃
Saturn	♄
Uranus	⛢
Neptune	♆

Solar System

	Mercury	Venus	Earth	Mars	Jupiter	Saturn	Uranus	Neptune
Mean radius (km)	2440	6052	6371	3397	71493	60267	25557	24766
Mass (kg)	$3.30 \cdot 10^{23}$	$4.87 \cdot 10^{24}$	$5.97 \cdot 10^{24}$	$6.42 \cdot 10^{23}$	$1.90 \cdot 10^{27}$	$5.69 \cdot 10^{26}$	$8.68 \cdot 10^{25}$	$1.02 \cdot 10^{26}$
Semi-major axis (au)	0.3871	0.7282	1	1.5237	5.2033	9.5826	19.2184	30.11
Orbital period (years)	0.240846	0.615198	1	1.88082	11.862	29.4571	84.0205	164.8
Orbital eccentricity	0.2056	0.0068	0.0167	0.0934	0.0484	0.0542	0.0472	0.0086

Physical constants

Name	Symbol	Value	Units
Speed of light	c	$2.9979 \cdot 10^{8}$	m s^{-1}
Universal gravitational constant	G	$6.67 \cdot 10^{-11}$	N $m^2 kg^{-2}$
Gravitational acceleration	g	9.81	m s^{-2}
Astronomical unit	au	$149.6 \cdot 10^{11}$	m
Parsec	pc	206265	au
Planck's constant	h	$6.63 \cdot 10^{-34}$	J s
Stefan-Boltzmann constant	σ	$5.67 \cdot 10^{-8}$	W $m^{-2}K^{-4}$
Boltzmann constant	k_B	$1.38 \cdot 10^{-23}$	J K^{-1}
Hubble constant	H_0	67.8	$kms^{-1}Mpc^{-1}$

Suggested resources

The following texts are suggested for an in-depth study of some of the topics presented in this book.

Theory

- Fundamental Astronomy, Hannu et al.
  This book presents many topics in a clear and balanced way, highlighting their physical basis. It requires a slightly more advanced knowledge of mathematics and physics.
- An Introduction to Modern Astrophysics, Carroll B., Ostlie D.
  This book contains all the astronomy you would need for a bachelor's degree. However, compared to the previous one, it can be dispersed and difficult to follow.
- Introduction to Classical Mechanics, David Morin.
  One of the best books on classical physics. With accessible explanations and hundreds of problems, this book is ideal for developing your problem solving skills.
- Concepts in Thermal Physics, Stephen and Katherine Blundell.
  This book covers the basics of thermal physics, thermodynamics and statistical mechanics, with applications in stellar astrophysics, condensed matter physics and climate change.
- Cosmology and Astrophysics Through Problems, Padmanabhan.
  This book is very advanced and requires good physics knowledge. Very recommended for those who want to study beyond the international competition and prepare for university.

Observation

- Stellarium.
  This software is one of the best resources to study the sky, interactively.
- Sky and Telescope's Pocket Sky Atlas, Roger W. Sinnott.
  It may be useful to have a reference on paper, by combining Stellarium with an atlas.

Index

Made in the USA
Las Vegas, NV
19 December 2022